International Conference on e-Health 2016

Held at the Multi-Conference on Computer Science and Information Systems (MCCSIS 2016)

Madeira, Portugal
1 – 4 July 2016

Editors:

Mario Macedo
Louis Rodrigues

ISBN: 978-1-5108-6524-2

Printed from e-media with permission by:

Curran Associates, Inc.
57 Morehouse Lane
Red Hook, NY 12571

Some format issues inherent in the e-media version may also appear in this print version.

Printed by Curran Associates, Inc. (2018)

For permission requests, please contact International Association for Development of the Information Society (IADIS) at the address below.

International Association for Development of the Information Society (IADIS)
Rua Sao Sebastiao Da Pedreira
No. 100, 30, 1050-209
Lisbon, Portugal

Phone: 351 21 3151373
Fax: 351 21 3151244

secretariat@iadis.org

Additional copies of this publication are available from:

Curran Associates, Inc.
57 Morehouse Lane
Red Hook, NY 12571 USA
Phone: 845-758-0400
Fax: 845-758-2633
Email: curran@proceedings.com
Web: www.proceedings.com

INTERNATIONAL CONFERENCE

E-HEALTH 2016

part of the

MULTI CONFERENCE ON COMPUTER SCIENCE AND INFORMATION SYSTEMS 2016

PROCEEDINGS OF THE INTERNATIONAL CONFERENCE

E-HEALTH 2016

Madeira, Portugal
JULY 1 - 4, 2016

Organised by
IADIS
International Association for Development of the Information Society

Sponsored by

TURISMO DE PORTUGAL madeira islands

Official Carrier

TAP PORTUGAL
with arms wide open
A STAR ALLIANCE MEMBER

Volume Editor:
Mário Macedo

Computer Science and Information Systems Series Editors:
Piet Kommers and Pedro Isaías

Associate Editor: Luís Rodrigues

TABLE OF CONTENTS

FULL PAPERS

SHORT PAPERS

REFLECTION PAPERS

POSTER

FOREWORD

These proceedings contain the papers of the International Conferences on e-Health 2016, which was organised by the International Association for Development of the Information Society, 1 - 3 July, 2016. This conference is part of the Multi Conference on Computer Science and Information Systems 2016, 1 - 4 July, which had a total of 606 submissions.

The use of ICTs (Information and Communication Technologies) in Healthcare Services is the main mechanism to improve efficiency and effectiveness. The e-Health (EH) 2016 conference aims to draw together information systems, practitioners and management experts from all quadrants involved in developing computer technology to improve healthcare quality.

Submissions were accepted under the following 3 main areas in the field of e-health within specific topics:

-Research Issues
-Management Issues
-Applications

The e-Health 2016 conference received 124 submissions from more than 28 countries. Each submission has been anonymously reviewed by an average of five independent reviewers, to ensure that accepted submissions were of a high standard. Consequently, only 20 full papers were approved which means an acceptance rate of 16%. A few more papers were accepted as short papers, reflection papers and posters. An extended version of the best papers will be published in the IADIS International Journal on Computer Science and Information Systems (ISSN: 1646-3692) and/or in the IADIS International Journal on WWW/Internet (ISSN: 1645-7641) and also in other selected journals, including journals from Inderscience.

Besides the presentation of full papers, short papers, reflection papers and posters the conference also included a keynote presentation from an internationally distinguished researcher. We would therefore like to express our gratitude to Dr. Malcolm J. Fisk, Senior Research Fellow, Centre for Computing and Responsible Research, De Montfort University, Leicester and Director, Telehealth Quality Group EEIG, UK, for accepting our invitation as keynote speaker.

This volume has taken shape as a result of the contributions from a number of individuals. We are grateful to all authors who have submitted their papers to enrich the conference proceedings. We wish to thank all members of the organizing committee, delegates, invitees and guests whose contribution and involvement are crucial for the success of the conference.

Last but not the least, we hope that everybody will have a good time in Madeira, and we invite all participants for the next edition of this conference.

Mário Macedo, Universidade Atlântica, Portugal
e-Health 2016 Program Chair

Piet Kommers, University of Twente, The Netherlands
Pedro Isaías, Universidade Aberta (Portuguese Open University), Portugal
MCCSIS 2016 General Conference Co-Chairs

Madeira, Portugal
July 2016

PROGRAM COMMITTEE

E-HEALTH CONFERENCE PROGRAM CHAIR

Mário Macedo, Universidade Atlântica, Portugal

MCCSIS GENERAL CONFERENCE CO-CHAIRS

Piet Kommers, University of Twente, The Netherlands
Pedro Isaías, Universidade Aberta (Portuguese Open University), Portugal
Heredina Fernandez Betancort, Universidad de Las Palmas de Gran Canaria, Spain

E-HEALTH CONFERENCE COMMITTEE MEMBERS

Abdel-Badeeh M.Salem, Ain Shams University, Egypt
Adel Taweel, Kings College, UK
Alain Pruski, Lcoms Université de Lorraine, France
Alexander Hoerbst, UMIT – University for Health Sciences, Medical Informatics and Technology, Austria
Amir Hajjam el hassani, University of Technology of Belfort Montbeliard, France
Andreas Schrader, University of Lübeck, Germany
Anja Henner, Oulu University of Applied Sciences, Finland
Antonio Lanatà, University of Pisa, Italy
Athina Lazakidou, University of Peloponnese, Greece
Bennoor Kazi, National Institute of Diseases of Chest & Hospital, Bangladesh
Bert-Jan van Beijnum, University of Twente, The Netherlands
Bobak Jack Mortazavi, Yale School of Medicine, USA
Bridget Kane, Karlstad University, Sweden
Carlos Fernández Llatas, Universidad Politécnica de Valencia, Spain
Cesar Caceres Taladriz, Rey Juan Carlos University, Spain
David Isern, University Rovira I Virgili, Spain
Edward Kacki, Imia, Poland
Elena Villalba, Universidad Politécnica de Madrid, Spain
Elske Ammenwerth, UMIT – University for Health Sciences, Medical Informatics and Technology, Austria
Emmanuel Andres, Université de Strasbourg , France
Enzo Pasquale Scilingo, University of Pisa, Italy
Eric Campo, Laas-cnrs, France
Farid Naït-Abdesselam, Paris Descartes University, France
Fernando Martin-sanchez, Weill Cornell Medicine, USA
Georg Duftschmid, Medical University of Vienna, Austria
George Georgiev, University of Wisconsin, USA
George Ghinea, Brunel University, UK
Guillaume Lopez, Aoyama Gakuin University , Japan
Jan Vejvalka, Charles University, Czech Republic
Jesuk Ko, Gwangju University, South Korea
Jose Luis Sanchez , Laboratorio de Computadoras y Control, Spain
Karla Navarro, Faculty of Engineering And Information Technology, Australia
Krzysztof Zielinski, Agh-University of Science And Technology, Poland

KEYNOTES LECTURE

PUTTING THE PATIENT MORE IN CONTROL: A CHANGING PERSPECTIVE FOR ICT AND TELEHEALTH

Dr Malcolm J. Fisk, Senior Research Fellow, Centre for Computing and Responsible Research, De Montfort University, Leicester and Director, Telehealth Quality Group EEIG, UK

ABSTRACT

Information and Communication Technologies are embedded as important tools of clinicians. ICT helps in the delivery of health services to their patients. 'Delivery' is, however, a one-way process and ICT can be used as a tool that disempowers and marginalises patients. There is an imperative, therefore, for ICT to be harnessed for the benefit of all of us … helping build health literacy; helping us to manage our behaviours and lifestyles; and helping us to become more equal partners in any health treatment. New approaches to ICT, particularly in the form of telehealth, can help to bring this about and put the patient more in control.

This presentation challenges some 'conventional' uses of ICT for health. It calls for a change in our understandings and attitudes. It even calls for a change in our language – promoting the notion of health service provision instead of 'delivery'. Service provision can, after all, include conventional approaches for the frailest and most vulnerable but it can also enable people (patients) to access and use services in relation to their needs and choices. ICT tools in the form of Internet links and mobile devices (both within the arena of telehealth) enable people to do this.

In the 'new world' for health and health services, therefore, ICT (including telehealth) will have an important part to play. But in playing that part ICT must be designed and configured with the needs of patients (people) as well as clinicians in mind. By doing this ICT has the potential to play a crucial part in helping improve people's lives, reduce disadvantage and empower us all … putting the patient more in control.

Full Papers

THE PORTUGUESE HEALTH DATA PLATFORM: AN ALTERNATIVE ARCHITECTURE APPROACH

Eduardo Pinto and António Carvalho Brito
INESC TEC, Faculdade de Engenharia, Universidade do Porto, Rua Dr. Roberto Frias, 4200-465 Porto, Portugal

ABSTRACT

Over the last years, the necessity for sharing clinical information among different healthcare institutions has been growing along with the growing population mobility. Nowadays, the ultimate goal is that clinical information is accessible regardless of context, place, time or where it was collected. In order to accomplish that, several countries have been launching projects usually known as Electronic Health Records (EHR), including Portugal in 2010. Those projects involve huge efforts since they gather challenges from business alignment to operational deployment and development. This paper describes and analyses the project that was launched in Portugal, in terms of its strategy and high-level architecture. We present a different architectural approach that we believe could help tackle some present and future problems the system faces. The approach follows the principles of Metropolis Model and advocates an effort distribution in development and information management with stakeholders through a focus on open and standardized interfaces.

KEYWORDS

e-Health Architectures; Electronic Health Record; Information Systems Architecture

1. INTRODUCTION

The health context is one of the most complex and critical, compelling the organizations to support several business processes and interacting with multiple stakeholders from clinicians to laboratory technicians, software houses or public and private healthcare institutions. With such a complex scenario it seems easy to understand the impact that the technology might have, automating processes or sharing information among all the interested parties. However, this complexity obviously raises the risks of failing for projects implementation, either because the products do not meet the requirements, the stakeholders' expectations or were built excessively monolithic what preclude future integration and evolution (Chu 2006). Over the last years, the society's objective of improving healthcare increased the necessity for sharing clinical information among different healthcare institutions (Haux 2006). This need represents enormous challenges both to software development and to business management areas.

In 2010, the Ministry of Health and the Faculty of Engineering of University of Porto (FEUP) signed a protocol of cooperation in this field. The goal was to find the best approaches that would led to the successful implementation of an EHR-like project in Portugal. Under the scope of that protocol, the research has been evolving in close collaboration between the Ministry's development team and the FEUP's research team. From the protocol's start until now, the project scope changed, growing from a "simple" EHR to a platform to foster clinical data sharing. *Plataforma de Dados de Saúde* (PDS - Health Data Platform, in English), launched by the Portuguese Ministry of Health, is a project that aims to allow, encourage and enforce clinical data sharing between the several institutions existing in Portugal.

PDS is composed by some projects with specific goals. This paper characterizes PDS and describes and analyses its components, the dependencies between them and their dependencies to external systems. Moreover, an alternative architecture approach is proposed arguing that it can improve the velocity with which more Portuguese health institutions and other stakeholders will be able to take advantage of PDS and all the advantages of accessing the right data at the right time, independently of its source.

Following this introduction, the Section 2 introduces some essential concepts in clinical data sharing field in general. The Section 3 briefly presents the Portuguese project (PDS): its components and dependencies. The next section is Section 4 and describes the alternative architecture approach proposed for the project future evolution. Finally, the article's conclusions are summed up at Section 5.

2. BACKGROUND CONCEPTS

The fully understanding of the work requires the clarification of some concepts that are explained in the next paragraphs. An Electronic Health Record (EHR) is a "longitudinal collection of electronic health information about individual patients and populations" (Gunter & Terry 2005). The main objective is to provide clinical information about a patient where it needs to be consulted, independently of its origin or location, helping to avoid clinical errors or duplication of efforts and resources. It is supposed to be a mechanism for integrating healthcare information with the purpose of improving care quality (Orszag 2008). The EHR concept stands as a natural evolution of the Paper-Based Patient Records. However, as argued by Tang and McDonald (Tang & McDonald 2006), the capacity the EHR have to be beneficial depends largely on factors like: comprehensiveness of information, duration of use and retention of data, degree of structure of data and ubiquity of access. Over the years, several countries have been launching initiatives to implement EHR projects. Some countries (Canada Health Infoway 2009; The Danish Ministry of Health 1996; Brennan 2005) have achieved more than others, but the common experience says it stands not only as a technological challenge, but way beyond that (Protti et al. 2009).

Often, these EHR projects aggregate other minor components. One of those components is usually the Patient Summary (PS)[1]: a set of information that allows a healthcare professional to have a quick and easy overview over a patient. The concept is used in the epSOS project (European Patients - Smart open Services) and a similar concept is used in the England's National Health Service, called Summary Care Records. Although the name may vary, the concept is practically the same and stands as an electronic record that will give healthcare staff faster and easier access to essential information about a patient. The epSOS PS contains the following data: 1) Demographic information (e.g. name, birth date, gender); 2) Most important clinical patient data (e.g. allergies, current medical problems, or major surgical procedures during the last six months); 3) Current medication including all prescribed medicines; 4) Meta-data about the PS itself (e.g. when and by whom was created or modified).Another interesting concept is the Personal Health Record (PHR) which is a system whereby individuals can "access, manage and share their health information, and that of others for whom they are authorized, in a private, secure, and confidential environment" (Connecting for Health 2003). As it is reported by some articles (Tang et al. 2006; Tang & Lee 2009), the approach of the system might differ from being a totally isolated one till being integrated with the national healthcare system. Also, there are several studies about the benefits that those systems can bring (Tang & Lee 2009; Fricton & Davies 2008).

3. THE PORTUGUESE PROJECT

In this section, we present the strategy followed by the Portuguese entity, responsible for the project that aims to foster clinical information sharing. *Plataforma de Dados de Saúde* (PDS) is the name of the project created by the Portuguese Ministry of Health, and managed by a company called Shared Services of Ministry of Health (SPMS), aiming to facilitate and leverage the share of clinical data among the several healthcare institutions, standing as a wider concept than a simple EHR, for instance. It started as a set of projects running in parallel with the goal of achieving value in the areas mentioned above. The projects are[2]:

- Patient Portal - My Health is an existing patient portal that supports the basic PHR functionalities;
- *Resumo Clínico Único do Utente (RCU2)* is the Portuguese implementation of the epSOS PS;
- International Portal is meant to allow a clinician in Portugal to access a foreign patient's PS;
- Professional Portal stands as the healthcare professionals' gate to the patient's clinical information. This portal gathers information from multiple sources, in order to provide the clinician with the most complete information about healthcare-related episodes.

---

[1] epSOS project: http://www.epsos.eu/

[2] Institutional information: http://spms.min-saude.pt/2013/11/pds-plataforma-de-dados-da-saude/

3.1 Internal Dependencies

The component projects/systems presented above have some information dependencies between them, as represented in Figure 1. The four interactions/dependencies stated are:

- Professional Portal has two: information introduced by the patient through the Patient Portal to be available to the clinicians (no. 1); the PS document of a patient from the PS repository system (no. 2);
- every PS document has some information that was originally introduced by the patient at the Patient Portal (no. 3);
- International Portal requests the PS document in order to send that information abroad the country (no. 4).

One important aspect to state is that the PS documents are HL7 Clinical Document Architecture compliant documents what is a great advantage in terms of fostering and facilitating the future integration of this information with other systems.

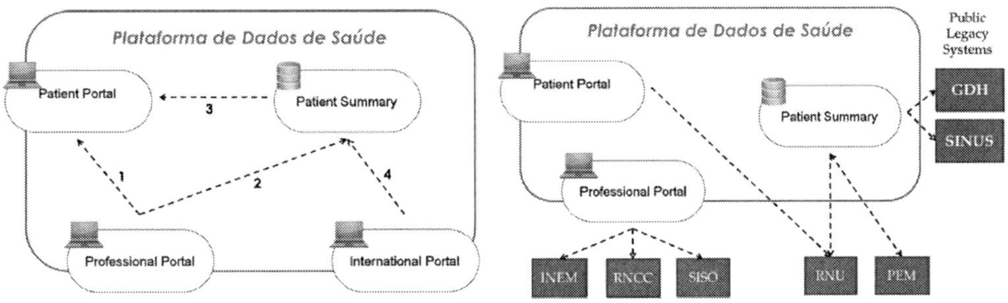

Figure 1. The projects that compose PDS and the internal (left) and external (right) information dependencies identified

3.2 External Dependencies

Thought the internal organization of PDS is already an interesting subject, the interest grows significantly with the complexity introduced by the dependencies existing with systems residing outside PDS scope, as represented in Figure 1. In this sense, the authors think the interactions can be classified in one of three types: information sources, information consumers and auxiliary systems. As expected, the information sources are several. Regarding the Professional Portal, some examples of information sources are: 1) INEM, which is the public institution responsible for the emergency episodes, 2) RNCC, which is the national network for continuity of care and 3) SISO, which stands as a central database related with the oral care services.

This data is used to build a sort of time-line with the different events related with that person and which can be filtered by some parameters. Concerning the RCU2, the information is provided essentially by the public legacy systems. That is, nowadays the information from hospitals is fetched from the national database of Diagnosis-related Groups (GDH). On the other hand, there is also information to be retrieved from the primary care institutions (SINUS), for instance patient allergies, medication and medical and nurse diagnoses. The electronic drugs dispensation project (PEM) is also a source of information to the PS.

Nowadays, the only external system acting as information consumer is PEM which uses the patient allergies to alert the clinician at the prescription act about eventual dangers or incompatibilities. In terms of auxiliary systems, the RNU is probably the most important one. The RNU is a national service that allows the patients unique identification and that own some demographic but important information about them, like the general practitioner or the primary healthcare institution associated.

3.3 Problems and Challenges

As described in the previous sections, the project depends on several interactions with other systems. The legacy systems have a crucial role as they are the source of information and the entry point where the professionals introduce the clinical data. An interesting study (Teixeira et al. 2013) evaluated the experience of the Portuguese healthcare professionals with their Health Information Systems. The EHR referred in that study is the Professional Portal described in the earlier sections. The two other systems evaluated in that paper are two legacy systems. The authors state that "aging health information systems are perceived as being fragmented, inadequate for professional's requirements and having severe usability problems" causing "significant disruptions during healthcare provision" and "decrease in clinical information quality".

When looking in detail to PDS evaluation (referred in the study as the EHR), among five experience requirements described, the system had a clearly positive evaluation in one (Perceived Usefulness), slightly positive in two (Information Availability and System Usability) and negative in the remaining two (Ease of Adoption and System Reliability). For someone that followed the project's implementation and knows the ecosystem it is possible to understand the reasons behind that:

- System Usability - when comparing two systems with a development time gap of 20 years it is expected that the newer system (PDS) is much more pleasant to use than the old ones (legacy systems). If we add the fact that some healthcare professionals were involved in the development process of that new software, the positive feedback is even more expected;
- Information Availability - the integration of information such as prescriptions or diagnoses is obviously acknowledged by the professional as very useful. Although, the current model allowed to integrate valuable information from the main sources (such as the biggest hospitals) there are still many other institutions with relevant information to be linked;
- Ease of Adoption - since this requirement is closely related to usability, at a first look it might seem odd why it had a negative evaluation when usability had a pretty positive one. There are two key reasons for that: 1) the lack of investment in training and formation of professionals when PDS was deployed 2) the fact that professionals are required to jump between the legacy system and the new system over the time, meaning there is not integration at a level that allows a doctor to use only one software for accessing all the patient's clinical information;
- System Reliability - definitely the worst rated experience requirement by the professionals due to frequent "reliability issues that propagate to other systems". As said, PDS relies on legacy systems and an old network that often fails causing the whole experience to be very frustrating.

The peer-to-peer integrations were a key strategy to deliver something valuable and working in a tiny time-frame but within a large and complex environment like this, it is unlikely it will be able to evolve towards a comprehensive ecosystem that needs to involve more than the legacy systems that are developed in-house by SPMS. The amount of different systems and institutions with relevant clinical data yet to integrate is very significant and a peer-to-peer strategy is certainly not the best choice for scalability and flexibility. Also, the integration done on a presentation level that was often used certainly allows the professionals to access more information but fail to provide an easy and pleasant experience.

Lastly, a system that collects information from several local repositories has to be capable of identifying unequivocally the professional. Currently, due to the existence of multiple IS it is very common that professionals share their access credentials with each other. These shares diminish the information quality accountability since it is not possible to guarantee who produced or accessed the information.

4. AN ALTERNATIVE ARCHITECTURE APPROACH

The problem of sharing clinical information between different institutions contains challenges that vary from high-level strategy to chosen standards (Pinto & Brito 2014). However, the focus of this article is on the architectural issues. Clements, Kazman and Klein advocate (Clements et al. 2001) that "architectures allow or preclude nearly all the system's quality attributes". In this sense, when creating a system where the quality attributes are critical, the architecture is certainly one of the most relevant aspects.

The concept of Systems-of-Systems (SoS) is closely related to this idea of achieving value through the connecting and integration between several peer systems. A system is classified as an SoS when there is an "assemblage of several components which individually may be regarded as systems". Maier suggests some architectural principles to deal with this kind of system like stable intermediate forms, policy triage, leverage at the interfaces and ensuring cooperation. Additionally in this field, the Ultra-Large-Scale Systems

(Northrop et al. 2006) and the Large-scale complex IT systems (Sommerville et al. 2012) are two recent terms that refer to the same type of systems. These are systems identified by extreme size in every imaginable dimension: lines of code, stakeholders, number of systems, etc. Northrop, et al. state some characteristics as decentralized control, "unknowable" and conflicting requirements, continuous evolution and others (Northrop et al. 2006).

If we lower the abstraction level, it makes sense to refer some architecture styles like Service-Oriented Architectures (SOA), Resource-Oriented Architectures (ROA) or Event-driven Architectures (EDA). SOA is an "architectural style that emphasizes implementation of components as modular services that can be discovered and used by clients" and that "emphasis on loose coupling between interacting services" (Srinivasan & Treadwell 2005). Over the years the client has been brought to the middle of the business, helping to create value. The Metropolis Model (Kazman & Chen 2009) appears as an attempt of describing very complex systems built from two basilar concepts: Open-source Software (OSS) and Community-Based Service Systems (CBSS). The Metropolis Model presents a new unified vision between the CBSS and the OSS, focusing deliberately in the crowd value generation through the definition of two levels: the kernel services and the periphery services.

4.1 Definition of the Kernel's Limits

The developments under the sign of PDS have led to the releasing of two main systems: the Patient Portal - My Notes and the Professional Portal, as described in Section 3. These two systems were developed almost as independent systems, with their own user interfaces, architecture and technologies.

In order to become a real platform for sharing clinical data, PDS should evolve with a well-delimited scope. Following the base principle of the Metropolis Model we suggest to: 1) delimit the scope of PDS to a platform for sharing data 2) split the logical domain from the presentation one. The suggestion is to update the applications so that the three existing interfaces became only the official clients to the clinical information repository (PDS). To achieve this, it is necessary to clearly separate the two domains, defining web services that are used to query and retrieve data, as represented in Fig. 3. This strategy meets the Metropolis Model main concept through the definition of a smaller but more robust kernel of the system. At the same time, it allows to transform those interfaces into open interfaces. By doing that, many stakeholders would have the opportunity to: 1) develop new applications using the existing services provided directly by the Ministry 2) integrate the information in the work-flow of the existing information systems and 3) create applications and services that use the available information in innovative ways.

The most significant part is that, the power and investment of the companies could create significant value to the patients, even with the Ministry spending no money unless the effort to provide and define the services. We believe that with this approach would be possible to take advantage of the several stakeholders, either from healthcare software houses, hospitals IT departments or patients and professionals individually. When talking about crowd-sourced where the information is the fundamental, it is not necessary to do high investments but to provide the right tools to the interested parties.

As we can see in Fig. 3, the system would have a reduced but much clearer scope. The existent portals would be progressively separated of PDS kernel. Also, they would start working as official clients supported by the Ministry of Health and using the same services available to any integration of every authorized institution. This approach incites IT departments to work towards full integration of PDS data. On the one hand, significant improvements in the user experience for professionals are expected as they are no longer required to jump between different applications to access the patient's clinical information regardless of where it was produced or stored. That would certainly increase the usability and ease of adoption for the system. On the other end, the scalability and flexibility of the system are improved as integrations peer-to-peer would no longer be needed and would rather happen in a much more agile and fast way.

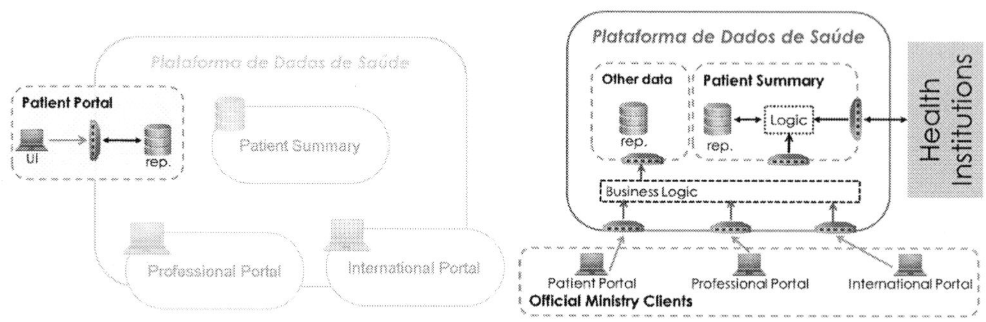

Figure 2. The transformation of the portals into official clients of an open-services layer that will also be available to other stakeholders

4.2 Model Premises

PDS is obviously a complex system and is expected to continue to evolve and became bigger as new services are needed. In that sense, the way it is internally organized is fundamental to allow new requirements and new services. The Service-oriented Architectures appear as an interesting solution to address some of the scalability and complexity issues. Some PDS services are easily identified as potential components, for instance:

- Data Anonymizer - receives some data and returns it without being possible to reveal the
- patient identity;
- Data Dispatch Manager - receives the data to be transferred to some institution and assures that it arrives the destination;
- PS Manager - important component to manage all the actions performed that are related with PS. It might be an example of a service composed by internal services itself;
- Audit Service - registry of all operations, information accesses and data flows.

The medium-term evolution of PDS depends on its capability to collect, merge and provide information from/to multiple institutions and contexts. Since the information we are talking about is very sensitive, it is absolutely essential the system to assure only authenticated users and/or systems have access to the clinical data which they are authorized to.

The security of the system relies on three vertices: patients, professionals and institutions. That means that when exchanging clinical information there must be no doubt about the patient to whom the data refers. Unfortunately, it is known that the local registries have sometimes wrong demographic patient information either because it was mistakenly introduced or because it is not up-to-date. In this sense, it is fundamental to validate and check the data against a patient national registry, which must be as updated as possible. In Portugal it exists and is called RNU. However, it is not just about having the records but having the needed services to take advantage of them. For instance, the existence of a national patient demographic database could allow an authenticated patient to schedule an appointment, insert personal health information, update the demographic registries, etc.

Regarding the professionals, their relation with the security question is expressed in two moments: the production and the consumption of the information. That is, when the information is produced at a local level, it has the signature of the professional who registered it, usually a clinician or a nurse. Thus, a central system that collects information from several local repositories has to be capable of identifying unequivocally the professional who produce it. Therefore, the creation of a Healthcare Professionals National Registry would have an enormous importance both for allowing the creation of unified authentication processes through different IS and for the administrative benefits that such knowledge would bring. It would be possible to avoid current common situations where healthcare professionals exchange their systems credentials. These shares would be less frequent if the professionals would not need to memorize several credentials to multiple IS. As a final consequence, the security of the whole system itself would raise significantly.

The institutions have the final role in this security triangle. The infrastructure that allows the clinical information to circulate through the network peers must ensure its integrity. Ideally, the central system and all the peers would share the same authentication and recognition process becoming extremely easy to add new peers to the network. This institutional layer appears in the lower level of the two previous, aiming to assure that the source and target physical infrastructures are the rightful ones and that the information carried out is delivered to its rightful destination.

In short, it is possible to identify three critical moments on which the information integrity and security depends: production, consumption and sharing. As it is stated above, patients, professionals and institutions have a fundamental role in guaranteeing the privacy, integrity and security of clinical data. Therefore, the necessity to identify them supports the construction of those three registries and the respective services. While the project's scope is restricted to the public institutions these questions of integrity and security may be wrongly overlooked because of the existence of a national private network. However, these concerns are absolutely essential when we think about opening the access to this clinical repository, that PDS intends to be, to the private entities (such as hospitals and clinics). Then, they will have a critical role of ensuring the credibility of the system. In that sense, it is important that current PDS processes start to be adapted prospecting a wider vision.

5. CONCLUSION

The Portuguese project (PDS) was designed as a compound of smaller projects based on well-known concepts, like the PS or the PHR. The existing political and social environment added an urgency to deliver short-term results, what happened to accelerate the process and avoided the creation of a megalomaniac project. However, as the project settles down there is a need to foresee it in a long-term perspective. PDS external dependencies and the way those dependencies are implemented determine the ability the project is able to scale and evolve. Currently, the professionals recognize the value added by PDS as well as its usefulness and usability. However, ease of adoption and availability are two examples of quality attributes being damaged.

The authors presented an Alternative Architecture Model intended to answer to some of the problems identified with the current state of the project. We argue that PDS should be presented, as the name suggests, as a platform to foster the clinical data sharing with the final objective of improving the patient healthcare services. In that sense, PDS should be responsible to guarantee the access to the data in a completely safe, authenticated and authorized way making use of three national registries that should exist and are fundamental to support this model. PDS might have a role as the catalyst of healthcare stakeholders' engagement, from professionals to software houses, aiming the development of software solutions that improve the healthcare services, either by public or private institutions. That strategy would allow the clinical information provided by PDS to be integrated by IT departments directly into legacy systems or new systems used by the healthcare professionals. That fact would improve the ease of adoption of PDS as well as its usability. The reduced efforts on SPMS could be directed towards the upgrading of legacy systems increasing the availability performance of all the system. Regarding security concerns, we believe it is not possible to assure integrity and security without being able to authenticate the three entities involved in the process: patients, professionals and institutions. Therefore, the model requires the existence of three essential blocks: Patients National Registry, Healthcare Professional National Registry and Institution National Registry. Those blocks are indispensable for assuring that the clinical data is only access by authorized parties. They also have a significant role as they can be used to build different services that allow both patients and professionals to have a superior user experience.

We believe that the developments now must be focused on providing the means and not the ends. This change of strategy will allow distributing the efforts through the interested stakeholders. A hospital either public or private should be able to integrate the existing PDS data directly into its information systems. A software developer should be able to develop a mobile application that uses Patient Portal to registry his daily weight records. The model essentially would enable the stakeholders to work in parallel creating new integrations and innovative services that could certainly improve the healthcare services requiring less effort from the Ministry of Health. Projects such this one are unlikely to reach a state of completeness since improvements will always be required. However, we believe this model shall allow not only accelerating the process of sharing clinical information but also to foster the appearance of innovation and ultimately benefit the healthcare patient services.

ACKNOWLEDGEMENT

Thanks to SPMS for the support and FCT for funding this research (SFRH/BD/90838/2012).

REFERENCES

Brennan, S., 2005. *The NHS IT project: the biggest computer programme in the world - ever!*, Radcliffe Publishing Ltd; 1 edition (April 2005).

Canada Health Infoway, 2009. *Building a Healthy Legacy Together - Annual Report 2008/2009*

Chu, S.C., 2006. From component-based to service oriented software architecture for healthcare. *Proceedings of 7th International Workshop on Enterprise networking and Computing in Healthcare Industry 2005 HEALTHCOM 2005*, 4(1), pp.96–100.

Clements, P., Kazman, R. & Klein, M., 2001. *Evaluating Software Architectures: Methods and Case Studies*, Boston, MA, USA: Addison-Wesley.

Connecting for Health, 2003. *The Personal Health Working Group Final*,

Fricton, J. & Davies, D., 2008. Personal health records to improve health information exchange and patient safety. *in Patient Safety: New Directions and*, pp.1–10.

Gunter, T.D. & Terry, N.P., 2005. The emergence of national electronic health record architectures in the United States and Australia: models, costs, and questions. *Journal of Medical Internet Research*, 7(1).

Haux, R., 2006. Health information systems - past, present, future. *International journal of medical informatics*, 75(3-4), pp.268–81.

Kazman, R. & Chen, H.-M., 2009. The metropolis model a new logic for development of crowdsourced systems. *Communications of the ACM*, 52(7), p.76.

Maier, M.W., 1998. Architecting principles for systems-of-systems. *Systems Engineering*, 1(4), pp.267–284.

Northrop, L. et al., 2006. *Ultra-Large-Scale Systems: The Software Challenge of the Future*, Pittsburgh, PA: Software Engineering Institute, Carnegie Mellon University.

Orszag, P.R., 2008. Evidence on the Costs and Benefits of Health Information Technology. In *Testimony before Congress*.

Pinto, E. & Brito, A.C., 2014. Architectural Key Dimensions for a Successful Electronic Health Record Implementation. In *ICEIS 2014 - 16th International Conference on Enterprise Information Systems*. pp. 136–143.

Protti, D., Johansen, I. & Perez-Torres, F., 2009. Comparing the application of Health Information Technology in primary care in Denmark and Andalucía, Spain. *International journal of medical informatics*, 78(4), pp.270–83.

Sommerville, I., Cliff, D. & Calinescu, R., 2012. Large-scale complex IT systems. *Communications of the ACM*, 55(7), pp.71–77.

Srinivasan, L. & Treadwell, J., 2005. An overview of service-oriented architecture, web services and grid computing. *HP Software Global Business Unit*, 2.

Tang, P. & McDonald, C., 2006. Electronic health record systems. *Biomedical Informatics*.

Tang, P.C. et al., 2006. Personal Health Records-Definitions, Benefits, and Strategies for Overcoming Barriers to Adoption. *Journal of the American Medical Informatics Association*, 13.

Tang, P.C. & Lee, T.H., 2009. Your doctor's office or the Internet? Two paths to personal health records. *New England Journal of Medicine*, 360(13), pp.1276–1278.

Teixeira, J.G. et al., 2013. Healthcare professionals as customers: A service perspective on Portuguese primary care health information systems. *2013 IEEE 15th International Conference on e-Health Networking, Applications and Services, Healthcom 2013*, pp.377–381.

The Danish Ministry of Health, 1996. *A Danish Health Care Data Network In Two Years*

SECURE DISTRIBUTED BACKUP MANAGEMENT OF PERSONAL HEALTH RECORDS

Divyashikha Sethia[1], Daya Gupta[1], Huzur Saran[2], Arpit Goel[1] and Radhika Kuchhal[1]
[1]*Department of Computer Science and Engineering, Delhi Technological University, India*
[2]*Indian Institute Of Technology, Delhi, India*

ABSTRACT

A patient can visit numerous hospitals or clinics in his lifetime, which therefore disseminates his medical data among multiple Hospital Information Systems (HIS) in the form of Electronic Health Record (EHR), managed by each hospital independently. The integrated Personal Health Records (PHR) can be kept on portable equipment such as USB or mobile device, and can be retained with the patient as he moves from one HIS to another. The backup of the PHR can be maintained on the cloud for refurbishing PHR in case of theft or remote access. The proposed system, maintains backup of the PHR consisting of references to each HIS and not entire data, for security reasons. In case of theft or remote access to PHR, a distributed health record management system queries individual HISs for clinical data of a particular patient and then can integrate all the responses together in standardized way to generate the aggregated health record. The responses received from different HIS may be in different formats, so the system translates all responses into a common format such as HL7 format, that represents the integrated PHR of the patient. The backup of records can be stored on distributed file system such as Hadoop in an encrypted form using Attribute Based Encryption (ABE) for fine grained access control. We observe that sequence files can provide reduced access time, highly reliable and available system.

KEYWORDS

Distributed health record, HL7, EHR, PHR, HIS, healthcard

1. INTRODUCTION

A Personal Health Record (PHR) is loosely defined as a patient data repository that can serve individuals, encouraging self care. It allows an individual to keep track of his complete medical history including current and past diagnoses, medications, lab results and much more (Sprague 2006). PHRs can also be used for providing efficient communication between patients and providers and even allow patients to setup future appointments (Tang 2006). It enables consolidation of medical data from patients as well as physicians. Patient collected information may include various physiological parameters that can be gathered through body sensors and physicians may specify clinical information such as lab tests results, prescribed medication or identified allergies. Apart from clinical information, PHRs can also store other information like patient's insurance information for billing purposes and emergency information that can be used in crisis or during disasters (Raymond 2007).

There are various types of PHRs that are being implemented today. In a broad view, they can be classified as either standalone PHRs or integrated PHRs (Raymond 2007; Detmer et al. 2008). *Standalone PHRs* are generally under the physical control of the individual and they can be carried on devices like USB drives or smart cards. The patient himself may create and update these records. Such PHRs can provide portability, security and confidentiality of records but the information cannot be trusted by a healthcare provider as patients may not be able to report exact test results making them inadequate (Wuerdeman et al. 2005). *Integrated PHRs,* on the other hand, include records from various sources such as Hospital Information Systems (HIS), pharmacies or insurance companies where patient data is generally kept in a centralized repository and made available to patient through web or email. Although, integrated PHRs have huge potential, they are difficult to realize in practice (Raymond 2007; Detmer et al. 2008). To reap benefits from both architectural styles, a hybrid PHR can be used which consists of a patient controlled health record, termed as healthcard, containing patient's vitals, insurance information, clinical observation, lab reports, prescriptions and medications. The healthcard can exchange data with various healthcare stakeholders such as doctor, nurse, pharmacist, lab technician etc in a standardized manner, promoting both self care and communication among patients and healthcare providers (Kaelber et al 2008).

Medical information for patients can be distributed among numerous healthcare facilities as the patient may move between different hospitals. In certain situations, patient data may be tethered to an insurance plan or a particular facility but he may move from one medical plan to another, disseminating the medical records. Although, the decentralization of records may deter healthcare services due to unavailability of complete medical account, it still outweighs a centralized strategy as it raises many legal, security and privacy concerns. The National Programme for IT (NPfIT) system that has been developed in U.K uses Spine to hold the complete patient record centrally (Spine Services). The system has met with severe criticism regarding confidentiality and misuse of patient data, summarized by Campion-Awwad et al. (2014). In contrast, Netherlands developed an Electronic Patient Dossier (EPD) system, that decentrally stores patient records into GBZs, which are information systems maintained by each facility individually (Noordende 2010; Force 2010). Fragmented records do help in alleviating confidentiality issues but it must deal with the incorporation of dispersed data and keeping it in a standard format.

This work assumes that patient retains PHR on a portable device such as a mobile device (Sethia et al. 2014) and proposes an integrated health record management solution in which records are decentrally stored on different HIS. Each record stored in HIS usually represents a visit or an encounter with a healthcare professional. A backup of the healthcard is retained on the cloud in a digital health vault consisting of references with HIS visits along with medical history files such as bulky media files related to lab tests or medical procedures. To maintain confidentiality of records, we do not keep all the records together at any other location, except that with the patient over which he has complete control. Although this mitigates the privacy issue, it makes the job of backing up the healthcard all the more difficult (Detmer et al. 2008). We propose a distributed query mechanism that keeps reference pointers to all these visit records at a central location which can be used to pull all the scattered EHRs required for restoration of the patient's complete medical account in case of damage to or theft of healthcard. All HISs may not share the same data model and hence the data dispersed may be in heterogeneous forms. To keep the records in a regular format, we use HL7 framework in which we translate EHRs retrieved from different sources using a common HL7v2 template. These HL7 based records encapsulated together represents the healthcard of the patient. Though the system is able to retrieve the PHR of an individual, keeping references to disseminated EHRs necessitates high security and availability. To provide high availability, we use Hadoop Distributed File System that allows replication of data while being highly fault tolerant and can be deployed on inexpensive hardware (Borthakur 2008). The references files are kept on HDFS in an encrypted format using Attribute based encryption. In this way, we are able to provide a novel method to securely aggregate distributed health records that can be used as backup for the PHR of the patient.

2. RELATED WORK

In the past years, people's interest in keeping track of their medical information has catapulted. They want access to their health records in a convenient way while without distressing about the confidentiality and security of the records (Detmer et al. 2008; Krohn 2007). This has led to exploration of PHRs and their different models, research in the area of integration of medical information and development of medical standards for transfer of information and interoperability. Adoption of PHR is evident in various parts of the world (Sprague 2006; Detmer et al. 2008; Halamka et al. 2008).

A simple standalone PHR may not require to integrate with other sources, but to provide better healthcare services it is essential to interact and incorporate with multiple healthcare facilities. A clinical data exchange system using Ensemble Integration platform has been developed in China that collects clinical data from various medical institutions and transforms and uploads it into a centralized data repository using HL7 (Wang et al. 2014). A hybrid architecture for sharing of health records has been suggested that keeps the actual patient records at different locations but maintains patient identifying information at a central component that is to be used for retrieval of distributed records (Baldock et al. 2014). The Dutch EPD system also stores patient records into decentralized component known as GBZs managed by each hospital locally. To access the dispersed records, a reference index for each patient is maintained at a central component storing references to all the particular patient's records and the idea was extended by the author that involves elimination of centralized component to reduce privacy and security risks by explicit transfer of references, when required, under the control of the patient or the doctor (Noordende 2010; Noordende 2011). The work proposed in this paper provides a novel manner of storing integrated records on the healthcard that is highly available along with a backup on a digital vault in the cloud which can refurbish the details of the card through distributed query mechanism.

3. LITERATURE REVIEW

3.1 Hospital Information Systems

A hospital or a healthcare information system (HIS) can be seen as automated system that not only provides electronic storage of patient health records but also satiates administrative and financial needs of a healthcare facility. A HIS may comprise of a data model that stores the patient record in a structured format electronically, authorization and security mechanisms for protection of patient data, information communication tools for promoting interoperability, data analysis and visualization tools for enhancing care services, and a sophisticated user interface for making the system simple to operate. Some popular open source HIS applications are OpenMRS, OpenEMR and GNU Health.

3.2 HL7 Framework

Different HISs store information in diverse formats such as in relational database, structured documents like XML or JSON, or even digitized copy of physical records making it very difficult to exchange information among various facilities (Eichelberg et al. 2005). This leads to requirement of standard ways for exchanging data to make systems interoperable. Some of such standards are HL7, openEHR, Continuity of Care Records (CCR) and Continuity of Care Documents (CCD). HL7 is an ANSI certified set of international standards used for medical data exchange. HL7v2 medical standard is the most popular interoperability standard in use. It provides a framework for sharing and integration of medical information. The standards dictate how information should be structured and transferred among similar or dissimilar systems. It is based on the assumption that health data is exchanged among various applications due to occurrence of some event such as patient admit or discharge, lab tests ordered or lab tests completed, pharmacy order or some medical procedure scheduled.

3.3 Mirth Connect

Mirth Connect is an open source tool that is widely used for transforming non standard data into standard format such as HL7. It allows routing of standard messages between multiple systems over various transfer protocols like TCP/IP, HTTP and SMTP. The software helps in building HL7 interfaces to non standard compliant applications. It connects two systems using a channel that can fetch data from some source, transform it according to rules set for the channel, filter information and then send it to the destination. All transactions in Mirth Connect can be stored in an internal database so as to group multiple data requests or to archive messages.

3.4 Hadoop

Apache Hadoop is an open source project for reliable, scalable, distributed processing and storage. Hadoop Distributed File System (HDFS) and MapReduce are the core services of Hadoop. HDFS is one of the file systems of Hadoop written in Java which divides a large file into small blocks and distribute them to various nodes called DataNodes. Redundant copies of each block can be kept in HDFS to avoid loss of data in case of failures. The master of HDFS cluster is called NameNode which maintains a database containing mapping from logical files to physical blocks in the DataNodes. Block replication is also managed by the NameNode (Cohen and Acharya 2014). MapReduce is data processing paradigm to get useful aggregated results from large volumes of data. In MapReduce first the records are processed by map task in a completely parallel manner and then output of map task becomes input to reduce task which merges results from different mappers to get an aggregated output (Yahoo Hadoop Tutorial).

3.5 Attribute Based Encryption

The patient sensitive data is stored on HDFS which needs to be secured as patient's privacy is recognized in laws such as health insurance portability and accountability act (HIPAA) (Narayan et al. 2010). To provide security we can store data in encrypted format so that if storage is compromised not much of data is lost. Attribute based Encryption (ABE) is a public key encryption technique which is useful when exact identities of people who can decrypt the data is unknown but only the descriptive attributes or credentials are known (Goyal et al. 2006). In Ciphertext-Policy ABE (CP-ABE) the keys have a set of attributes and ciphertext contains an access policy. The user can decrypt a particular ciphertext only if the user's key attribute satisfies the access policy contained in the ciphertext (Bethencourt et al. 2007).

4. DESIGN AND IMPLEMENTATION

The work assumes that the primary PHR of the patient resides on a portable device such as a mobile device (Sethia et al. 2014). In this architecture we provide overview of how backup can be retained on the cloud which can offer to reconstruct the healthcard in case of theft and also access some of the medical information in a secure manner using Attribute Encryption and using Distributed File system such as Hadoop. The section describes the overall architecture, different components involved along with their tasks and an overview of implementation of our work.

4.1 System Architecture

The entire system, as shown in Figure 1, has three main components: a data query and integration system, a middle layer of HL7 interface, and individual hospital information systems storing records in a distributed manner. The system can be used by clinical administrator to provide a backup for the complete medical record of an individual that can then be handed over to the person through a portable device.

Each local HIS stores medical data about patients who have an encounter with a physician affiliated with that hospital. This medical data may encompass information like patient's demographics, current vitals, medication history, prescriptions, diagnosis information, lab test results etc. They keep health records in an electronic format according to their information model and may provide an external interface for retrieval and update of records over the network. Each HIS is responsible for maintenance and availability of health records and it also ensures correctness of medical information stored.

Since these HISs may not share a common data model, health information of a patient gets distributed into heterogeneous data sources. The records pulled from these source needs to be translated into a common format so that they can be assimilated together while maintaining interoperability standards so as to support easy communication and data exchange with different healthcare professionals. HL7 is a feasible medical standard as it provides a common structure to medical information and separate HL7 string messages assembled together can represent complete medical history of the patient, while keeping the storage requirement to bare minimum. To retrieve all the medical records of a patient, we maintain a list of reference pointers at a centralized component. These pointers are used by the data query and integration system to request data from each HIS having patient's record. These pointers are maintained in reference file created for each patient registered in the system identified by an external patient identifier. As data received in response from different sources may not be in a standard format, we use HL7 interface layer to forward each query and then get back responses in standardized HL7 layout. All of the HL7 messages received by the system can be encapsulated together into a container format like JSON, representing the integrated EHR, and forwarded to the administrator. The integrated EHR with some pre-parsing of HL7 data can be used to realize the integrated PHR of the patient. The administrator then yields control of the record to the individual who keeps the record on his personal device ensuring confidentiality of records while enabling him to keep track of his medical information.

4.2 Reference Storage and Management

Reference pointers to all the HIS along with files such as lab reports, prescriptions and medications are stored on Hadoop Distributed File System to provide scalable, reliable storage in case of theft or damage to healthcard. The data stored in HDFS is encrypted to maintain patient privacy using Ciphertext-Policy Attribute Based Encryption as this method is closer to Role based access control (Bethencourt et al. 2007). While encrypting data we define the access policy consisting of attributes of people who can decrypt that data. Depending upon the role different parts of the data can be accessed by different people like an administrator have access to reference pointers to refurbish lost card whereas a pharmacist has access to only medication prescribed to the patient and nothing else.

4.3 Implementation

For implementation of the system in our local environment, we have used open source hospital information systems: OpenMRS, OpenEMR and GNU Health for simulating storage of health records at different locations but the system can use any HIS that supports export of patient's medical data. All of these three systems are built on different technologies and use different data models to store patient records. Both OpenMRS and GNU Health provide REST web services for retrieval of data in non HL7 format but there is no such provision in OpenEMR. We use Mirth Connect to build external HL7 interfaces to these data sources, enabling us to transform raw data into standard HL7 data. We connect to OpenEMR using a Database Connector so as to retrieve raw data as requested and then transform into HL7 records. Mirth Connect can also send these transformed records over the network to a desired destination. Each of these transformed records may signify an encounter or a visit to a healthcare professional and they are represented as a combination of ADT A28 and ORU R01 messages in HL7.

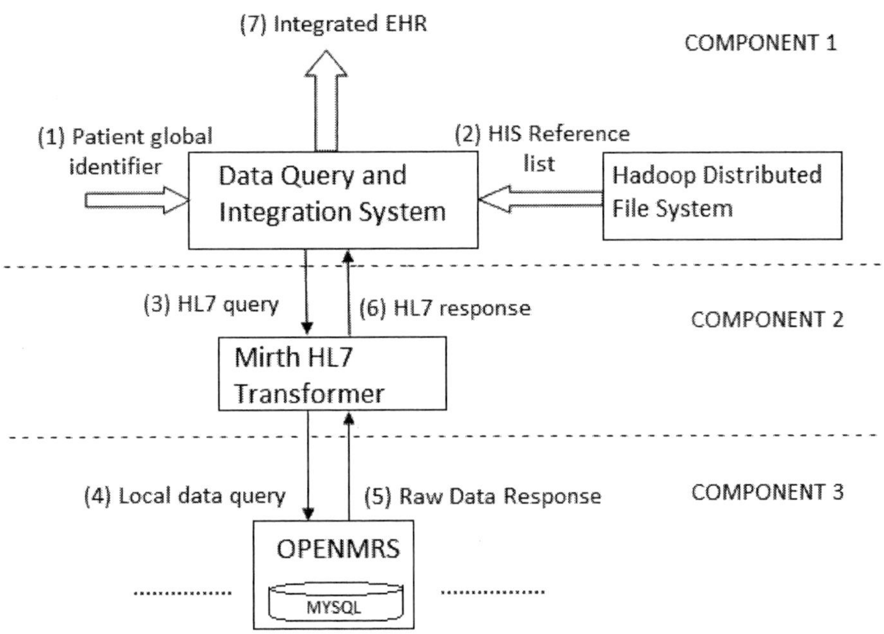

Figure 1. Overall Workflow with components used in the system

An overall workflow is given in Figure 1, in which the system first takes patient identifier as an input along with administrator username and password. The data query system then retrieves the encrypted reference file stored on HDFS and decrypted it using ABE. These references are then forwarded to HL7 interfaces built using Mirth Connect. Each interface then retrieves data from the mentioned source and translates it into appropriate format. The transformed message is then routed back to the system and integrated together inside a structured JSON. This JSON file is then handed back as the integrated healthcard of the individual.

5. RESULTS

HDFS is not geared up to efficiently access small files; it is primarily designed for access to large files. In Hadoop each map task processes a block of input at a time. If we have many small files, then each map task processes very little input and there are lot of map task each of which imposes extra bookkeeping overhead. So we need to merge small files into one large file. Merging will also reduce NameNode memory footprint and startup time, so to improve the performance we used Sequence files. In sequence files, small files are stored in the form of key value pair where filename of each small file is used as key and its content as value

We tested the availability of HDFS by reading decrypted references through Java multithreaded application created with Java version 1.7. We took files of size 1MB for testing the availability and response time of reference files when retrieved in a pseudo concurrent environment. Our hadoop cluster is established on cloud named as Baadal. It is an IITD cloud computing environment for high performance computing, an indigenously developed cloud orchestration and virtualization management software that can work with multiple virtualization technologies like KVM, Xen and VMware. The Hadoop cluster has 4 nodes, one NameNode and 3 DataNodes. Each node has hadoop 2.6.0 version and connected with 10Gbps Ethernet network.

Table 1. Local File System Vs HDFS Vs Sequence Files in HDFS

No. of files accessed	**Time taken by Local File System (sec)**	**Time taken by Hadoop DFS (sec)**	**Time taken by Sequence files (sec)**
1	0.02	0.09	0.001
10	2.56	1.2	0.008
100	123.07	25.28	0.07

Table 1 shows the time taken to access patient references when files of size 1 MB are stored on local file system, HDFS and in the form of sequence file on HDFS. Time taken to access references stored in Sequence file is less compared to those stored in Local file system and those stored as small files not merged into one large file on HDFS.

6. CONCLUSION AND FUTURE WORK

The personal health record of a person allows him to actively monitor his health while giving him direct control over his clinical information. A Patient visiting different healthcare facility will have his data distributed among multiple heterogeneous HISs while he keeps a copy of the record that gets consolidated into his PHR. Although the hybrid PHR model allows confidentiality of data but if, somehow, the healthcard is lost it becomes very complex to restore the PHR. The distributed mechanism discussed provides a secure way to manage and create a backup of the PHR with the help of trusted HISs. The reference model allows integration of information when required and the use of ABE to encrypt the references provides security to the backup system. Keeping these references on the HDFS allows for their high availability even on economical hardware.

Through implementation and testing in simulated environment, we can see that the proposed system successfully recreates the PHR of the individual in a reliable manner but realizing it in a real environment do pose certain challenges. Some of the major issues include trustworthiness of different components used and compliance of medical standards by the different healthcare organizations. In a country like India where private sector governs the healthcare industry and sharing of medical data is limited among competing organizations, it becomes essential that the control over the information is in the hands of the patient and he can allow sharing of data with other health professionals while also enabling him to watch over his complete medical history.

We plan to add pushing of references to the central system either by the PHR or by individual HISs. The system also faces certain technical challenges like appending of records in sequence files and improving response times for large media files. We plan to use Apache HBase for improving performance for reading and writing Big data and also to provide better response times for small files in addition to allowing appending of data.

REFERENCES

Apache HBase. Available at: http://hbase.apache.org/ [Accessed May, 10, 2015].

Baldock, D.J. et al, 2014. *Electronic health record sharing using hybrid architecture*. U.S. Patent No. 8650045.

Bethencourt, J. et al, 2007. Ciphertext-policy attribute-based encryption. *IEEE Symposium on Security and Privacy, SP'07,* Berkeley, CA, pp 321-334.

Borthakur, D. 2008. *HDFS Architecture Guide*. Available at: https://hadoop.apache.org/docs/r1.2.1/hdfs_design.html [Accessed May, 10, 2015]

Campion-Awwad, O. et al, 2014. *The National Programme for IT in the NHS, A Case history*. Available at: https://www.cl.cam.ac.uk/~rja14/Papers/npfit-mpp-2014-case-history.pdf

Cohen, J. C. and Acharya, S., 2014. Towards a trusted HDFS storage platform: Mitigating threats to Hadoop infrastructures using hardware-accelerated encryption with TPM-rooted key protection. *In Journal of Information Security and Applications*, Vol. 19, No. 3, pp 224-244.

Daglish, D. and Archer, N., 2009. Electronic personal health record systems: a brief review of privacy, security, and architectural issues. *IEEE World Congress on Privacy, Security, Trust and the Management of e-Business, CONGRESS'09,* Saint John, NB, pp 110-120.

Detmer, D. et al, 2008. Integrated personal health records: transformative tools for consumer-centric care. *In BMC medical informatics and decision making*, Biomed Central, Vol 8, No 1.

Dinh, T.D., 2009. *Hadoop Performance Evaluation.* Unpublished thesis. Ruprecht-Karls University at Heidelberg.

Force, G. E. T., 2010. *Electronic health records: A global perspective Second Edition.* Technical report, HIMSS Enterprise Systems Steering Committee.

Goyal, V. et al, 2006. Attribute-based encryption for fine-grained access control of encrypted data. *Proceedings of the 13th ACM conference on Computer and communications security*, pp 89-98.

GNU Health – Freedom and Equity in Healthcare. Available at: *http://health.gnu.org/* [Accessed May, 1, 2015]

Halamka, J.D. et al, 2008. Early experiences with personal health records. *Journal of the American Medical Informatics Association*, Vol. 15, No. 1, pp 1-7.

Kaelber, D.C. et al, 2008. A research agenda for personal health records (PHRs). *Journal of the American Medical Informatics Association,* Vol. 15, No. 6, pp 729-736.

Krohn, R., 2007. The consumer-centric personal health record—it's time. *Journal of Healthcare Information Management,* Vol. 21, No. 1, pp 21-23.

Mirth Connect. Available at: http://www.mirth.com/Products-and-Services/Mirth-Connect [Accessed March, 10, 2015]

Noordende, G.V., 2010. Security in the Dutch electronic patient record system. *Proceedings of the second annual workshop on Security and privacy in medical and home-care systems*, ACM, pp 21-32.

Noordende, G.V., 2011. Controlled dissemination of electronic medical records. *Proceedings of the 2nd USENIX conference on Health security and privacy*, USENIX Association, pp1-4.

Narayan, S. et al, 2010. Privacy preserving EHR system using attribute-based infrastructure. *Proceedings of the 2010 ACM workshop on Cloud computing security workshop*, ACM, pp 47-52.

OpenEMR Project. Available at: *http://www.open-emr.org/*[Accessed April, 22, 2015]

OpenMRS. Available at: *http://openmrs.org* [Accessed April, 28, 2015]

Raymond, B., 2007. Realizing the Transformative Potential of Personal Health Records. *Kaiser Permanente Institute for Health Policy, In Focus*.

Sethia, D. et al, 2014. NFC based secure mobile healthcare system. *IEEE Sixth International Conference on Communication Systems and Networks (COMSNETS)*, Bangalore, pp 1-6

Spine Services-Health and Social Care Information Centre. Available at: *http://systems.hscic.gov.uk/spine*[Accessed March, 10, 2015]

Sprague, L., 2006. Personal health records: the people's choice. *NHPF Issue Brief,* No. 820, pp 1-13

Tang, P. C. et al, *2006*. Personal health records: definitions, benefits, and strategies for overcoming barriers to adoption. *Journal of the American Medical Informatics Association,* Vol. 13, No. 2, pp 121-126.

The Small files problem. Available at: *http://blog.cloudera.com/blog/2009/02/the-small-files-problem/* [Accessed July, 1, 2015]

Wang, Y. et al, 2014. Design and Development of a Clinical Data Exchange System Based on Ensemble Integration Platform. *Frontier and Future Development of Information Technology in Medicine and Education*, Springer, Netherlands, pp 385-392.

Wuerdeman, L. et al, 2005. How accurate is information that patients contribute to their electronic health record*? AMIA Annual Symposium Proceedings,* pp 834-838.

Yahoo! Hadoop tutorial. Available at: *http://developers.yahoo.com/hadoop/tutorial* [Accessed July, 22, 2015].

INTERNATIONAL DATA TRANSFER FROM THE EUROPEAN UNION TO THE UNITED STATES

Nathalie Devillier
Grenoble Ecole de Management, France

ABSTRACT

The international transfer relies on the Data Protection Directive. It requires special precautions when personal data is transferred to countries outside the European Economic Area (EEA, which includes all EU countries and in addition, non-EU countries Iceland, Liechtenstein and Norway) that do not provide EU-standard data protection. This article describes the different available exceptions or derogations with a special focus on transfer from the EU to the United States. Data transfer to the US used to rely on a Safe Harbor Agreement adopted by the European Commission in 2000. In October 2015, the European Court of Justice ruled this decision was invalid which deprived the transfer of its legal basis. On Feb. 2, 2016 the European Commission announced it has reached a political agreement with the US. This article aims at highlighting the terms and conditions of such transfers in this new context.

KEYWORDS

Data protection, European Regulation, Law, Privacy Shield, United States.

1. INTRODUCTION

The international data transfer raises both legal and political issues. The United States share the goal of improving the protection of data of their citizens, but do not address this issue in the same way as in the European Union. While the Common Law of each US State recognizes the right to privacy, federal statutes offer a protection of personal data in some sectors (health, finance, child protection…). From an international law perspective, in 1980, the OECD issued recommendations for protection of personal data in the form of 7 principles. The international transfer of personal data from the European Union to countries outside the European Economic Area (EEA, which includes all EU countries and in addition, non-EU countries Iceland, Liechtenstein and Norway) relies on the Data Protection Directive.[1] According to the Directive, companies operating in the European Union are not permitted to send personal data to "third countries" outside the EEA, unless they guarantee adequate levels of protection and require special precautions when the recipient country does not provide EU-standard data protection. The Safe Harbor Privacy Principles were developed from 1998 to 2000 and were designed to prevent private organizations within the European Union or United States which store customer data, from accidentally disclosing or losing personal information. Concretely, US companies could opt into a program and be certified if they adhered to seven principles. In July 2000, the European Commission (EC) decided that US companies complying with the principles and registering their certification that they met the EU requirements, the so-called "safe harbor scheme", were allowed to transfer data from the EU to the US. This is referred to as the Safe Harbor Decision. The European Court of Justice (ECJ) recently declared this decision invalid since "legislation permitting the [US] public authorities to have access on a generalized basis to the content of electronic communications must be regarded as compromising the essence of the fundamental right to respect for private life." As a result, the EC and the US needed to negotiate a new Safe Harbor Agreement. European and U.S. negotiators agreed a data pact called "Privacy Shield" on Feb. 2, 2016 that should prevent European Union regulators from restricting data transfers by companies such as Google and Amazon across the Atlantic. Meanwhile, the EC put forward its EU Data

---

[1] Directive 95/46/EC of the European Parliament and of the Council of 24 October 1995 on the protection of individuals with regard to the processing of personal data and on the free movement of such data, *Official Journal* L 281 , 23/11/1995 P. 0031 – 0050.

Protection Reform in January 2012 to make Europe fit for the digital market. The General Data Protection Regulation (GDPR) will be passed by the European Parliament during spring 2016.[2]

This arises the complex political context of counter terrorism, the Transatlantic Trade and Investment Partnership ("TTIP") negotiations, alongside with technological and health expansion (big data and Quantified self).

Previous works included the analysis of : Health data Confidentiality and Security in the European Union (Ferraud-Ciandet, 2009), Health the legal background of eHealth services in the European Union (Ferraud-Ciandet, 2010), the analysis of eHealth and telemedicine French legal framework (Ferraud-Ciandet, 2011), the Electronic Health Record at the EU level (Devillier, 2014), the GDPR analysis (Devillier, 2016-2) and the consequences of the Safe Harbor invalidation (Devillier, 2016-1).

5554 U.S. companies in particular tech giants like Google, Microsoft, and Facebook, rely heavily on the easy flow of data. They are therefore primarily concerned with reestablishing a legal framework for data transfers; still the real issue for Europeans is mass surveillance by government. U.S. spying has been a contentious issue for European citizens ever since Edward Snowden revealed the extent of NSA surveillance. The purpose of this study is to clarify the legal framework of international health data transfer from the EU to the US jointly with the Directive, the Privacy Shield and the GDPR. This study will concretely assess the Privacy Shield's content with regard to ECJ conclusions: does it really offer an adequate level of protection? Is the Privacy Shield providing EU residents effective judicial redress options in case of privacy violations?

This paper is therefore an up-to-date contribution to the topic and offers a comprehensive legal perspective also intelligible for non-specialist of this field of study.

2. INTERNATIONAL DATA TRANSFER FROM THE SAFE HARBOR DECISION TO THE PRIVACY SHIELD AGREEMENT

From a European law perspective, both the Directive and the GDPR offer options to allow international data transfer from the EU to non-EEE states and therefore the US (2.1). The Safe Harbor Decision adopted in 2000 by the EC provided a legal basis to such transfer, including for health data; still the ECJ ruled it was invalid and was compromising the essence of the fundamental right to respect for private life (2.2). Now, the Privacy Shield replaces the Safe Harbor: what's its content? Does it offer adequate guarantees to protect EU citizens' privacy? (2.3)

2.1 Available Options for Data Transfer from the EU to the US

Special precautions need to be taken when personal data is transferred to countries outside the EEA that do not provide EU-standard data protection. Without such precautions, the high standards of data protection established by the Data Protection Directive would quickly be undermined, given the ease with which data can be moved around in international networks.

2.1.1 Options under the Directive

The Directive 95/46/EC is the reference text, at European level, on the protection of personal data. It sets up a regulatory framework which seeks to strike a balance between a high level of protection for the privacy of individuals and the free movement of personal data within the EU (Bennet, 1997 and Wong, 2012). This Directive applies to data processed by automated means (e.g. a computer database of customers) and data contained in or intended to be part of non automated filing systems (traditional paper files).[3] The Directive aims to protect the rights and freedoms of persons with respect to the processing of personal data by laying down the key criteria for making processing lawful and the principles of data quality.

---

[2] European Commission - Press release, Agreement on Commission's EU data protection reform will boost Digital Single Market, IP/15/6321, Dec. 15, 2015.

[3] It does not apply to the processing of data by a natural person in the course of purely personal or household activities or in the course of an activity which falls outside the scope of Community law, such as operations concerning public security, defence or State security.

One of the **principles of data quality**, is that personal data must be processed **fairly and lawfully**, and collected for specified, explicit and legitimate purposes. They must also be adequate, relevant and not excessive, accurate and, where necessary, kept up to date, must not be stored for longer than necessary and solely for the purposes for which they were collected. Special **categories** of processing are subject to specific requirements, in particular data concerning health. This provision comes with certain qualifications concerning, for example, cases where processing is necessary to protect the vital interests of the data subject or for the purposes of preventive medicine and medical diagnosis (art. 8).

Transfers of personal data from a Member State **to a third country** with an adequate level of protection are authorized. However, although transfers may not take place when an adequate level of protection is not guaranteed, there are a number of exceptions to this rule listed in the Directive, e.g. the data subject himself agrees to the transfer, in the event of the conclusion of a contract, it is necessary for public interest grounds, but also if Binding Corporate Rules (BCRs) or Standard Contractual Clauses (SCCs) have been authorized by the Member State (art. 25 – 26). The Commission may find that a third country ensures an adequate level of protection within the meaning of art. 25 paragraph 2, by reason of its domestic law or of the international commitments it has entered into, particularly upon conclusion of negotiations, for the protection of the private lives and basic freedoms and rights of individuals. Member States shall take the measures necessary to comply with the Commission's decision. Every person shall have the right to a **judicial remedy** for any breach of the rights guaranteed by national law applicable to the processing in question. In addition, any person who has suffered damage as a result of the unlawful processing of their personal data is entitled to receive compensation for the damage suffered (art. 22).

As a result, there were different options available to allow the international data transfer from the EU to the US, the EC decision being the widest channel to articulate EU and US views on privacy.

2.1.2 Options under the GDPR

The GDPR is a Regulation in the making by which the European Commission intends to strengthen and unify data protection for individuals within the EU. It also addresses export of personal data outside the EU. The Commission's primary objectives of the GDPR are to give citizens back the control of their personal data and to simplify the regulatory environment for international business by unifying the regulation within the EU. When GDPR takes effect in 2018, it will replace the data protection directive.

The proposal for the European Data Protection Regulation contains the following key changes: the Regulation applies to organizations based inside or outside the European Union provided they process personal data of EU residents. It gives a wide definition of personal data which must be read jointly with the Article 29 data Protection Working Party letter and annex of Feb.5, 2015 on Health data in apps and devices.[4]

Transfers of personal data to a third country or an international organization must ensure that the level of protection of individuals guaranteed by the Regulation is not undermined (art. 40). They may take place where the Commission has decided that the third country, or a territory or one or more specified sectors within that third country, or the international organization in question ensures an adequate level of protection (art. 41). When assessing the adequacy of the level of protection, the Commission shall, in particular, take account of effective and enforceable data subject rights and effective administrative and judicial redress for the data subjects whose personal data are being transferred (art. 41-2-a). The third country should ensure effective independent data protection supervision, provide for cooperation mechanisms with the European data protection authorities, and the data subjects should be provided with effective and enforceable rights and effective administrative and **judicial redress.** Indeed, the GDPR also creates a right for each data subject to an **effective judicial remedy** if they consider that their rights under the Regulation have been infringed as a result of the processing of their personal data in noncompliance with this Regulation (art. 75).

In the absence of such a decision, a controller or processor may transfer personal data to a third country or an international organization only if it has adduced **appropriate safeguards**, and on condition that enforceable data subject rights and effective legal remedies for data subjects are available (art. 42) such as BCRs and SCCs, an approved Code of conduct or an approved Certification mechanism. Other **derogations** similar to the directive are still valid such as the necessary transfer in order to protect the vital interests or for the conclusion or performance of a contract concluded in the interest of the data subject (art. 44).

[4] Letter of Feb.5, 2015 [http://ec.europa.eu/justice/data-protection/article-29/documentation/other-document/files/2015/20150205_letter_art29wp_ec_health_data_after_plenary_en.pdf] and annex [http://ec.europa.eu/justice/data-protection/article-29/documentation/other-document/files/2015/20150205_letter_art29wp_ec_health_data_after_plenary_annex_en.pdf].

As a consequence, the international data transfer from the EU to the US, may rely on ane EC decision such as the Safe Harbor 2.0 or Privacy Shield provided it meets the RGDP above requirements.

2.2 International Data transfer under the Safe Harbor Decision

2.2.1 Safe Harbor's Content and Functioning

The EC decided that US companies complying with the principles and registering their certification that they met the EU requirements, the so-called "safe harbor scheme", were allowed to transfer data from the EU to the US (Long, 2002 and Cline, 2006). This is referred to as the Safe Harbor Decision.[5] The seven principles are the following: Notice - Individuals must be informed that their data is being collected and about how it will be used, Choice - Individuals must have the option to opt out of the collection and forward transfer of the data to third parties, Onward Transfer - Transfers of data to third parties may only occur to other organizations that follow adequate data protection principles, Security - Reasonable efforts must be made to prevent loss of collected information, Data Integrity - Data must be relevant and reliable for the purpose it was collected for, Access - Individuals must be able to access information held about them, and correct or delete it if it is inaccurate, Enforcement - There must be effective means of enforcing these rules.

Only U.S. organizations regulated by the Federal Trade Commission or the Department of Transportation may participate in this voluntary program. After opting in, an organization must have appropriate employee training and an effective dispute mechanism in place, and self re-certify every 12 months in writing that it agrees to adhere to the U.S.-EU Safe Harbor Framework's principles, including notice, choice, access, and enforcement.[6] It can either perform a self-assessment to verify that it complies with the principles, or hire a third-party to perform the assessment. Companies pay an annual $100 fee for registration except for first time registration ($200).

The U.S. government does not regulate Safe Harbor, which is **self-regulated through its private sector members** and the dispute resolution entities they pick. The Federal Trade Commission manages the system under the oversight of the U.S. Department of Commerce to comply with the commitments can be penalized under the Federal Trade Commission Act by administrative orders and civil penalties of up to $16,000 per day for violations. If an organization fails to comply with the framework it must promptly notify the Department of Commerce, or else it can be prosecuted under the *False Statements Act*[7].

The Safe Harbor agreement allowed companies to self-certify when transferring data to their U.S. servers, which could then be accessed by U.S. law enforcement. For instance, in Microsoft's lawsuit against the U.S. government, the government said that because it's a U.S. company, it doesn't matter where the user is from -- it can just get that data.

This scheme has been criticized in a 2002 review by the European Union which found "a substantial number of organizations that have self-certified adherence to the Safe Harbor do not seem to be observing the expected degree of transparency as regards their overall commitment or as regards the contents of their privacy policies" and that "not all dispute resolution mechanisms have indicated publicly their intention to enforce Safe Harbor rules and not all have in place privacy practices applicable to themselves".[8] In 2004, the Commission services expressed concerns about the number of self-certified organizations that have not published a privacy policy or that have published a policy that is not compliant with the Principles.[9]

Again in 2013, the EC noted that 'safeguards that are provided under US law are mostly available to US citizens or legal residents' and that, '[m]oreover, there are no opportunities for either EU or US data subjects

---

[5] Commission Decision of 26 July 2000 pursuant to Directive 95/46/EC of the European Parliament and of the Council on the adequacy of the protection provided by the safe harbor privacy principles and related frequently asked questions issued by the US Department of Commerce (notified under document number C(2000) 2441) 2000/520/EC, *Official Journal L 215 , 25/08/2000 P. 0007 – 0047.*

[6] This might be checked on the U.S.-EU Safe Harbor Overview official website: http://www.export.gov/safeharbor/eu/eg_main_018476.asp.

[7] The False Claims Act (FCA), 31 U.S.C. §§ 3729 – 3733, online [www.justice.gov/.../C-FRAUDS_FCA_Primer.pdf].

[8] European Commission (2002) The application of Commission Decision on the adequate protection of personal data provided by the Safe Harbor Privacy Principles, SEC(2002)196, Feb. 13, 2002, [http://web.archive.org/web/20060724174359/http://www.ec.europa.eu/justice_home/fsj/privacy/docs/adequacy/sec-2002-196/sec-2002-196_en.pdf].

[9] European Commission, The implementation of Commission Decision on the adequate protection of personal data provided by the Safe Harbor Privacy Principles, SEC(2004)1323, Oct. 20, 2004, [http://web.archive.org/web/20060724173657/http://www.ec.europa.eu/justice_home/fsj/privacy/docs/adequacy/sec-2004-1323_en.pdf].

to obtain access, rectification or erasure of data, or administrative or judicial redress with regard to collection and further processing of their personal data taking place under the US surveillance programmes'.[10] Also, the Commission noted that the data subjects had no administrative or judicial means of redress enabling, in particular, the data relating to them to be accessed and, as the case may be, rectified or erased.

Moreover, an report to the President Obama raised the question whether US law sufficiently respects the legitimate privacy interests of non US citizens. It also pointed that "the United States should be a leader in championing the protection by all nations of fundamental human rights, including the right of privacy, which is central to human dignity".[11] It therefore recommended that in the absence of a specific and compelling showing, the US Government should follow the model of the Department of Homeland Security, and apply the Privacy Act of 1974 in the same way to both US persons and non-US person which means equal access to judicial redress.

2.2.2 ECJ ruling's Impact on Safe Habor

The ECJ Case. Maximilian Schrems issued a complaint to the Irish supervisory authority (Data Protection Commissioner), after the Snowden revelations in 2013, that his Facebook data was being processed on U.S. servers where it's not adequately protected. The complaint was rejected because in a decision from July 2000, the European Commission believed that under the Safe Harbor agreement, the U.S. ensures sufficient protection for European citizens' data. Therefore, the Irish body believed it can't further investigate his complaint. However, the Irish High Court later questioned whether the European Commission can override a national government from investigating such complaints. Now, the European Court of Justice has ruled that a European Commission decision "cannot eliminate or even reduce the powers available to the national supervisory authorities under the Charter of Fundamental Rights of the European Union and the directive." The ECJ also noted that the Safe Harbor agreement didn't have strong limitations that would prevent the U.S. government from accessing EU citizens' data unless it was for strict national security purposes and added that a legislation permitting the public authorities to have access on a generalized basis to the content of electronic communications must be regarded as compromising the essence of the fundamental right to respect for private life." The Court said that the Safe Harbor agreement didn't provide any way for EU citizens to pursue legal remedies to access, rectify or erase data about them, which "compromises the essence of the fundamental right to effective judicial protection." Finally, the Court ruled that the Irish supervisory authority must examine Mr. Schrems' case to decide whether to suspend the transfer of European Union subscribers' data to the United States, on the grounds that the data can't be guaranteed adequate protection.[12]

The ECJ judgment means that the 2000 **Safe Harbor law could not supersede scrutiny at the national level** (Malgieri, 2016). Such agreements "cannot eliminate or even reduce the national supervisory authorities' powers… if the national supervisory authorities receive individual complaints, that does not (…) prevent them, by virtue of their investigative powers and their independence, from forming their own opinion on the general level of protection ensured by a third country and from drawing the appropriate conclusion when they determine individual cases."[13] It also reiterated that the absence of administrative or judicial means of redress "does not respect the essence of the fundamental right to effective judicial protection, as enshrined in Article 47 of the Charter" (par.95).[14].

The G29 declared: "If by the end of January 2016, no appropriate solution is found with the US authorities and depending on the assessment of the transfer tools by the Working Party, EU data protection

---

[10] EC Communication, Limitations and redress possibilities, Nov. 27, 2013, COM(2013) 847 final, point 7.2, [http://eur-lex.europa.eu/resource.html?uri=cellar:551c0723-784a-11e3-b889-01aa75ed71a1.0001.01/DOC_1&format=PDF].

[11] Liberty and Security in a Changing World, Report and Recommendations to the President's Review Group on Intelligence and Communications technologies, Dec. 12, 2013, p.155 [https://www.whitehouse.gov/sites/default/files/docs/2013-12-12_rg_final_report.pdf].

[12] Case C-362/14, Maximillian Schrems v. Data Protection Commissioner, Judgment of the Court, Oct. 6, 2015, online [http://curia.europa.eu/juris/document/document.jsf?docid=169195&doclang=EN].

[13] Case C-362/14, Maximillian Schrems v. Data Protection Commissioner, Sept. 15, 2015, Opinion of Advocate General Bot, par. 61, online [http://eur-lex.europa.eu/legal-content/EN/TXT/?uri=CELEX%3A62014CC0362].

[14] The first paragraph of Article 47 of the Charter requires everyone whose rights and freedoms guaranteed by the law of the European Union are violated to have the right to an effective remedy before a tribunal in compliance with the conditions laid down in that article. The very existence of effective judicial review designed to ensure compliance with provisions of EU law is inherent in the existence of the rule of law (see, to this effect, judgments in *Les Verts* v *Parliament*, 294/83, EU:C:1986:166, paragraph 23; *Johnston*, 222/84, EU:C:1986:206, paragraphs 18 and 19; *Heylens and Others*, 222/86, EU:C:1987:442, paragraph 14; and *UGT-Rioja and Others*, C-428/06 to C-434/06, EU:C:2008:488, paragraph 80).

authorities are committed to take all necessary and appropriate actions, which may include coordinated enforcement actions. (...) transfers that are still taking place under the Safe Harbor decision after the CJEU judgment are unlawful."[15] On 6 November, 2015 the Commission issued guidance for companies on the possibilities of transatlantic data transfers following the ruling until a new framework is put in place.[16] On 2 December, 2015 the College of Commissioners discussed the progress of the negotiations and Commissioner Jourová received a mandate to pursue the negotiations on a renewed and safe framework with the US.

2.3 International Data Transfer under the Privacy Shield Agreement

2.3.1 Privacy Shield Content

To date, the full version of the negotiations was not made public. The EC just published a press release stating the EU-US Privacy Shield "reflects the requirements set out by the European Court of Justice in its ruling on 6 October 2015." [17] It is supposed to provide stronger obligations on companies in the U.S. to protect the personal data of Europeans and stronger monitoring and enforcement by the U.S. Department of Commerce and Federal Trade Commission, including through increased cooperation with European Data Protection Authorities in order to prevent a generalized access.

The new arrangement will include three key elements. First, **strong obligations on companies handling Europeans' personal data and robust enforcement:** U.S. companies wishing to import personal data from Europe will need to commit to robust obligations on how personal data is processed and individual rights are guaranteed. The Department of Commerce will monitor that companies publish their commitments, which makes them enforceable under U.S. law by the US. Federal Trade Commission. In addition, any company handling human resources data from Europe has to commit to comply with decisions by European DPAs.

Second, **clear safeguards and transparency obligations on U.S. government access. F**or the first time, the US has given the EU written assurances that the access of public authorities for law enforcement and national security will be subject to clear limitations, safeguards and oversight mechanisms. These exceptions must be used only to the extent necessary and proportionate. The U.S. has ruled out indiscriminate mass surveillance on the personal data transferred to the US under the new arrangement. To regularly monitor the functioning of the arrangement there will be an annual joint review, which will also include the issue of national security access. The European Commission and the U.S. Department of Commerce will conduct the review and invite national intelligence experts from the U.S. and European Data Protection Authorities to it.

And third, **effective protection of EU citizens' rights with several redress possibilities.** Any citizen who considers that their data has been misused under the new arrangement will have several redress possibilities. Companies have deadlines to reply to complaints. European DPAs can refer complaints to the Department of Commerce and the Federal Trade Commission. In addition, Alternative Dispute resolution will be free of charge. For complaints on possible access by national intelligence authorities, a new Ombudsperson will be created.

The Vice-President **Ansip** and Commissioner **Jourová** are to prepare a draft adequacy decision in the first weeks of 2016, which could then be adopted after obtaining the advice of the Article 29 Working Party and after consulting a committee composed of representatives of the Member States. In the meantime, the U.S. side will make the necessary preparations to put in place the new framework, monitoring mechanisms and new Ombudsman. The final agreement will have to undergo the EU process of "comitology" — review by committees of data protection regulators, EU member states and others before it can take effect. And then it will face inevitable legal challenge in European courts and the Court of Justice again.

The legal strenght of these exchange of letters and verbal agreement is quite low: it is even not an agreement, and even less a formal decision as required by both the directive and the future RGDP. There is no official agreement yet.

---

[15] Statement of the Article 29 Working Party, Oct. 16, 2015, [http://www.cnil.fr/fileadmin/documents/Communications/20151016_wp29_statement_on_schrems_judgement.pdf].

[16] EC, Communication on the transfer of personal data from the Eu to the United States (Schrems), COM(2015)566 final, Nov. 6, 2015, [http://ec.europa.eu/justice/data-protection/international-transfers/adequacy/files/eu-us_data_flows_communication_final.pdf].

[17] EC, EU Commission and United States agree on new framework for transatlantic data flows: EU-US Privacy Shield, feb. 2, 2016, IP/16/216.

2.3.2 Privacy Shield in the Light of the Schrems' Case

The ECJ judgment ade clear it is necessary to respect the essence of the fundamental right to effective judicial protection enshrined in Article 47 of the Charter. The Privacy Shield affirms that any citizen who considers that their data has been misused under the new arrangement will have several redress possibilities still it doesn't explicitly says that EU residents will have the same redress options as US citizens (i.e. available under the Privacy Act). Instead, an Ombudsperson will be created which is an alternative dispute resolution system (ADR).

Mediation is not a judicial redress. It is the fastest growing ADR method. Unlike litigation, mediation provides a forum in which parties can resolve their own disputes, with the help of a neutral third party. Mediation depends upon the commitment of the disputants to solve their own problems. The mediator, also known as a facilitator, never imposes a decision upon the parties. Rather, the mediator's job is to keep the parties talking and to help move them through the more difficult points of contention. To do this, the mediator typically takes the parties through several stages. At the end, when the parties agree upon the broad terms of a settlement, they formally reaffirm their understanding of that settlement, complete the final details, and sign a settlement agreement. As a consequence, the agreement is nothing else but a contract, not a judgment. One may question the opportunity of having such dispute regarding privacy solved by mediation. Will a citizen be able to face a tech giant such as Google ?

This creates an arbitrary discrimination between data subjects. In the same situation, US citizens are entitled to sue US companies with a class action whereas EU citizens will only have a mediation system. Moreover, EU law recognizes US citizen have the same rights to sue businesses at EU courts which reinforces the discrimination. This might change since on Jan. 2, 2016, the **Judicial Redress Act** was placed on US Senate Legislative Calendar. This bill authorizes the Department of Justice (DOJ) to designate foreign countries or regional economic integration organizations whose natural citizens may bring civil actions under the Privacy Act of 1974 against certain U.S. government agencies for purposes of accessing, amending, or redressing unlawful disclosures of records transferred from a foreign country to the United States to prevent, investigate, detect, or prosecute criminal offenses. DOJ, with the concurrence of the Department of State, the Department of the Treasury, and the Department of Homeland Security, may designate countries or organizations whose citizens may pursue such civil remedies if the person's country or organization has appropriate privacy protections for sharing information with the United States.[18] Still Republican senators are looking to insert a provision that would oblige the Attorney General to certify whether a country whose citizens will have redress don't have policies that endanger U.S. national security. Therefore, the proposition of having effective judicial redress looks iffy.

Again, details about those "robust obligations" in the Privacy Shield have not been revealed. The concrete compliance of the Privacy Shield with the Schrems judgment will therefore be possible only once it will be published. There's also a suspicion that the verbal agreement, along with the annual certifications it contains, is more intended to keep European courts from getting involved than to lock in any real improvement in data privacy. Adding to the complexity of the agreement between the EU and the U.S. is the problem that it must be consistent with the RGDP independent of the Privacy Shied agreement.

3. CONCLUSION

The Privacy Shield offers a temporary legal basis to maintain international data transfer from the EU to the US. Still, since its content is not fully made public the concrete advantages and risks are difficult to identify. From the EC press release and official sources, it looks like the judicial redress options required by the ECJ ruling are not available. Instead, the Privacy Shield proposes mediation which is a completely different dispute settlement system. In the absence of a complete transparency from the EU and US authorities on the Privacy Shield's content in terms of obligations and remedies, it remains unclear whether it complies or not with the Schrems judgment. Civil society, citizens, companies should therefore pay particular attention to the forthcoming agreement, just like lawyers do.

---

[18] H.R.1428 - Judicial Redress Act of 2015, 114th Congress (2015-2016), [https://www.congress.gov/bill/114th-congress/house-bill/1428].

REFERENCES

Book

Ferraud-Ciandet N., 2009. *Protection de la santé et sécurité alimentaire en droit international*. Larcier, Bruxelles.

Ferraud-Ciandet N., 2011. *Droit de la télésanté et de la telemedicine.* Heures de France, Paris.

Book Chapter

Ferraud-Ciandet N., 2013. *Éthique et droit de la télémédecine en Argentine* In *Les systèmes informatisés complexes en santé*, Dalloz, Paris.

Journal

Bellman S., Johnson E. J., Kobrin S. J. & Lohse G. L., 2004. International Differences in Information Privacy Concerns: A Global Survey of Consumers, *The Information Society: An International Journal,* vol. 20, No. 5, pp 313-324.

Bennet C. J. & Raab C. D., 1997. The Adequacy of Privacy: The European Union Data Protection Directive and the North American Response, *The Information Society: An International Journal,* vol. 13, No. 3, pp 245-264.

Cline J., 2006. Roadmap for an International Safe Harbor framework, *International Review of Law, Computers & Technology*, vol. 20, No. 3, pp 361-376.

Devillier N., 2016-1. Des négociations UE–États-Unis tendues autour des données personnelles. *The Conversation*, online [https://theconversation.com/des-negociations-ue-etats-unis-tendues-autour-des-donnees-personnelles-53753] retrieved on Feb.6, 2016.

Devillier N., 2016-2. Les citoyens européens prêts à contrôler leurs données personnelles face aux géants du web. *The Conversation*, online [https://theconversation.com/les-citoyens-europeens-prets-a-controler-leurs-donnees-personnelles-face-aux-geants-du-web-52666] retrieved on Feb.6, 2016.

Devillier N., 2014. De la gestion à l'optimisation des données de santé : L'état d'avancement des dossiers médicaux électroniques, *Journal International de Bioéthique,* vol. 25, No. 3, pp 81-96.

Ferraud-Ciandet N., 2010. L'Union européenne et la télésanté, *Revue Trimestrielle de Droit Européen*, vol. 3, No. 3, pp 537-561.

Ferraud-Ciandet N., 2009. Health Data's Security and Confidentiality in the European Union: The Legal Framework of Medical records', *European Law Journal,* vol. 1, No. 1, pp 12-14.

Long W. J. & Quek M. P., 2002. Personal data privacy protection in an age of globalization: the US-EU safe harbor compromise, *Journal of European Public Policy,* vol. 9, No. 3, pp 325-344.

Malgieri G., 2016. Trade Secrets v Personal Data: a possible solution for balancing rights, International Data Privacy Law, Jan. 29, 2016, online, doi: 10.1093/idpl/ipv030.

Wong R., 2012. The Data Protection Directive 95/46/EC: Idealisms and realisms, International Review of Law, *Computers & Technology,* vol. 26, No. 3, pp 229-244.

A COMPETENCE BASED MSC IN HEALTH INFORMATICS FOR RWANDA

Frank Verbeke [a,b], Graham Wright [c] and Marc Nyssen [a,b]
[a] *Department of Biostatistics and Medical Informatics, Vrije Universiteit Brussel, Brussels, Belgium*
[b] *Regional e-Health Center of Excellence, University of Rwanda, Kigali, Rwanda*
[c] *Centre for Health Informatics Research and Development, University of Fort Hare, South Africa*

ABSTRACT

A MSc in Health Informatics program (MSc HI) has been running at the Regional e-Health Center of Excellence of Rwanda (REHCE) since 2011. A review of the program was ordered in February 2014, when 2 cohorts of students had completed the post-graduate certificate and diploma courses and most of them had started preparatory activity for their master thesis.

The study developed a method for mapping course content on health informatics competences and knowledge units, enabling identification and measurement of knowledge gaps and content redundancy. Using this method, the analysis of regulatory and program documents combined with stakeholder interviews demonstrated that the existing MSc HI curriculum did not completely address the needs of the Rwandan health sector and teaching strategies did not always match student's expectations. Based on a detailed Rwandan health informatics needs assessment, IMIA's *Recommendations on Education in Biomedical and Health Informatics* and the *IMIA Health Informatics Knowledge Base*, a new curriculum was developed, providing a better match with the specifics of healthcare in the Central African region. The new curriculum has been put into practice in the 2014/2015 academic year and options for regional extension of the program to Eastern DRC (Bukavu) and Burundi (Bujumbura) are being investigated.

KEYWORDS

Education, Medical Graduate, Teaching, Curriculum, Knowledge

1. INTRODUCTION

A MSc in Health Informatics program (MSc HI) has been running at the Regional e-Health Center of Excellence of Rwanda (REHCE) since 2011. It was originally hosted at the Kigali Institute of Science and Technology and then moved to the Kigali Health Institute that was integrated in 2014 in the College of Medicine and Health Sciences (CMHS) of the University of Rwanda. The main objective of this 2-year part-time program is to prepare students graduating from the MSc HI program to provide leadership in the Health Informatics field. The program combines modules in computer science, health informatics, and health management. The core Health Informatics curriculum includes modules in Health Information Systems, Information Storage, Data retrieval and Management, Research Methods and Outcomes Analysis. Special areas of concentration include Electronic Health-care Records, Health Information Systems and Health Enterprise Architecture.

A review of the program was ordered in February 2014, when 2 cohorts of students had completed the post-graduate certificate and diploma courses and most of them had started preparatory activity for their master thesis.

2. MATERIALS AND METHODS

The review activities consisted of:

- A study of national regulations and documents relevant to the MSc HI program
- A study of the MSc HI program documents

- A study of international standards, recommendations and literature on health informatics post-graduate education programs which have preferably been endorsed by the International Medical Informatics Association (IMIA)
- A mapping of existing MSc HI curriculum content on international standards and regulations
- A field study of needs and competences in Rwanda in the field of health informatics, mainly through a number of interviews with relevant stakeholders, including the Ministry of Health (MoH) staff, NGOs, public and private health facilities, education & training institutions, health software development companies and research organizations.
- An evaluation of how the actual MSc HI curriculum was perceived by students, teaching staff and CMHS faculty.
- Development of a modified program which fills detected competence and content gaps, provides a suitable curriculum to fill those disparities and remedies identified program delivery issues.
- Analysis of managerial & financial aspects of the REHCE and the MSc HI curriculum in particular.
- Discussion of the modified program during a stakeholder workshop.
- Integration of stakeholder feedback in a final review report.

3. RESULTS

3.1 Analysis of Regulations and Program Documents

According to the MSc HI documentation, the program has the following high-level aims:

- Develop health informatics skills to the professionals and managers in the healthcare sector with hands-on skills required to cope with the ever increasing use of sophisticated health information systems.
- Equip students with generic tools to enable them to respond to future developments in healthcare and information technology while grounding them in the "real world" of current initiatives in information policy, technology and management.
- Attract healthcare professionals from the Eastern & Central African Regions, to provide knowledge and skills that will help them maximize the use of ICT tools in the healthcare sector in the region.
- Prepare professionals to respond to the changing dynamics of healthcare by providing them with the skills necessary to develop, apply and evaluate the use of information technology in the healthcare arena.
- Produce professionals who can lead organizations in the application of information technology to improve the delivery of healthcare.

In addition, the program has the following educational aims:

- To develop proficiency in analytical skills, procedures and methods relevant to the field of health and clinical services delivery
- To develop skills relevant to clinical or public health practice
- To develop enhanced qualitative and quantitative data analysis capabilities
- To develop awareness of the major ethical issues associated with the health informatics domain and appreciation of social determinants and cultural sensitivities related to these matters
- To acquire proficiency in the concepts, language and fundamentals of the field of study
- To support the development and application of evidence based practice
- To contribute to the development of professional knowledge and practice through education, research scholarly debate, presentation and publication
- To develop skills in research methodology and teaching practice with options to expand experience in each of these domains.

Learning outcomes had been defined as follows:

A. Knowledge and Understanding

1. The capabilities of various data, networking, and enterprise architectures in the healthcare system
2. Managerial issues and technologies related to interoperability in the healthcare system
3. The product strategies of major hardware, software and telecommunications vendors
4. National and global architectural standards, regulations and future developments in healthcare systems
5. Interaction of health system components with a competitive environment
6. The role of Enterprise Resource Planning (ERP), Supply Chain Management (SCM), and Customer relationship management (CRM) systems as components of the enterprise architecture
7. The impact of automation on work practices
8. Strategic implications of emerging technologies
9. Tools and techniques of project planning and management
10. The organizational change process
11. The strategic use of information technology from a business perspective at the enterprise level

B. Cognitive/Intellectual Skills/Application of Knowledge

1. Examine issues related to the acquisitions and ongoing management of products, services and contracts in healthcare systems
2. Recognize and improve processes to achieve efficiency and compliance objectives.
3. Examine the current and potential impact of information and information technology on the healthcare industry.
4. Examine the characteristics of the information age and explore the implications of emerging ethical concerns such as information privacy, accuracy, property and accessibility.
5. Examine the issues which constitute a safe digital environment.

C. Communication/ICT/Numeracy/Analytic Techniques/ Practical Skills

1. Implement and manage security and disaster recovery plans and business continuity from an overall organizational perspective
2. Develop detailed project plans, schedules and budgets; estimate project resources; allocate/coordinate resources and interface with management.
3. Use project management software
4. Identify stakeholders, assess potential impacts of projects and develop strategies of overcoming resistance, politics and other human issues.

D. General transferable skills

1. Present both technical and managerial ideas in a variety of formats (written, verbal etc.).
2. Demonstrate an effective leadership
3. Efficiently manage time, human, technical and financial resources

Examination of the stated objectives highlighted a lack of synergy with international recommendations for Health Informatics Specialists. In addition analysis of the module outcomes as learning outputs showed further gaps between defined learning outcomes and those subsequently included in modules. In particular many aspects of research appeared to have been lost in the curriculum documentation.

3.2 Student Interviews

Student feedback on the running MSc HI was obtained from a series of 4 individual student interviews and 1 student group discussion. Comments were provided on curriculum content, opportunities to develop practical skills, tutor and student evaluation systems and program staffing.

Interviews with students highlighted that a number of modules were well received and students thought that they had a major impact on their progress in Health Informatics. Such perceptions of course may have been influenced by different factors such as teaching style, teacher relationships and difficulty of the subject. When asked "*which modules had the biggest impact on you*" the responses included:

- E-health Enterprise Architecture
- Introduction to Health Care management and organization
- Software based clinical decision making and support systems
- Applications of Mobile phone technology in Health Informatics
- Introduction to Health Care management and organization

A number of modules were thought to be "*just theory*", "*not practical*" and "*not important in the Rwandan situation*". The former included the modules "*Introduction to ICT*" and "*Introduction to management Information Systems*" which could have been delivered with more practical examples and demonstrations. A module thought to be "*not important*" was "*Introduction to GIS Systems in Health Informatics including Epidemic Monitoring and Control*". This module also had comments about the high number of exercises during the teaching activity.

A comment made regarding a number of modules was that "*student engagement was missing in much of the delivery*". This was particularly highlighted with the video conferencing sessions which were said to be "*not interactive*"

Students also commented on the apparent lack of sequencing of the modules, which seems to have been an issue of who was available to teach rather than curriculum design.

3.3 Rwandan Health Informatics Needs Assessment

Interviews have been organized with a number of major stakeholders, including Ministry of Health (MoH) staff, NGOs, public and private health facilities, education & training institutions, health software development initiatives and research organizations. The main focus of these interviews was to inventory competences needed for the Rwandan health informatics setting and to identify future (short and long term) employment opportunities for MSc HI graduates.

Information obtained from the MoH clearly showed a growing need for people with synergic skills covering Information Technologies (IT), Information Systems (IS), Information Management (IM) and Health System (HS) knowledge. Generally speaking, there is a shift from IT and IS towards IM and HS needs: internal MoH software and systems development activities are progressively phasing out in favor of private sector outsourcing. Consequently, the MoH requires new management skills relating to contracting, people management and project management.

Health facilities identified competences related to health care research (mainly for university teaching hospitals), data mining and statistical and epidemiological reporting based on Electronic Health Record (EHR) content. Public and private hospitals described the urgent need for Chief Information Officer (CIO) profiles who again must be capable of combining IT, IS and IM knowledge and skills.

At the moment of the review, only few local software and systems development companies existed in Rwanda; most of them were subsidiaries of international actors whose R&D activity has not been relocated. Locally required competences are related to system integration, customization and training activities involving predominantly IS and IM skills.

3.4 International Standards

Studying international health informatics post-graduate education standards and recommendations was another part of the review process. The reviewers have mainly studied documents from the International Medical Informatics Association (IMIA) and the American Medical Informatics Association (AMIA). Both organizations being well in line according to health informatics education standards, IMIA documents have been used as a reference.

3.5 IMIA Recommendations

The "Recommendations of the International Medical Informatics Association (IMIA) on Education in Biomedical and Health Informatics" [4] have been used as a core reference document. An essential part of this document was the list of weighed core competences for Health Informatics specialists. This competences list served as one of the benchmarks for evaluating the Rwandan MSc HI curriculum. It should be noted that all of the necessary competences identified by the Rwandan stakeholders were covered by the IMIA recommendation.

3.6 IMIA Knowledge Base

In 2011, IMIA also published a knowledge base structuring the knowledge of the Health Informatics domain into 14 topics which were subdivided into 245 knowledge units [5]. This unit list served in the review process as a tool for measuring the coverage of Health Informatics knowledge by the MSc HI curriculum.

3.7 Development of a Reference Curriculum

Based on the results obtained from the Rwandan Health Informatics needs assessment, the evaluation of the existing curriculum and against the background of international health informatics education standards, the reviewers started the process of developing a reference curriculum for Rwanda. This reference curriculum was then compared to the existing one resulting in a competence- and content gap analysis, which lead to a series of recommendations for reaching the target curriculum.

In a first exercise, 40 competences identified in the "IMIA recommendations for health informatics education" have been mapped on 0 or more knowledge units from the IMIA knowledge base. This resulted in a total of 400 mappings, clearly demonstrating the knowledge overlap for many of the competences. Only for 1 competence, no corresponding knowledge unit could be found: *"2.1 Fundamentals of human functioning and biosciences (anatomy, physiology, microbiology, genomics, and clinical disciplines such as internal medicine, surgery etc.)"*.

Each module of the existing MSc HI curriculum has then been mapped on a number of IMIA competences. This showed that a total set of 38 IMIA competences (out of 40) were covered by the existing curriculum, which was a remarkably good score. Only the following competences remained uncovered:

2.1 Fundamentals of human functioning and biosciences (anatomy, physiology, microbiology, genomics, and clinical disciplines such as internal medicine, surgery etc.)

2.2 Fundamentals of what constitutes health, from physiological, sociological, psychological, nutritional, emotional, environmental, cultural, spiritual perspectives and its assessment

It was concluded that in terms of competences, the actual MSc HI curriculum responded well to the Rwandan stakeholder expectations.

Based on the mapping between existing MSc HI modules and IMIA competences combined with the previously performed mapping between IMIA competences and IMIA knowledge units, an automatic mapping between existing MSc HI modules and IMIA knowledge units was generated in order to evaluate the coverage of the Health Informatics knowledge domain by the existing program. This demonstrated that 26 (out of 245) knowledge units were not covered by the existing program.

The automatically generated mapping table between existing curriculum modules and IMIA knowledge units was then cleaned for redundancy (assigning each knowledge unit only to 1 single module) and for completeness (modifying module definitions in order to cover previously missed knowledge units). This provided a set of modules with 100% knowledge coverage whilst avoiding any redundant teaching.

In a final phase, a number of modules have been (partially) merged in order to come to an equally distributed module weight of 10 credits for each module, with 6 modules in the post-graduate certificate year and 6 modules in the post-graduate diploma year.

All modules of the developed curriculum have been completed with references to recent literature. Some of the referenced papers may then be chosen by the module manager to be mandatory literature during lectures, assignments or examinations and other papers may remain optional yet useful for developing critical scientific reading skills. It is encouraged that such reading skills be peer-reviewed within each module.

3.8 Post-Graduate Certificate Program

The post-graduate certificate program essentially focuses on Information Technology (IT) and Information Systems (IS) after an introduction to the Health Informatics domain. The development of practical skills and essential knowledge of the Rwandan health informatics context are core components of the program. The following modules constitute the proposed post-graduate certificate program:

Introduction to Health Informatics: is an introduction to the discipline of Health Informatics and the scope of the program. It covers a number of basic concepts in Health Informatics and ensures that the students from different background have the opportunity to share perceptions of Health and Medical Informatics.

Health care management and organization: covers the concepts of Healthcare Management in relation to Health Informatics. The health Informatics concepts underpinning management, organizational culture and socio-technical aspects are explored.

Knowledge management in healthcare delivery: covers the Health Informatics topics of the use of knowledge within a number of areas, in particular the processing of Data into Information, and then into Knowledge. The use of such knowledge is explored in Decision Making and Education.

Management information systems: explores the connection between information systems (IS) and business performance. It also explores the issues of security, transparency, traceability and Return on Investment.

Electronic health records management and hospital & health information Systems: assist students to understand the complexities of managing individual and community based health information. They will be able to understand different approaches of health record modeling; purpose based structuring of health information and the principles of integrated health information management.

Health Informatics applications including PACs, MIT, Telemedicine and mobile technologies: explores the concepts of ubiquitous computing technologies to provide assertive support in the population. It explores the use of telemedicine and e-Health solutions for medical support. This is underpinned by an exploration of wireless technologies.

3.9 Post-Graduate Diploma

The post-graduate diploma year is supposed to add IM knowledge and skills, leading the student to a broader comprehension of the complete Health Informatics domain and its place in the health system ("*the big picture*"). It also adds essential project management skills and elements of critical thinking enabling well-considered evaluation of different options. Research skills and scientific reasoning have also been formally included in the program in order to better cope with some research related shortcomings discovered in the existing curriculum. The following modules are part of the diploma program:

Software based clinical decision making and support systems: explores the concepts of information to support health professionals in their decision making and development of therapeutic strategies. It covers clinical pathways and guidelines and biomedical modeling and simulation as well as the principles of data representation and cognitive aspects of information processing.

Public health informatics including patient information kiosks, websites and public health systems for epidemiology, epidemic control and GIS: explores the concepts of data representation and information analysis. It will explore information tools to support education for the public and health professionals, epidemiology and public health.

Social-cultural, legal and economic impact of health informatics: explores ethical, legal, and social issues arising in the use of computer-based technology and information systems in the delivery of health care. It also includes health informatics ethics and regulatory frameworks.

Project management and IT introduction in health care delivery, case studies in health informatics: in this module students will learn the principles of project management and of using information management within projects. Through exploration of health information projects, students will gain a real-world understanding of how to manage biomedical informatics projects. They will take into account the socio-technical aspects of implementation.

Research, monitoring and evaluation in Health Informatics: is an introduction to research in the areas of health informatics and covers a wide range of methods and techniques. These range from epidemiology data and analysis and mapping to measures of patient outcomes following medical interventions. Evaluation and monitoring of programs of care and clinical audit are also included.

E-health Enterprise Architecture: explores the details of healthcare information technology (HIT) interoperability and standards. The evolution of technology in healthcare, along with the impact on clinical information systems, will be studied. The benefits of integrating healthcare information systems will be investigated, as will the challenges of integrating systems across disparate organizations, healthcare disciplines, and technologies. The value proposition of a standards-based approach to integration will be presented.

The thesis still accounts for 60 credits in the reviewed curriculum. Student and teaching staff interviews revealed that thesis development has been problematic for most of the students. Therefore it has been foreseen to better prepare students for this part during the certificate and diploma years by:

- Explicity including research methods and scientific writing skills in the module on *"Research, monitoring and evaluation in health informatics"*
- Including mandatory participation in seminars and conferences in health informatics (from ordinary attendance to submission and presentation of a paper)
- Programming 1 or 2 training sessions on drafting a thesis proposal for obtaining ethical clearance

4. DISCUSSION

The reviewed program learning outcomes have been completely aligned with IMIA recommendations (see above), with the exception of *"2.1 Fundamentals of human functioning and biosciences (anatomy, physiology, microbiology, genomics, and clinical disciplines such as internal medicine, surgery etc.)"*. This new curriculum was put into practice in the 2014/2015 academic year. The new MSc HI program has been tailored to the Central African context and options for regional extension of the program to Eastern DRC (Bukavu) and Burundi (Bujumbura) are being investigated.

The reviewers have not been able to analyze any previously existing evaluation documentation mostly because an external examiner had not been appointed for the program. The program content gaps and teaching strategy issues discovered during the curriculum review however demonstrated the importance of such periodical external examination.

Less than 40% of the students from the first 2 cohorts were able to find a job related to their HI degree. Recently, the Sustainable Development Goals have brought international pressure for implementing universal health coverage on a global scale. Also in low income countries, this goes with the development of large scale health insurance information systems that must connect insurers with government bodies and the point of care. The number of job opportunities in Africa that require the skills developed in this master program, has been constantly growing since the second half of 2015.

REFERENCES

1. Hersch W, Margolis A, Quirós F, Otero P, Building a health informatics workforce in developing countries. *Health Affairs,* 29, 274-277, 2010
2. Luna D, Otero C, Marcelo A, Health Informatics in Developing Countries: Systematic Review of Reviews. Contribution of the IMIA Working Group Health Informatics for Development. *Yearbook of medical informatics,* 8, 28-33, 2012
3. Verbeke F, Karara G, Nyssen M, Evaluating the Impact of ICT-tools on Health Care Delivery in Sub-Saharan Hospitals, *Stud Health Technol Inform*, August 22, 2013
4. Marin H, Martin-Sanchez F, Recommendations of the International Medical Informatics Association (IMIA) on Education in Biomedical and Health Informatics. *Acta Informatica Medica,* 18, 2010
5. Wright G, Development of the IMIA Knowledge Base, *SA Journal of Information Management*, Vol 13, No 1, 2011
6. Verbeke F, Contributions to the Development of Health Informatics to address e-Health Challenges in Rwanda, *Rwanda Medical Journal*, Vol. 69, No. 4, 2012
7. Berner E, Informatics Education in Healthcare: Lessons Learned, Springer, London, ISBN 978-1-4471, 2014
8. Verbeke F, Shamashanga P, Amisi C, Karara G, The National Plan for Health Informatics Development in DRC, CEDESURK, Kinshasa, ISBN 978-99951-69-03-7, October 20, 2014

Address for correspondence

Frank Verbeke, MD, PhD

Department of Biostatistics and Medical Informatics, Vrije Universiteit Brussel, Laarbeeklaan 103, B-1080 Brussels, Belgium

A ROBOT-AVATAR: EASIER ACCESS TO EDUCATION AND REDUCTION IN ISOLATION?

Jorun Børsting and Alma Leora Culén
Department of Informatics, University of Oslo, P. Boks 1080, 0316 Blindern, Oslo, Norway

ABSTRACT

This paper presents a qualitative study of deployment and use of a robot as a sick child's avatar at school. Many children and youth suffer from a range of chronic illnesses that make them, often for long stretches of time, deprived of normal education and social life. The participants in our study are adolescents who have been diagnosed with, and suffered from, Myalgic Encephalomyelitis (ME/CFS) for at least one year. They attend school typically between 1 – 3 hours per week, and the robot-avatar is intended as a way for them to extend the time spent in the classroom and to increase their social presence. The paper discusses tools made to help us understand these young people and their relation to technology (concerning their physical and mental condition) and their connectedness to others (friends, family, teachers and other relevant connections). Further, the paper reports on experiences with the use of robot-avatars by participants, their parents, and their schoolteachers.

KEYWORDS

e-health; e-learning; robot-avatar; adolescents; ME/CFS

1. INTRODUCTION

This article is concerned with how well can the technology support access to school or alleviate isolation for adolescents who are, due to a chronic illness, forced to stay at home or in the hospital. The chronic illness is defined as in (Mokkink et al., 2008) and entails 1) medical conditions lasting longer than six months, 2) are not (yet) curable, 3) affect the child's daily functioning and well-being, and 4) require a disproportionate amount of health care resources. In the case of children and adolescents, the latter point also implies resources of parents and others in the immediate family, as well as those of the educational system.

The case explored in this paper is that of adolescents suffering from Myalgic Encephalomyelitis, also known as the Chronic Fatigue Syndrome (ME/CFS) and their use of a robot-avatar as a tool that could help increase school attendance and, perhaps, decrease social isolation that is often a consequence of their condition. The use of robot-avatars with adolescents suffering from ME/CFS is part of a larger study concerned with avatar's use by children and adolescents suffering from other chronic conditions, such as cancer, heart diseases, asthma and others. The aim of the larger study is to see how different illnesses affects the use of the robot and if there are special needs related to some conditions that require additions, modifications, customizations, etc. of the robot. ME/CFS, as a multi-system ailment, certainly has aspects that are different from other chronic conditions, for example, the energy modulation mentioned in the literature review. Increased care regarding ethical concerns while doing research with youth suffering from this condition is also needed. While they do not necessarily show visible signs of illness, their efforts to participate may have health-related consequences. The condition is also often misunderstood and stigmatized.

Ten robot-avatars used in the study with ME/CFS adolescents are high-fidelity prototypes, currently under development by a startup *No Isolation,* see ("No Isolation," 2015). For the duration of our qualitative use study (from September 2015 – August 2016 when the robots are to be launched as commercial products), they may be considered as research products (Odom et al., 2016), placed in the context of everyday lives of adolescents with ME/CFS. The specific research concern discussed in this paper is how well robot-avatars meet school needs, and how they affect the social arena related to schools. So far, the robots were given to nine families with youth suffering from ME/CFS, and arrangements have been made with their schools around the use of the robot in the classroom. These arrangements entailed school's permission (principal's), agreement with teachers, and last, but not least, consent of other children and their parents to the use of the robot in the classroom. For this paper, because of space limitations, we report on findings from two cases that stand out as complementary and representative of findings otherwise. The first case focuses on the use of the very first prototype made and placed with an adolescent whom we call Jon (age 14). This case alone illustrates best diverse usability issues that have had an impact on how the robot-avatar did at school. The second case is that of a boy whom we call Peter (age 13) and his robot-avatar. Peter's case is representative of several other participants who could attend school very seldom during this school year.

The paper is structured as follows. In the next section, the literature review on adolescents with ME/CFS is presented. The subsequent section briefly describes robot-avatars and then addresses the research design. Experiences with the robot, from both the participant's and the school's perspective, are described in Section 4. Section 5 presents our findings. Conclusions and future work end the paper.

2. THE LITERATURE REVIEW

This literature review is divided into three parts. The first part offers an insight into what ME/CFS is and how are adolescents suffering from it affected. In recent years, the increase in chronic diseases, in particular among youth, has been significant (Perrin et al., 2014). Among these diseases, ME/CFS has become a growing concern (Farmer et al., 2004), not only for those suffering from it and their families, but also for the medical science, health management, and society at large. It is no longer a rare disease, and it affects children from about the age of 11 at the same rate as adults, which is well over 3% of the population by some definitions of the disease (Jason et al., 2012). Understanding this is crucial for understanding adolescents with ME/CFS, their abilities, and expectations, also with and of technology. The second part relates the condition with the issue of social isolation. The last section relates previous research on ME/CFS and technologies designed, or used, to help sufferers.

2.1 What is ME/CFS and how it Affects Adolescent Sufferers?

ME/CFS is a debilitating multisystem illness resulting in a number of symptoms that include severe physical and cognitive exhaustion, confusion, difficulties with memory, concentration, sensitivity to light and noise among others, see (Afari and Buchwald, 2003; Bringsli et al., 2014). In order for someone to be diagnosed with ME/CFS, the symptoms need to be present to such a degree that they clearly limit person's ability to carry out ordinary daily activities (National Collaborating Centre for Primary Care (UK), 2007). Sufferers are classified in four groups: mild (an approximate 50% reduction in pre-illness activity level), moderate (mostly housebound), severe (mostly bedridden) or very severe (totally bedridden and in need of help with basic functions) (Carruthers et al., 2011).

The origin of the disease remains an enigma. Treatments are highly individualized, often symptom-based, and include both pharmacological and behavioral strategies. The most common form of help to those suffering from the illness in Norway is provided through self-management courses. These courses teach people how to stabilize symptoms, balance rest and activity etc. (Bringsli et al., 2014). This is in line with a more general trend to promote self-management, especially for patients with chronic diseases (Holman and Lorig, 2004). Self-management generally implies medical management, behavioral management, and emotional management (Corbin and Strauss, 1988; Lorig and Holman, 2003). For ME/CFS sufferers, it also includes a so-called *energy modulation*, where a sufferer is considered to have a limited amount of energy to use per day and that amount needs to be distributed carefully. Research shows that energy modulation has an effect on the illness (Jason et al., 2009). Overuse implies extra exhaustion lasting between several hours to several days, leading to increased pain, sensitivity and overall worsened physical and cognitive condition. A person with ME/CFS, however, often does not know in advance effects of activities, such as meeting a friend. Thus, energy modulation is challenging and difficult. For us as researchers, it was crucial to learn about this and behave accordingly while doing our research at homes of these young people.

For children and adolescents, symptoms are often ascribed to other causes, making their position more difficult, as Dowsett and Colby reported already in 1997 (Dowsett and Colby, 1997) and still true today:

"Do children and adolescents suffer from ME/CFS? Simple common sense tells most parents, teachers and doctors that they do and often more severely than adults. Yet there remains a sizeable proportion of

professionals in Health Care, Education and Social Services who are still prepared to ascribe the numerous, disabling but seemingly unconnected symptoms of this illness in young people to anorexia, depression, school phobia or a dysfunctional family background. All are, at least, agreed that the illness presents a considerable economic, educational and social problem", (Dowsett and Colby, 1997).

2.2 ME/CFS and Social Isolation

Social isolation is often associated with people suffering from ME/CFS (Best and Butler, 2014; Drachler et al., 2009; Roche and Tucker, 2003). Drachler reviewed thirty-two quantitative and qualitative studies, including the views of over 2500 people with ME/CFS. They found seven issues in need of major support. One of these seven was the need to support ways to develop strategies to maintain/regain social participation. They further stated that "outside the family, education was the major focus of social participation for many young people" (Drachler et al., 2009). However, because of their illness symptoms they could not attend school, and therefore some reported having lost connection with friends and teachers (Roche and Tucker, 2003). Home tuition is an often used alternative to attending regular education for young people suffering from ME/CFS (Garralda and Rangel, 2004). Drachler found that home tuition "allowed a flexible schedule of learning within the limits of their condition, but reduced social participation at an important stage of social development" (Drachler et al., 2009). Asbring finds that for ME/CFS sufferers just attending school, or having a job for an adult, signifies living 'normally', and further, "providing purposeful activity, an opportunity for social interaction, a sense of achievement, self-value and social recognition, income and social security" (Asbring and Närvänen, 2002; Clarke and James, 2003; Schoofs et al., 2004) are important.

2.3 Adolescent ME/CFS and Technology

There is little in the literature on supporting children and adolescents with ME/CFS through the use of technology. As mentioned, ME/CFS sufferers have multiple challenges, at the physical and cognitive level, coupled with reduced tolerance to light and noise. These challenges affect their experiences with technologies. Special attention is required when introducing a new technology into to the life of these adolescents; one needs to take care not to introduce new problems. However, social and educational challenges are very important, as was also found by (Wadley et al., 2014), whose study with hospitalized children concludes that mediated communication between hospitals, homes and schools is beneficial. Very few other examples of previous research that we could find, are also related to educational and social arenas. General use of social media was described in (Grajales et al., 2014), where authors have analyzed and synthesized 76 articles, 44 websites, and 11 policies/reports and presented findings according to 10 different categories of social media: (1) blogs (e.g., WordPress), (2) microblogs (e.g., Twitter), (3) social networking sites (Facebook), (4) professional networking sites (LinkedIn), (5) thematic networking sites (e.g., 23 and Me), (6) wikis (Wikipedia), (7) mashups (HealthMap), (8) collaborative filtering sites (Digg), (9) media sharing sites (YouTube, Slideshare), and others (SecondLife). Some of their results, in particular, those related to the second life, are relevant for adolescents with ME/CFS and have been described in the work of Best and Butler (Best and Butler, 2013a, 2013b). Literature addressing education for adolescents with ME/CFS is scarce as well, and papers found mostly focus on e-learning, see (Sheridan et al., 2013). Diverse e-learning platforms are seen as an opportunity for children with ME/CFS to participate in classroom activities even though they are staying at home. However, the access to school and educational materials is only one of the challenges that sufferers meet, thus, it does not address, for example, the cognitive problems.

In summary, there is a great need for research exploring design and use of technology for ME/CFS adolescents that does not worsen their condition and supports, in particular, education and social presence.

3. ROBOT-AVATARS AND THE RESEARCH DESIGN

The participants for the study were recruited in collaboration with the Norwegian ME/CFS association ("Norges ME-forening," 2016). Nine adolescents, over 12 and under 16 years of age, were chosen to participate in the study. The recruitment choices were made based on the insight that ME association had on participant's condition (only mild or moderate cases were considered) and difficulties related to school attendance. Five boys and four girls were recruited. The tenth robot was given to Jon, replacing the very first prototype made. This paper then shows the trajectory of deployment of two robot-avatars at Peter and Jon's schools. Working with them has helped shape our methods, discover problems related to the classroom use and uncover some interesting leads for our current and future research. The work with *No Isolation,* discussing the research design, took about three months, while the work with participants started in early January of 2016. Jon got his robot in early February, and Peter shortly after. Jon and Peter could use their robots for about a week, and after that, the robots were placed into the classroom where, at the time of this writing, they have been for over four months. Peter lives close to school (ca. 2 km), but was capable of attending at most one school class per week. Jon lives far from school (ca. 24 km), but was capable of attending one class every second day. Both Peter and Jon live with their parents and a younger sibling.

3.1 Robot-Avatars

Robot-avatars are intended for a single person's use. They are to be controlled by their owners, allowing the avatars to become their eyes, ears, and voice. An avatar is controlled using touch on a smartphone or a tablet. It has a microphone, speakers and camera installed, but it does not have the ability to store any data, it simply streams, see Figure 1, without any recording capabilities, between a sick adolescent and their school. An app, installed on a personal mobile device such as a smartphone or a tablet, is used to control the robot in real-time, see Figure 1d). This control is simple and implies signaling presence when the contact between the adolescent and the classroom is established (the eyes light up), using touchpad to move the avatar's head in the desired direction and touching the light button, signaling that they would like to speak (the light on top of avatar's head lights up, Figure 1a)). A robot-avatar is using 4G so that it can be taken to class trips or places where there is no Wi-Fi, ensuring access everywhere.

Figure 1. a) Jon's old and new avatars. The old, black one, has a red sticker from the class trip to the museum, b) Peter's classmates made a smile for his avatar, c) the inside of an avatar, d) the app showing how Jon sees a remote location

An adolescent can see his peers and teachers. He/she is not visible. The intention is for others to see the robot as a physical representation of a sick person. While this is not crucial for ME/CFS adolescents (much more relevant for other illnesses where a child may be in a hospital), they also are often not in a shape they want others to witness.

3.2 Methods

As discussed in Section 2, we used the literature review to gain an understanding of what has been done in terms of design and use of technology for ME/CFS sufferers. To gain an understanding of the user group itself, we have organized an expert sense-making session on ME/CFS. The participants have a deep knowledge of ME/CFS, having daily contact with children, youth and adults with ME, and their families. The group consisted of representatives from the Norwegian ME organization, the organization that provides support to families of ME/CFS sufferers, the youth ME/CFS organization, and a medical doctor. The insights from this session were very important and reported in (Børsting and Culén, 2016). For present paper, it suffices to say that it has helped us understand better the concept of energy modulation and many other issues relevant to everyday living with ME/CFS.

Our choices of methods and tools to use with participants were guided by our previous experiences in working with vulnerable children and youth (Culén, 2015; Culén and Karpova, 2014; Culén and Velden, 2013), where vulnerability is to be understood as a set of risks that should be reduced by new technologies such as the robot-avatar. Thus, on the one hand, we needed to develop methods and tools to collect data on participant's habits around school and social connections related to school, enabling us to see changes in behavior and social patterns. Their condition was not to deteriorate due to increased school attendance and social activities via the avatar, or efforts to participate in this research. On the other hand, we needed to collect data on the robot use in the classroom, diverse usability issues and opportunities for re-design. The first part required great sensitivity to ethical issues. Firstly, as part of the research was done at participant's homes, if they got tired, or their pain increased, it was harder for them to tell us to leave, so we needed to observe them closely and ask several times if they are still ok with participation. Secondly, as mentioned, the energy modulation is challenging, and sometimes even the participants do not know how some activity will affect them. We have, thus, made every effort to be fast and efficient beforehand. Finally, some ethical concerns were related to other classroom peers and their parents. Even though they were thoroughly informed about the study, aware that robot-avatars do not store any information and that all data is encrypted, some still felt uneasy with the possibility that their child could be seen from a remote location. In one case, a request was made that the parents of a participant sign that only their child will ever use the robot-avatar and that no one else would ever observe the classroom with their child.

When delivering an avatar to participant's home, we first interviewed the participant and then helped them to install the app and bring the avatar to life and made sure that they would know how to use them later. When the robot-avatar was delivered to Jon, our first participant, this initial visit took a long time (approximately three hours) due to diverse technical problems. Visiting Peter, and especially more recent participants, it took much less time. Interviews typically lasted about 30 minutes.

Tools such as the cool wall (Fitton et al., 2012; Read et al., 2011), social mapping and the technology use mapping, described in (Culén, 2015), were used, see Figure 2. The interview was audio recorded, and the photos, as per consent allowed, taken.

The cool wall, shown in Figure 2 a), was used as an ice-breaker technique that at the same time allowed us to discuss what they consider cool in terms of technology and games, and why. Large magnets and the board allowed for easy movement of items and ease in changing categories, if needed. Some icons that we considered likely to be used by our participants were made a-priory. We left some magnets white, and participants easily add their own items. We also asked if any of these things that they consider as cool or neutral tire them.

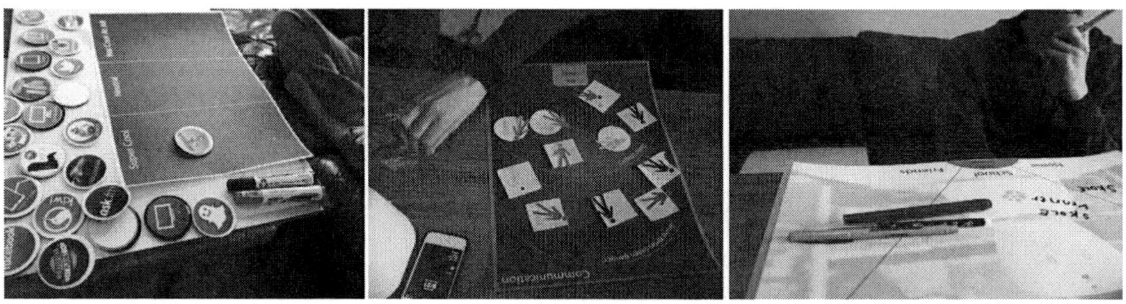

Figure 2. Tools used: a) the cool wall b) the communication map c) the future technology map

Figure 2 b) shows a social map that places the participant in the center (the green square marked as me), then a circle of close relations with daily contact was drawn, and then the circle with people with whom the communication is occasional. Small people cutouts, in different colours to separate people into categories such as friends, family, really close friends, doctors, teachers etc., were provided. Each participant could define their own categories, as well as choose colours for each category. A similar map was used to see what kind of technologies they use daily and for what (this map is not shown in Figure 2). Here, we also aimed to find out if some technology platforms had a negative impact on their health, that is if they increased their illness symptoms. The same icons as for the cool-wall, as well as the self-made ones, were used to discuss their habits around technologies such as Skype, smartphones, computers, TV and others. Finally, a map of envisioned use of the avatar was presented, see Figure 2c). Here, participants could write directly on the map any thoughts around how the robot-avatar will be used. The same procedure was used with all nine participants, but as mentioned, this paper focuses on Jon and Peter's case.

After this initial visit, data on Jon and Peter's cases was gathered through frequent email contacts with their parents. In total, 69 messages were exchanged. 37 of these handled organizational matters, while 32 were the actual feedback on use, technical problems and design and personalization suggestions. After a period of 6 weeks of use, an online interview was done with Jon and Peter. The online format was chosen in order to allow Jon and Peter to answer 17 open questions at the time of their own choosing. After ten weeks of use, 14 open questions interview was done with three schoolteachers, two from Jon's school and one from Peter's school. Thereafter, observations were carried out at both schools, while Peter and Jon attended their classrooms remotely. At the end of the class observations, we administered a nine questions questionnaire to their classmates. Lastly, a closing interview with Peter and Jon, using the same tools as the initial one, was done.

In summary, this is an interpretative study, where data was collected through face-to-face and online interviews with our participants, classroom observations, classmates' questionnaires, as well as feedback from parents in the form of email and teachers through online interviews.

4. FINDINGS AND DISCUSSION

It is appropriate to mention, prior to discussing findings, that there were some issues with these first prototypes. For example, one had to take the battery out of the robot and charge it daily, see Figure 1c). One also needed to charge the modem. Furthermore, 4G coverage was supposed to work everywhere, but this was not the case. These two problems have, naturally, affected the experiences with avatars. Coping strategies that emerged to address these issues were interesting for this research.

The amount of data collected in this study is large, and themes that emerged rather diverse, i.e., we could discuss design, personalization, technical issues and other issues based on the collected data. We focus only on access to school and social relations findings that emerged from interviews, classroom observations and student questionnaires regarding access to education and social connections at school.

4.1 Findings from Interviews

We start by reflecting on the usefulness of the tools for the purpose of interviews. The cool wall served well as an icebreaker and conversation starter. It was an easy exercise to do and allowed us to address a wide range of topics on technology use, and more importantly, their personal likes and dislikes. It was also useful when addressing topics that could be perceived as difficult to talk about, such as how long they can play before they get tired, or how many close or distant friends they have. We found out that the use of these tangible magnets, pens and maps enabled us to collect more and richer data than we would have gotten if we just asked questions, addressing these same topics directly. Jon and Peter found these exercises to be fun. Since Jon's mother (and both parents in Peter's case) was present during the interview, these tools enabled us to have a more direct dialogue with the interviewee only. They also enabled an easy overview of changes between the initial and final interviews.

Some concerns regarding computers were raised during the initial interview. Jon stated that he gets worse when using a stationary computer. Peter's parents reported that it was their impression that Peter gets worse with prolonged use of the stationary computer. In particular, that he gets worse if sound levels in games he plays are high. The adverse reaction to noise was an important issue at school also. The parents pointed out that lots of noise in the classroom could also be a problem when using the robot-avatar. Peter did not comment, rather, he said that he favored the computer for playing the game of Minecraft. That game connects him to the three friends he has. Minecraft and Skype, on the computer, were the primary routes to social contacts for Peter. Jon was also using his phone, SMS, Skype and diverse games to connect daily with his friends and actively keep close contacts. When using the future avatar use map, Jon was quick to point out that the categories we set should be mixed, see (Figure 2 c). We have separated home, school and friends, but Jon insisted that school friends are very important.

In online interviews, both Peter and Jon reported that they have been able to tend to their education much more frequently than before. Peter reported that he was able to attend remotely in one week, as much as he had been able to attend during the three months period prior to the use of the avatar. He also stated that in periods of increased illness symptoms, he was still able to attend school, something that would not be possible without the robot-avatar.

Jon wrote in the online interview *"I have had great pleasure from using the robot. I've learned something new and feel that I am more included".* Jon has used the robot to talk to some of his classmates during breaks at school when the teacher was not present. In addition, he could use it during the group work in the class.

All three teachers found the current prototype setup with daily charging to be problematic. It took time and it was an easily forgettable task. The mentioned 4G coverage problems affected negatively the answers from two teachers at Jon's school. They, however, still saw a potential for the robot to work well when these technical issues were eliminated. One of Jon's teachers stated that the robot had been mostly out of working order, but still, when asked if it was perceived to be useful for Jon, the answer was "Yes, exciting". The other teacher answered "Currently, not much" to this question. Peter's teacher, from school that had good 4G coverage, reported much more positive experiences. Her answer to the question related to robot's usefulness was *"Yes, I find it very useful. He gets a feeling of being a bit more included, although he is not physically present."*

Many issues addressed by teachers were also brought up by Jon and Peter, i.e., the need for a signal when they want to say something, or another signal if illness symptoms on that specific day were worse and they wanted to observe only, rather than participate. Peter's teacher said that it was easy to forget that Peter was present and that she is trying to be more aware and include him in the work just like his classmates.

4.2 Findings from Classroom Observations

We attended a lesson in English at Peter's school. Fifteen students were present. After the greetings and explanation as to who we are, the focus shifted to the avatar. There was a lot of confusion about whether Peter was present via his avatar, or not. This unsureness was something that Peter reported previously in his online interview. When it was confirmed that the robot was on, the teacher greeted Peter: "Can you hear me? Hi!" "Yes", answered Peter. Then the class proceeded as usual. The teacher gave some examples of what prepositions in English are, and proceeded to do exercises from the book. She also directly addressed Peter: "Peter, do you know where we are?" Peter confirmed. The teacher asked him if he could read from the book. He did. "Perfect, totally correct, perfect!" Then just like everyone else was called to answer questions, so was Peter. His responses were always correct. The teacher also made sure that Peter had no problem seeing the text on the blackboard. The dialogue went well, and we felt that Peter was included in line with other students. However, Peter did not interact much with the robot-avatar. That is, the robot never moved, it did not turn to follow who spoke in the class. Peter did not request to speak on his own. He spoke low when he spoke. The robot was positioned on a school bench, where just one other student was sitting. At some point, the student asked the teacher if she could move to sit with a friend saying "It is the most boring thing to sit alone". Another student offered to move 'Peter' to his desk at the start of the class, but this was not approved.

Figure 3. a) Peter's classroom, b) Jon's classroom at the start c) Jon's classroom at the end of the semester

The experience in Jon's classroom was in strong contrast with this one. Jon's mother has picked up the avatar the day prior to our visit and charged it overnight. She delivered it fully charged to school the next morning, in order to make sure that everything works. She has experienced problems with charging routines at school before. The following little digression illustrates this point: *several times both Jon and Peter had substitute teachers. Since substitute teachers had no training in handling the robot-avatar and they did not know that it had to be charged, the communication was not possible. One of the substitute teachers in Jon's class refused to have the robot – something that was not expected, so again, Jon could not take part in the class as scheduled. This happened at a point when the avatar has been used for some time already and when Jon started to expect flawless access to teaching. So, it was very frustrating that he could not attend school through the robot as expected. The outcome at Jon's school was to take the robot away from the classroom and keep it locked in a closet. At Peter's school, the routines around charging were changed, and responsibility for charging the robot delegated to students. This has worked perfectly.* On the morning of our visit, the principal walked us to Jon's classroom. We attended a math lecture. The avatar was not there. We greeted the class and the teacher, while the principal went to pick up the avatar from the closet (even though the mother returned it just some minutes earlier, specifically so that it could be ready). The avatar was placed next to the blackboard, see Figure 3c), where the 4G coverage was weak. The teacher did not greet the student but proceeded with solving an equation and explaining principles behind what he did. Avatar's eyes light up when it is on. Sitting at the back, we could see the light (the connection) fluctuating. One of us went to the teacher and asked if he could make sure that the student is actually present. The teacher checked and Jon answered. The teacher then told students to do exercises from their book. The avatar's connection continued to flicker. One researcher got up and placed the modem in a different location, where the connection was decent. Not much more happened in terms of interaction with Jon during the hour.

4.3 Findings from Questionnaires

There were six out of nine questions that are relevant for this paper (the remaining three questions addresses the personalization and design issues). These were the following: 1) Does it feel like Jon (Peter) is present in the classroom when he uses the robot-avatar? 2) What is good with the robot-avatar in the classroom? 3) What is it that you do not like with the robot-avatar in the classroom? 4) Have you used the avatar to talk to Jon (Peter) at breaks? 5) What do you think, how is it to have the robot-avatar in the classroom? 6) If the robot-avatar were yours, how would you use it? The first and the fourth questions were yes/no questions, while the fifth was a 5 point smiley scale. The remaining two questions were open.

In Jon's classroom, 23 students filled the questionnaire. In Peter's classroom 15. The first question in Jon's classroom was answered in positive 8 times and 15 times in negative, while in Peter's 5 were negative, 7 positive and 3 undecided.

So, in Jon's classroom, most students did not feel like he was there. Answers to other questions clarified why this was so. For the second question, students wrote: "That Jon can attend the school", "When he is not here, he can be here using the avatar", "It is good for him, he can get some teaching and be social", "I do not think over it, but it is nice that he can attend the school from home", "Nothing special, he does not use it much", "It is special, not many classrooms have a robot", "It is cool to have something nobody else has", "Nice for him". The negative aspects were: "It is scary", "It should be used more", "It is missing hands!", "Noise from his home is heard", "Black and boring". However, six friends reported that they did talk to Jon during breaks (before the robot went to live in the closet).

One student found the robot to be very negative, the majority (11 students) gave it the biggest smile, and 5 a smile. As for the last question, they had many ideas, some along the line of current use (go to school when sick), and others different: "Pretend to work", "Watch others learn", "Talk", "Joke", "To cheat on tests", "Be with people when they are sick".

Many in Peter's classroom gave very similar answers as to what is good with the robot. For the negative aspects, three students added that the robot is fragile (it actually fell to the ground and the shell broke, but the functionality was not affected), it needs to be handled with care. One student wrote that the voice is too low and it is hard to hear Peter talk.

Nine students answered the fifth question with the biggest smile, 5 with a smile, one person did not care (neutral), and one was mildly frowny.

In summary, here too, technical problems with robot-avatars were mentioned, its fragility, noise it can be transmitted from Jon or Peter's home, the noise the robot makes when it turns from. It was not experienced as a true representation of Jon or Peter. In spite of all this, teachers, participants, and parents were very positive towards the future possibilities of robot-avatars, once all technical problems were eliminated. Most positive statements pertained to the exact purpose for which avatars were made.

5. DISCUSSION

This paper focus on how the robot-avatar provides access to education, social participation at school and some important ethical considerations when doing research with youth suffering from ME/CFS.

As mentioned in the introduction, increased care regarding ethical concerns while doing research with youth suffering from ME/CFS is needed. Special considerations need to be taken to ensure that the use of the robot in everyday lives does not worsen their condition. When introducing or developing technology for youth with ME/CFS, it is especially important to be aware of and sensitive towards the challenges related to the energy modulation in order to avoid post-exertional malaise. Any sensory inputs, cognitive or physical activities related to the use of the robot, which exceed individual boundaries, can worsen their condition, often experience over several days post-exertion. Therefore, it is also important to know that youth with ME/CFS could easily surpass their own limits, in order to comply with demands from others or, simply because the robot is fun to use. This highlights the importance of carefully evaluating the design and use of the robot and how it affects the user. We now recommend the youth and their parents to be careful and design strategies for use that suit their child.

Peter has had days of increased illness symptoms during this period. Neither he nor his parents attribute these health-related problems to the robot. They said that Peter periodically gets better or worse. The robot-avatar continued to be used on the days that he could. Jon, on the other hand, has designed a use strategy in line with the energy modulation strategy, consisting of 'small doses' of use. However, to be sure that it is not the robot that causes an increase in illness symptoms, it would be wise for Peter also to adopt a strategy for use, like Jon's, based on energy modulation (Jason et al., 2009).

If robot-avatars are to provide access to schools successfully, they need to be in perfect functioning order, easy to use and useful. The level of "being finished" that our avatars had, could not work for a commercial product. But they were absolutely amazing as research products (Odom et al., 2009). We got a lot of great data, in particular, related to design and personalization because avatars were indeed seen as something that can be improved, and thinking about suggestions worthwhile. However, when it comes to access and socializing, they were somewhat impeded by this "prototype" status. In fact, we heard the teacher at Jon's school tell the students just before they started filling the survey: "Remember, this is just a prototype, so you need to be VERY honest!"

As mentioned in the literature survey, there were attempts to use tablets and similar to ensure access to school (Sheridan et al., 2013). However, considering tablets as representatives of a child were not the issue brought up. The emotional aspect that physical representation of a child by avatar opens up for was important to explore. The stronger the emotional connection with the avatar as a representative of the child, the better outcomes can be expected from its use in terms of social connections. A good example of this is the fact that Jon's teachers, and in particular, the substitute teacher mentioned earlier, do not see the robot as Jon. If they did they most certainly would not lock him in the closet whenever Jon did not attend a lesson. Jon's mother did make an emotional connection with the avatar, and voiced her opinion as "it is like locking him (Jon) in a closet", identifying Jon's ability to access friends and teaching with that of the avatar. Jon has increased the number of closest friends by one and added a few more to the group of occasional contacts, and he thinks that this is due to the use of the robot, despite the fact that lately, it has been living in the closet. This increase he attributes to the ability to talk to them more often and participate in social interactions with peers in breaks. Jon was much more positive towards the robot when he could use it to talk to friends in the breaks, than after this possibility was eliminated. When Jon was able to participate in breaks with the robot, he described that he felt more included socially in his class (as was reported through the online interview). His parents also noted that, in general, he seemed more emotionally satisfied and happy.

For teachers, it was naturally easier to see the robot as a prototype, because they were the ones having to do extra work, where charging the robot, or placing the modem where there is reception are only one aspect of it. They also had to think how not to forget the child, as Peter's teacher pointed out. They had to find natural ways to include the child properly in a conversation, also when what they do in the classroom is really not easy to attend remotely. How willing they are to address these issues determines how well the robot can be used. In our cases, at one school, teachers used a workaround, where students performed the charging activities, while at the other school the robot-avatar was used less and less.

Do robot-avatars enable easier access to education and reduce isolation? They need to be completely functional, but all our participants believe in the potential they promise.

6. CONCLUSION

We argue that focusing on the needs of youth suffering from ME/CFS is needed. Our study does so by researching how a technology could support these youth in their everyday lives. Access to education and social connectedness are identified as particularly relevant part of the everyday life for this user group. The technology in our study is a robot-avatar, which is intended as means to reduce social isolation and increase possibilities of getting access to education. It acts as an avatar for children at risk of isolation and represents them in social contexts at school and within friends and family. Already at this stage of prototyping, the robot-avatar represents the solution that solves the problems of access to education for ME/CFS sufferers better than any of the technical solutions discussed in the literature section. In addition, the first trials in real life show huge potential of the avatar, both in terms of ease of use, and because lessons learned from these two cases most certainly generalize to many of those suffering from ME/CFS. This because the participants had many problems common to ME/CFS, and the feedback from the study is that using the avatar has worsened none of the symptoms. On the contrary, very positive feedback from participants, school and parents were received. It is still important to adopt strategies for use that are in line with the need for energy modulation related to the ME/CFS illness. Being aware of and sensitive to issues related to the energy modulation is highly important through the whole process of doing research or developing technologies for and with people suffering from ME/CFS. Our research indicates that the robot-avatar, the added work tasks for teachers and organizational issues still needs more work, but the technology has the potential to support youth suffering from ME to access school and reduce their exposure to social isolation.

ACKNOWLEDGEMENT

Immense gratitude is hereby given to *No Isolation* for robot-avatars dedicated to our ME/CFS study, fruitful cooperation and great technical support along the way. Thanks are given to ME organization for helping with participant recruitment process, and providing valuable advice anytime we needed it. We also gratefully acknowledge the cooperation of teachers and principals for supporting the project and facilitating the use of the robot.

REFERENCES

Afari, N., Buchwald, D., 2003. Chronic Fatigue Syndrome: A Review. Am. J. Psychiatry 160, 221–236.

Asbring, P., Närvänen, A.-L., 2002. Women's experiences of stigma in relation to chronic fatigue syndrome and fibromyalgia. Qual. Health Res. 12, 148–160.

Best, K., Butler, S., 2014. Virtual Space: Creating a Place for Social Support in Second Life. Space Cult.

Best, K., Butler, S., 2013a. Second life avatars as extensions of social and physical bodies in people with Myalgic Encephalomyelitis/Chronic Fatigue Syndrome. Continuum 27, 837–849.

Best, K., Butler, S., 2013b. The Necessity of Control for Computer and Internet Users with Myalgic Encephalomyelitis/Chronic Fatigue Syndrome. Inf. Commun. Soc. 16, 1042–1060.

Børsting, J., Culén, A.L., 2016. Sense-Making in Complex Healthcare Domains: The Role of Technology in Every Day Lives of Youth Suffering from ME/CFS. Presented at the eTELEMED 2016, The Eighth International Conference on eHealth, Telemedicine, and Social Medicine, pp. 236–242.

Bringsli, G.J., Gilje, A., Wold, B.K.G., 2014. The Norwegian ME Association National Survey Abridged English Version. http://me-foreningen.com/meforeningen/innhold/div/2014/05/ME-Nat-Norwegian-Survey-Abr-Eng-Ver.pdf

Carruthers, B.M.et.al., 2011. Myalgic encephalomyelitis: International Consensus Criteria. J. Intern. Med. 270, 327–338.

Clarke, J.N., James, S., 2003. The radicalized self: the impact on the self of the contested nature of the diagnosis of chronic fatigue syndrome. Soc. Sci. Med. 1982 57, 1387–1395.

Corbin, J.M., Strauss, A., 1988. Unending work and care: Managing chronic illness at home, The Jossey-Bass health series and The Jossey-Bass social and behavioral science series. Jossey-Bass, San Francisco, CA, US.

Culén, A.L., 2015. Later Life: Living Alone, Social Connectedness and ICT, Digital Human Modeling - Applications in Health, Safety, Ergonomics and Risk Management: Ergonomics and Health. Springer 2015, pp. 401-412

Culén, A.L., Karpova, A., 2014. Designing with Vulnerable Children: a Researchers Perspective, in: Human-Computer Interfaces and Interactivity: Emergent Research and Applications Book, IGI Global. IGI Global, pp. 118–136.

Culén, A.L., Velden, M. van der, 2013. The Digital Life of Vulnerable Users: Designing with Children, Patients, and Elderly, in: Aanestad, M., Bratteteig, T. (Eds.), Nordic Contributions in IS Research, Lecture Notes in Business Information Processing. Springer Berlin Heidelberg, pp. 53–71.

Dowsett, E.G., Colby, J., 1997. Long-Term Sickness Absence Due to ME/CFS in UK Schools. J. Chronic Fatigue Syndr. 3, 29–42.

Drachler, L.M., Leite, C.C.J., Hooper, L., Hong, C., Pheby, D., Nacul, L., Lacerda, E., Campion, P., Killett, A., McArthur, M., Poland, F., 2009. The expressed needs of people with Chronic Fatigue Syndrome/Myalgic Encephalomyelitis: A systematic review. BMC Public Health 9, 458.

Farmer, A., Fowler, T., Scourfield, J., Thapar, A., 2004. Prevalence of chronic disabling fatigue in children and adolescents. Br. J. Psychiatry 184, 477–481.

Fitton, D., Horton, M., Read, J.C., Little, L., Toth, N., 2012. Climbing the cool wall: exploring teenage preferences of cool. ACM Press, p. 2093.

Garralda, M.E., Rangel, L., 2004. Impairment and coping in children and adolescents with chronic fatigue syndrome: a comparative study with other paediatric disorders. J. Child Psychol. Psychiatry 45, 543–552.

Grajales, F.J., Sheps, S., Ho, K., Novak-Lauscher, H., Eysenbach, G., 2014. Social media: a review and tutorial of applications in medicine and health care. J. Med. Internet Res. 16, e13.

Holman, H., Lorig, K., 2004. Patient self-management: a key to effectiveness and efficiency in care of chronic disease. Public Health Rep. 119, 239–243.

Jason, L.A., Brown, A., Clyne, E., Bartgis, L., Evans, M., Brown, M., 2012. Contrasting case definitions for chronic fatigue syndrome, Myalgic Encephalomyelitis/chronic fatigue syndrome and myalgic encephalomyelitis. Eval. Health Prof. 35, 280–304.

Jason, L., Benton, M., Torres-Harding, S., Muldowney, K., 2009. The impact of energy modulation on physical functioning and fatigue severity among patients with ME/CFS. Patient Educ. Couns. 77, 237–241.

Lorig, K.R., Holman, H., 2003. Self-management education: history, definition, outcomes, and mechanisms. Ann. Behav. Med. Publ. Soc. Behav. Med. 26, 1–7.

Mokkink, L.B., van der Lee, J.H., Grootenhuis, M.A., Offringa, M., Heymans, H.S.A., Dutch National Consensus Committee Chronic Diseases and Health Conditions in Childhood, 2008. Defining chronic diseases and health conditions in childhood (0-18 years of age): national consensus in the Netherlands. Eur. J. Pediatr. 167, 1441–1447.

National Collaborating Centre for Primary Care (UK), 2007. Chronic Fatigue Syndrome/Myalgic Encephalomyelitis (or Encephalopathy): Diagnosis and Management of Chronic Fatigue Syndrome/Myalgic Encephalomyelitis (or Encephalopathy) in Adults and Children, National Institute for Health and Clinical Excellence: Guidance. Royal College of General Practitioners (UK), London.

No Isolation [WWW Document], 2015. URL http://www.noisolation.com/en/ (accessed 3.2.16).

Norges ME-forening [WWW Document], 2016. URL http://me-foreningen.com/meforeningen/ (accessed 6.3.16).

Odom, W., Pierce, J., Stolterman, E., Blevis, E., 2009. Understanding why we preserve some things and discard others in the context of interaction design, in: Proceedings of the SIGCHI Conference on Human Factors in Computing Systems, CHI '09. ACM, New York, NY, USA, pp. 1053–1062.

Odom, W., Wakkary, R., Lim, Y., Desjardins, A., Hengeveld, B., Banks, R., 2016. From Research Prototype to Research Product, in: Proceedings of the 2016 CHI Conference on Human Factors in Computing Systems, CHI '16. ACM, New York, NY, USA, pp. 2549–2561.

Perrin, J.M., Anderson, L.E., Cleave, J.V., 2014. The Rise In Chronic Conditions Among Infants, Children, And Youth Can Be Met With Continued Health System Innovations. Health Aff. (Millwood) 33, 2099–2105.

Read, J., Fitton, D., Cowan, B., Beale, R., Guo, Y., Horton, M., 2011. Understanding and designing cool technologies for teenagers. ACM Press, p. 1567.

Roche, J., Tucker, S., 2003. Extending the Social Exclusion Debate An Exploration of the Family Lives of Young Carers and Young People with Me. Childhood 10, 439–456.

Schoofs, N., Bambini, D., Ronning, P., Bielak, E., Woehl, J., 2004. Death of a lifestyle: the effects of social support and healthcare support on the quality of life of persons with fibromyalgia and/or chronic fatigue syndrome. Orthop. Nurs. 23, 364–374.

Sheridan, A.K., Scott, L., MacDonald, N., Murray, L., Holt, S., Kat, A., 2013. Exploring E-learning provision for Children with ME in Scotland. Educ. 2, 78–80.

Wadley, G., Vetere, F., Hopkins, L., Green, J., Kulik, L., 2014. Exploring ambient technology for connecting hospitalised children with school and home. Int. J. Hum.-Comput. Stud. 72, 640–653.

ADOLESCENTS SURVEILLANCE SYSTEM FOR OBESITY IN EUROPE. A PIONEERING PROJECT TO PREVENT OBESITY IN ITALY USING E-TECHNOLOGY: THE ASSO* PROJECT * ADOLESCENCE SURVEILLANCE SYSTEM FOR OBESITY PREVENTION

Monèm Jemni[1], João Viana[2] and Garden Tabacchi[3]
[1]*Qatar University, College of Arts and Sciences, Department of Sport Science, Doha, Qatar*
[2]*Research Centre in Sports Sciences, Health Sciences and Human Development, CIDESD, University Institute of Maia, ISMAI, Portugal*
[3]*Department of Sciences for Health Promotion and Mother Child Care "G. D'Alessandro", University of Palermo, Italy.*

ABSTRACT

Adolescents' obesity is a major concern in our modern life; it could lead to significant increase in the rate of obese in the future generations and consequently the health budget. The ASSO Project in Italy (Adolescence Surveillance System for Obesity prevention) is tackling this new pandemic through multi facets monitoring system using the new e-technology (life style, diet, smoking, drinking and exercise habits). The project is in process of being implemented in Italy after a pilot in the South of the country. The objective of this short communication is to provide an overview of the project, the method applied, its implementation and the very first exercise related results in more than 800 teenagers.
78% of the assessed teenagers are found to be sedentary. More than three quarters of the 22% active adolescents are being such because they undertake sports at competitive or non-competitive levels. The cardiorespiratory endurance of these adolescent' sample is found to be very weak with no difference between the under 16 and the over 16 years old, thing which will raise many concerns.

KEYWORDS

Obesity, fitness testing, cardiorespiratory assessment, body composition

1. INTRODUCTION

We all know the world's population are facing an unprecedented obesity pandemic. Not only are the rich and occidental countries struggling with alarming numbers of obese people but also poorer countries are also reporting an increasing number of deaths directly or indirectly related to obesity. History shows also that humans are becoming less active when compared to previous eras.

Longitudinal studies in all countries since the early 2000s have found increasing rates of overweight in children (Chen et al., 2002). There is at least a moderate correlation between weight in childhood and adulthood, as well as obesity in childhood and later health problems (Hills et al., 2007). The World Health Organization showed that the main cause for mortality is cardiovascular disease (WHO 2005), with obesity one of the major causes. Obesity also causes reduced work productivity (Olshansky et al, 2005). Higher governmental health authorities are requesting scientist and researchers how could they contribute in tackling this pandemic, each from different procspective.

Europe is a vastly heterogeneous continent where the rates of obesity vary significantly from one country to another. Even within one country, obesity varies between ethnicities. Levels of physical activity have long been known to be lower in South Asian people than the general UK population, with heterogeneity within sub-groups, with Bangladeshis the least physically active.(Hayes L, White M, Unwin N, Bhopal R, Fischbacher C, Harland J, et al. Patterns of physical activity and relationship with risk markers for cardiovascular disease and diabetes in Indian, Pakistani, Bangladeshi and European adults in a UK population. Journal of Public Health. 2002 September 1, 2002;24(3):170-8.) Heart disease prevalence is higher within the South Asians communities living in the UK when compared to other ethnicities driven probably by a) their lower level of physical activity; b) their lower socioeconomic position and c) their higher prevalence of diabetes. (Zaman et Jemni 2011).

Leave Europe and one observes interestingly different cultural attitudes to obesity. A recent discussion with my postgraduate students centred around the social scenario within central African people some of whom expect that a married man should put on some weight in order to show his prosperity, being thus well fed at home and hence his new wife being congratulated on her choice of spouse, overweight being a sign of wealth and good health. Similarly, my Chinese students mentioned that there is no shame in being obese in certain areas in China.

The European Commission is taking the prevalence of adolescence's obesity very seriously and has put together a long-term strategy to tackle the pandemic from different prospectus. Several EU parliamentary reports highlighted the extent of the issue. Studies of populations from the EU's 28 Member States showed that the majority (39% to 42 %) never exercise nor engage into regular or non-regular sport/physical activities, while only 8% to 9% exercise regularly (Special Eurobarometer 412). Italy in particular, amongst the Southern European countries, seems to embrace the highest prevalence of people who never exercise. Further investigations across the EU state members have shown that children's engagement with sport and/or exercise is even worse: 68% of the under 15 years old never exercised at the end of their education cycle. This figure is also significantly high between the ages of 16-19 years (45%) (Special Eurobarometer 412).

EU experts have led several consultancy programmes to highlight how physical activity could effectively contribute with an action plan that could change lives. Consultations revealed that 'Health-enhancing physical activity' is a top priority amongst the consensus topics in the EU agenda for sport (The European Dimension in Sport).

The EU has set out actions based on priority themes by mid 2013; amongst these themes are the following:

- Put together a monitoring process for children and adolescents obesity;
- Explore ways to promote health enhancing physical activity and participation in grassroots sport
- Identification of a strategy when the prevalence of obesity has risen to a significant number

(European Union Work-Plan for Sport for 2011-2014, 2011/C 162/01)

The EU Commission ACKNOWLEDGES that: *"physical activities and sport can contribute to the achievement of the objectives of the Europe 2020 strategy for smart, sustainable and inclusive growth."* (European Union Work-Plan for Sport for 2011-2014, 2011/C 162/01)

The above cooperation plan puts forward **Physical activity** as one of the most important health determinants in modern society. Regular physical activities/sport can make a major contribution to the reduction of weight and obesity and the prevention of a number of related serious diseases. They have a strong potential to contribute to smart, sustainable growth through their positive effects on education and public health. They help limit the rise in health expenditure by improving the health and productivity of the population and by ensuring a higher quality of life through old age (EU White Paper on Sport COM/2007/0391 final). The EU has set out an **Actions based on priority themes**, amongst these themes are the following:

Explore ways to promote health enhancing physical activity and participation in grassroots sport	*Expert Group 'Sport, Health and Participation'*	*Identification of measures by mid-2013*

(European Union Work-Plan for Sport for 2011-2014, 2011/C 162/01)

In this context, part of aftermath highlights the importance of an early intervention to tackle children's obesity from different sides. Experts agreed that multi-faceted interventions are more successful than single facets. In this context, projects that tackle child obesity from the medical, physical, psychological, social, life style management and also physical education have shown better results than only one or two-sided interventions.

Apart from the above-mentioned benefits, physical fitness is a powerful health marker; it influences many life aspects. It has several emotional, individual, intellectual, financial, and social impacts on persons (Bailey et al, 2013). Schools and colleges represent a perfect environment to collect data related to the current generation's health. They are also ideal sites to undertake pioneering interventions on a wider scale that may impact the future of a country. Many countries have actually recognized the importance of assessing physical fitness in schools. They have included assessment strategies in their education syllabus (Cvejić and Ostojić 2013).

Physical activity, in this context, could be a key health marker in any culture or ethnic minority. There is evidence showing that home, indoors or outdoors activities vary between adults and youngsters, men and women in any cultural context.

The objective of this short communication is to provide an overview of the project, the method applied, its implementation and the very first exercise related results in more than 800 teenagers.

As the project is in process of being implemented in Italy, the below results show the baseline data only.

2. THE ASSO PROJECT

2.1 Method Overview

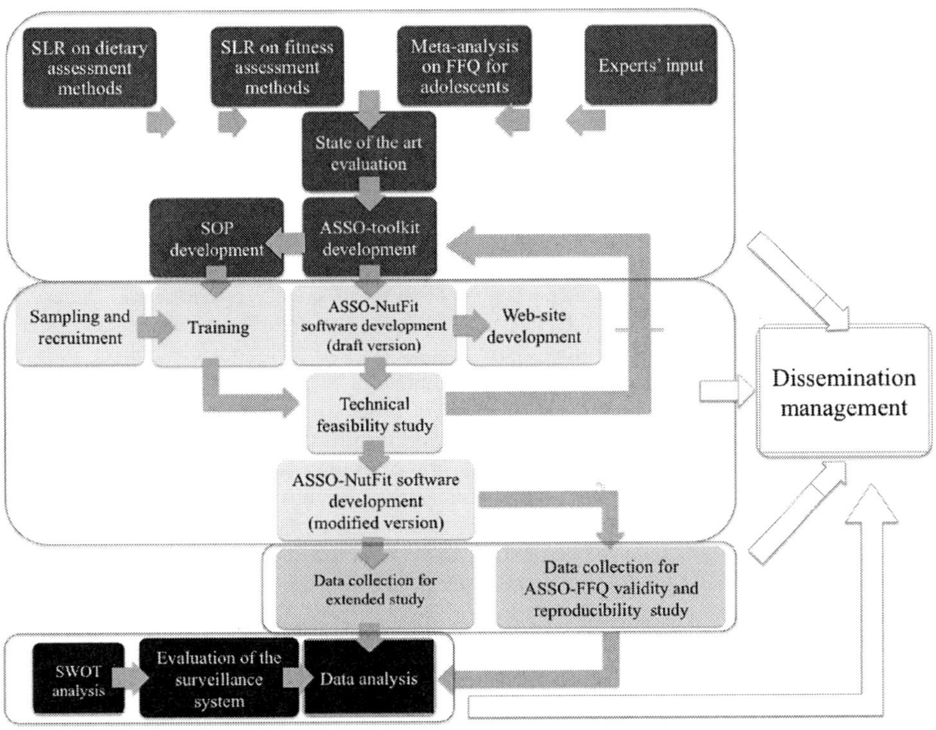

Figure 1. Overall structure of the Adolescents and Surveillance System for the Obesity prevention (ASSO) Project
SLR. Systematic Literature Review; FFQ. Food frequency questionnaire; WP. Work Packages; SOP. Standard operating procedures; SWOT. Strengths, weaknesss, opportunities, threads.

The ASSO Project is inspired by and from the EU policies and guidelines (www.assoproject.info). It addresses key issues related to monitoring scholarly adolescent's health variables from different aspects (body composition, physical fitness, diet, drinking and smoking habits) whilst applying the latest e technology for effectiveness. An important purpose is to monitoring adolescents' behavior towards a more active and healthy life style. The overall aim of the ASSO Project is to create an online database, where prevention tools could be applied when necessary. Monitoring process was put together according to the state of the art literature. There are four habits being monitored in order to give an overall accurate picture on the adolescents' health:

- Dietary habits
- Smoking habits
- Drinking habits
- Exercise habits

All these variables are monitored via a guided system during different periods: at school, at home with the parents/guardians, with relatives or grand parents, during holiday, week days and weekends, as they could vary according to the context, hence providing a holistic picture of the adolescents life style behaviour change in different environments.

2.2 Design and Procedure

The ASSO Project was designed and implemented according to the Figure 1. In order to come up with a relevant and a scientifically based assessment tool for each of the above habits, an analytic review of the current state of the art literature and/or a meta-analysis has been undertaken by a team of experts including local Italian specialists (Bianco et al 2015a; Filippi et al 2014; Tabacchi et al, 2014, 2015, 2016). The teams have then drafted four toolkits relevant to monitoring each area as shown in Figure 1 (from top to bottom):

ASSO-PASAQ: Physical Activity, Smoke and Alcohol Questionnaire;

ASSO-FHQ: Food Habits Questionnaire;

ASSO-FFQ: Food Frequency Questionnaire;

ASSO-BFMT: Body and Fitness Measures Tool.

All the above toolkits have several sections related to the nature of the data to collect. Some are descriptive and others contain photos and short video clips demonstrating the requirements, such as the case for the food portion sizes. The ASSO-FFQ, for example, is structured in three sections (foods, beverages and supplements) (Tabacchi et al 2016). The ASSO-BMFT includes two sections: body and fitness measures, which include the description of the anthropometric parameters and the Fitness Tests Battery (FTB) (Bianco et al 2015a; Bianco et al 2015b); and the Body and Fitness Measures Form, where all collected data can be reported by the operators. Full description of the procedures and techniques to collect body weight, height, waist circumference and for fitness tests were included in the Standard Operating Procedures (SOPs), properly developed to provide standardised methods in order to prevent systematic errors in the collection and reporting of data. The FTB includes five tests: Hand grip test; Standing broad jump test; Sit up test; 4 x 10 m shuttle run test; 20 m shuttle run test. A total time of 30 min per student was needed for the collection of the anthropometric measurements and the performance of all fitness tests; some of them could be administered to more students at the same time, thus reducing the total time needed for the whole class.

A total time of 75 min was required per student to perform the entire ASSO-toolkit, this being considered a reasonable time to promote the introduction of the ASSO in schools as a continuous surveillance system strategy.

Together with the research team, IT engineers have worked closely to create a relevant, easy and user friendly Software called the ASSO-NutFit enabling the users (mainly adolescents and teachers) to enter their data according to each section. Naturally, several software drafts have been piloted before its launch amongst a sample of teenagers.

In order to enable data entry, the ASSO team has run several training days within the schools involved in the research. These have included not only the software data entry but also data collection. Science and physical education teachers have been trained to assist in the questionnaires completion and also to collect the body and fitness variables.

2.3 Subjects

The ASSO Project has been piloted within the city of Palermo, Sicily, Italy. Public and private high schools of every type (college, technical institute and professional institute) were all represented with the aim of including all socio-economic categories of the city. A multistage sampling was applied for the schools and subjects' selection. A total sample size was calculated for an estimate precision of $\pm 3\%$ and CI 95%, evaluated in accordance to the frequency of overweight/obese subjects of 25% in Southern Italy (HBSC Study 2011). Relevant formula to calculate the required sample size was applied (more details could be found in Tabacchi et al, 2015, 2016).

An overestimated sample size of 1096 was required in order to count for the attrition and to get a representative sample size. Hence, seven schools have been selected and around 160 students per school were required. All students were provided with an informed consent to be signed by their parents/guardians and an information sheet. Disabled students and all those students for whom a proper collection of data was not possible, were initially included in the study but not included in the database's elaboration, as well as, any students under 14 years old or over 17 years old. An identification code and a password were attributed to each student in order to respect their privacy and to allow him/her to enter their online data.

2.4 Implementation and Application

The Project was conducted over a period of three years (2011-2014). More than 1000 students provided the requested data, which have been collated in the ASSO-NutFit database, and are at the moment under analysis and elaboration. A SWOT analysis has been elaborated discussed between the team members and recently published to give the readers the finest details about the procedure, the method used and the overall evaluation of the project (Tabacchi et al 2016).

Although, the system has been piloted in a small area of the Italian territory, the first analysis of the data, shows that is it has the potential to be extended to the whole territory and abroad, as it provides a complete framework on all the territory and the possibility of regional data comparisons of the adolescent lifestyles.

3. RESULTS

This section provides the very first insight on the exercise related results in 811 teenagers.

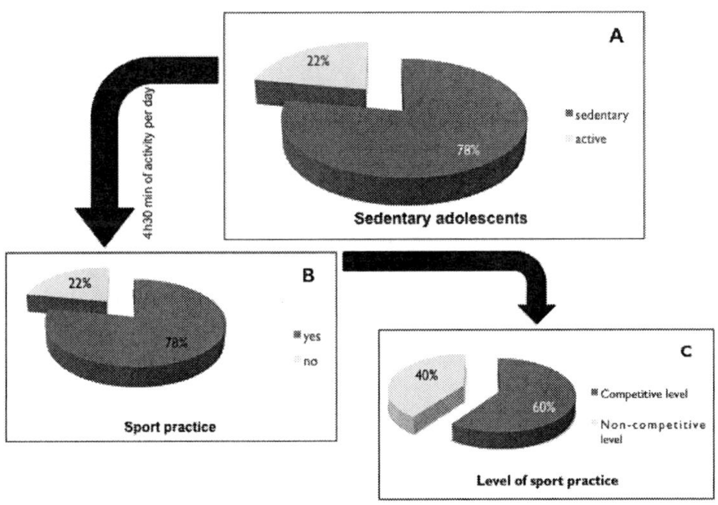

Figure 2. Exercise habits in a sample of 811 Italian adolescents (504 males and 307 females)

Figure 2 provides an overview on the exercise habits in a sample of 811 Italian adolescents (504 males and 307 females). The analysis of the lifestyle questionnaires showed that only 22% amongst the 811 teenagers are active; however three quarters of these 22% are being active because they undertake sports at competitive or non-competitive levels.

One of the most important fitness related result is the cardiorespiratory endurance of these adolescents' sample assessed by the 20 m shuttle run test. It was indeed very weak with no difference between the under 16 and the over 16 years old. This result is indeed quite significant and could trigger some quick action plans to tackle the trend (Figure 3).

Unexpectedly, when divided into age groups, the over 16 years old boys did not significantly out perform the under 16 yrs (6.33 ± 2.41 V 5.91 ± 2.47 stages respectively; $p>0.05$; Fig 3) and similarly when compared the under 16's females to their peers over 16 years (3.90 ± 1.74 V 3.85 ± 1.45 stages, respectively; $p>0.05$, Fig 3).

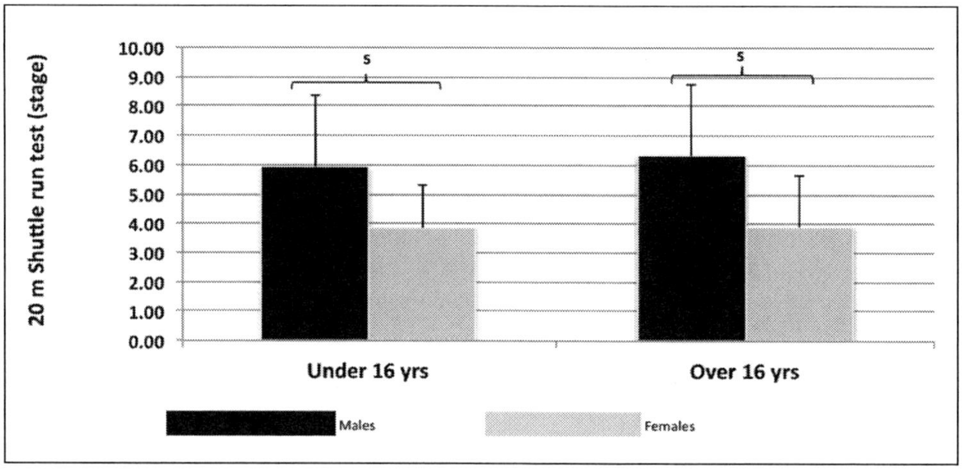

Figure 3. Males and females results of the 20 m shuttle run test per age and gender groups

4. CONCLUSION

The ASSO web-based system design was based on the state of the art research in the life style assessments specifically for teenagers. The ASSO tool kits provide ways to collect live information about the: Physical Activity, Smoke and Alcohol consumption, the Food Habits and frequency not only at home but also in different environment and times of the year. The ASSO Body and Fitness Measures Tool provides an overall assessment of the teenagers fitness level. The very first results of the fitness related results showed only 22% amongst the 811 teenagers were active. The cardiorespiratory endurance of these adolescents was very weak with no difference between the under 16 and the over 16 years old. Such result could be taken seriously by the health authorities to adopt necessary solutions.

Finally, the ASSO e-tool kit was designed to be administered by a leading institution that could give directions to the local government. It could provide live and current evidence-based conclusions and facilitate a network between the Ministry of Health and schools. Accurate reports could be drawn and relevant measures could be applied when necessary, all within a synergic way to achieving the government, the EU objectives and possible saving future lives.

REFERENCES

Bailey R, Hillman C, Arent S, Petitpas A (2013) Physical Activity: An Underestimated Investment in Human Capital? Richard Bailey, Charles Hillman, Shawn Arent, and Albert Petitpas. J Physical Activity and Health, 2013, 10, 289-308.

Bianco A, Jemni M, Ramos J, Thomas E, Tabacchi G. A systematic review to determine reliability and usefulness of the field-based test battery for the assessment of physical fitness in adolescents. The ASSO project. Int J of Occupational Medicine and Environmental Health, (2015a): Vol 28(3): 445-78. Doi: 10.13075/ijomeh.1896.00393

Bianco A, Palma A, Jemni M, Filippi A.R, Patti A, Thomas E, Mammina C, Tabacchi G. A fitness index model for Italian adolescents living in Southern Italy. The ASSO project. J Sports Med Phys Fitness. (2015b) Oct 16. [Epub ahead of print -PMID: 26472604]

Cvejić D PT, Ostojić S. Assessment of physical fitness in children and adolescents. Physical Education and Sport. 2013;11(2):135-45.

European Union SPECIAL EUROBAROMETER 412 - Sport and physical activity

European Union White Paper on Sport COM/2007/0391 final.

European Union Work-Plan for Sport for 2011-2014, 2011/C 162/01

Filippi AR, Amodio E, Napoli G, Breda J, Bianco A, Jemni M, Censi L, Mammina C, Tabacchi G. (2014) The web-based ASSO-food frequency questionnaire for adolescents: relative and absolute reproducibility assessment. Nutr J. 17;13(1):119.

HBSC Italia. Stili di vita e salute dei giovani in età scolare - Rapporto sui dati regionali HBSC 2009 -2010. Istituto Superiore di Sanità 2011. (http://www.hbsc.unito.it/it/images/pdf/hbsc/sicilia_report_hbsc_2010.pdf) Accessed April 2014.

Olshansky SJ1, Passaro DJ, Hershow RC, Layden J, Carnes BA, Brody J, Hayflick L, Butler RN, Allison DB, Ludwig DS. (2005) A potential decline in life expectancy in the United States in the 21st century. N Engl J Med. 17;352(11):1138-45.

Special Eurobarometer 412 - Sport and physical activity. EU commission, Brussels

Tabacchi G, Bianco A, Alessi N, Filippi A.R, Napoli G, Jemni M, Censi L, Breda J, Schumann N.L, Firenze A, Vitale F, Mammina C. Design, Implementation, and Evaluation of the Adolescents and Surveillance System for the Obesity Prevention Project. Medicine, (2016), 95(12) p e3143. Open Access 1 March 2016. doi: 10.1097/MD.0000000000003143

Tabacchi G, Filippi A. R., Amodio E, Jemni M, Bianco A, Mammina C. A meta-analysis of the validity of food frequency questionnaires targeted to adolescents. Public Health Nutrition. (2015), 10:1-16. [Sep 2015. Doi: 10.1017/S1368980015002505)

Tabacchi G, Amodio E, Di Pasquale Ma, Bianco A, Jemni M, Mammina C. Validation and reproducibility of dietary assessment methods in adolescents: a systematic literature review. Public Health Nutrition. (2014), 17(12):2700-14. doi: 10.1017/S1368980013003157

Zaman J, Jemni M. (2011) South Asians, physical exercise and heart disease. Heart. 97(8):607-609.

ESTIMATION OF ACCURACY OF RECOMMENDED DIAGNOSTIC AND TREATMENT ACTIONS BASED ON PRECEDENT APPROACH

Vladimir L. Malykh[1], Igor N. Kononenko[2] and Sergey V. Rudetskiy[1]
[1]Medical Informatics Research Center of the Ailamazyan Program Systems Institute of RAS
Pereslavl-Zalessky, Yaroslavl Region, Russia, 152021
[2]Medical Diagnostical Center 'D+'
Ailamazyan Program Systems Institute of RAS, Pereslavl-Zalessky, Yaroslavl Region, Russia, 152021

ABSTRACT

The paper is devoted to estimation of accuracy of recommended diagnostic and treatment actions based on precedent approach. For a given clinical case, nearest case search is performed in a database of clinical cases according to defined metrics. The case found is used as a diagnostic and treatment recommendation. Based on representative samples, the accuracy of such recommendations was statistically evaluated for seven mostly widespread nosologies.

KEYWORDS

Medical Informatics; Clinical Decision Support; Precedent Approach

1. INTRODUCTION

The review of Butko (1990) gives an overview of materials of the 5th International Congress for Medical Informatics and discusses the problem of medical knowledge formalization. Medical knowledge was supposed to be scientific or empirical. It was stated that models based on scientific medical knowledge have both generalized and limited character, but empirical knowledge cannot be representative and adequate for population as a whole. Thus, formalized medical knowledge consists of scientific theories and medical practice fixed in databases (DB) and knowledge bases (KB). Over the last few years, both components of medical knowledge continue developing. Due to widespread automation of diagnostic and treatment process (DTP), a new class of systems has risen – Medical Information Systems (MISs). Implementation of MISs has allowed enhancing the value of the empirical component of medical knowledge. There is a considerable body of empirical knowledge in databases of MISs that describes and formalizes millions clinical cases in the form of electronic medical records (EMRs). Recently, there is a growing trend towards the creation of national clinical data banks (Herrett 2015). Scientific components of medical knowledge continue evolving rapidly. New diagnostic approaches as well as new methods of treatment and new drugs are emerging. Recently, personalized medicine is rising that uses the patient's genetic composition to tailor strategies for patient-specific disease detection, treatment, or prevention. According to the paper of (Kohane 2009), personalized medicine is the best practice of clinical decision-making such that the decisions made maximize the outcomes that the patient most cares about and minimize the expenses that the patient fears the most, on the basis of as much knowledge about the individual's state as is available. Such points of view upon personalized medicine focus on clinical decision making and reemphasize the urgency and importance of research and development in this field. Clinical practice is, on the one hand, conservative and traditional, it is based on accumulated and verified experience (evidence-based medicine) and, on the other, there is a clear need to continuously harness new medical knowledge. Decision support systems intended to help physicians are threatened by obsolescence and the knowledge they contain is at risk of being outdated, and it is therefore essential to ensure continued formalization of knowledge and its constant updating. According to (Butko 1990), this is one of the reasons for extremely narrow dissemination and few practical implementations of such systems.

Attempts to overcome critical difficulties in creating clinical decision support systems have led to an expansion of alternative approaches.

The first of them is to supply healthcare professionals with relevant sources of information that could help them to independently take decisions. Frankly speaking, a decision support system of such a kind generates no ready solutions but instead recommends that physicians study the information and get their many questions answered. One example of this approach is the UpToDate system [Evidence-Based Clinical Decision Support at the Point of Care, http://www.uptodate.com/home].

The second approach is based on developing highly specialized decision support systems. Examples are Decision Support Systems of (Nazarenko 2015, Palchunov 2015).

The third approach pretends to being global and aims at developing a cognitive system that is able to self-learning and accumulating knowledge from unformalized text sources. One example is the IBM Watson system [IBM Watson, http://www.ibm.com/smarterplanet/us/en/ibmwatson/]. At the XVII Data Analytics and Management in Data Intensive Domains International Conference, DAMDID/RCDL'2015, the presentation (Arutyunyan 2015, Gavrilov 2015) devoted to the Watson system showed that the results of cognitive approach to creating a clinical decision support system appear to be very poor. Designing a highly specialized decision support system (oncology) led to a long-term training of the system by a team of experts. The hope that a Decision Support System (DSS) could be able to adopt medical knowledge and effectively use it is not materializing.

All of the above-mentioned approaches have shortcomings as well, such as the need of involving experts, regular updating of knowledge bases, and they are highly specialized.

In this paper, we consider a universal precedent-based approach to designing DSS. It is based on a large volume of formalized and verified clinical data (Big Databases). The main idea of precedent-based approach is quite simple – to find a most similar case in a database and use it to support clinical decision making. Precedents of clinical cases could be filtered taking into account the credibility of medical organizations in which precedents were created, the credibility of general practitioners who authored those precedents, and the relevance of precedents to the modern medical technologies.

To the success of precedent-based approach, there is a need for representative databases of clinical cases. (Herrett 2015) gives an example of the primary care database of anonymized medical records (EMRs) from general practitioners, with coverage of over 11.3 million patients from 674 practices in the UK.

The authors of this publication were involved in designing the INTERIN PROMIS MIS [http://www.interin.ru/] which is widely used in major healthcare organizations of Moscow, Russian Federation. Our estimates suggest that nowadays there is about 3 million EMRs in INTERIN PROMIS. It means that it is possible to create widespread representative nationwide DBs of clinical cases. Precedent-based approach focuses on practical use of knowledge from such DBs. The Institute for System Programming of the Russian Academy of Sciences (ISP RAS) (Karpov 2010) pioneered the use of precedent-based approach for clinical decision support in Russia. Independently, experts of Ailamazyan Program Systems Institute of the Russian Academy of Sciences (Ailamazyan PSI of RAS) carried out R&D in the field of implementation of precedent-based approach in clinical practice (Malykh 2014, Malykh 2015). The research was supported by the Ministry of Education and Science of the Russian Federation and by the Russian Foundation for Basic Research. As a conception, precedent-based approach was represented at the 15th World Congress on Health and Biomedical Informatics (Malykh 2015). Ailamazyan PSI of RAS had a certain advantage because the specialists of this Institute had an access to representative Big Data Bases of clinical cases of the INTERIN PROMIS MIS. In 2015, investigations on statistical evaluation of accuracy of recommended diagnostic and treatment services generated with precedent-based approach were performed using clinical data of the Clinical hospital of the President Administration of the Russian Federation.

In this paper, the authors present the first results of their research. These results are of critical importance as they allow to evaluate both effectiveness and accuracy of precedent-based approach in decision making systems.

2. PRECEDENT APPROACH TO DECISION MAKING IN CLINICAL PROCESSES

2.1 Model and Methods

For further and deeper understanding of the target formulation, we need to quote the research conducted by (Malykh 2015). Modern medical information systems keep electronic medical records (EMRs) and contain information on millions of various clinical cases formalized to a certain degree. The degree of formalization of the stored clinical data in various MISs may vary. Up-to-date medical information systems contain models of DTP as a time sequence of management events (which include ordering of diagnostic tests and administering of treatment) as well as monitoring events (which describe the patient's state). Management events are better formalized; hospitals keep statistical and economic records on them which include maintaining registers of services rendered, invoicing, planning and centralized control. Medical data associated with monitoring and diagnostic tests are insufficiently formalized.

One of our previous studies describes the possibility of modeling DTP using controlled stochastic Markov processes (Malykh 2014). This model is based on the assumption that DTP is a controlled process. The model sets forth a management notion, u, and a state notion, x. Management means decisions made by a physician and implemented in the future. Management is limited by prescription of various diagnostic and treatment measures by a physician, including diagnostic testing, drug and treatment administration, surgical treatment as well as manipulation, etc. The physician's choice is based on the accumulated medical knowledge about how a certain disease should be treated and on the physician's individual experience. Rather than taking into account all subsets of DTP elements as management options, the physician considers only those subsets that were previously used in similar situations and proved their clinical effectiveness. Management is explicitly case-based reasoning.

Management in the current situation (x^i, u^i) will be determined not only by the state (x^i) but also by the whole history of the process and management at earlier steps of DTP $\{i, i-1, i-2, ...\}$. This is determined by the specific nature of the treatment process. The physician makes decisions based on the life history, disease history, family history, history of allergies and the entire course of the given clinical case. To take account of the DTP memory effect, it was proposed that the data from the above-mentioned histories are included in discrete states of the process, with management integrated. Each control element of the DTP may be associated with a certain integral characteristic of application of this element in DTP. For example, the total dose consumed by a patient will be considered the integral characteristic for a drug and a total radiation dose will be considered the integral characteristic for radiation therapy. Frequency of application of a certain element can also be considered as an integral characteristic (e.g., the number of electrocardiographic examinations).

Figure 1 represents a controlled discrete process with memory where each state **x** contains information about the integral management characteristics **u**, observed values **v** and history values **A** (anamnesis) relevant to a process.

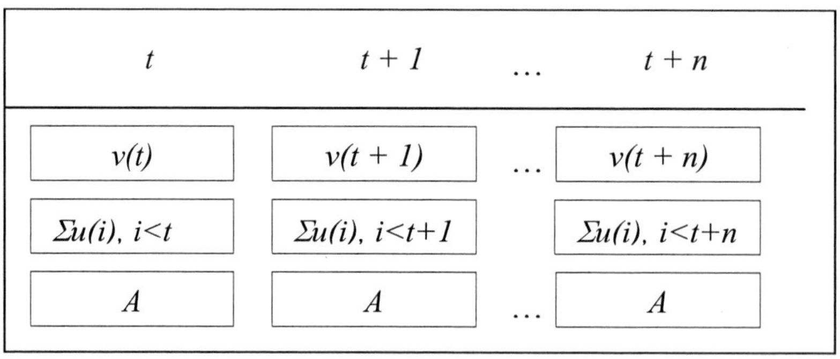

Figure 1. The DTP model

Modeling the DTP with controlled stochastic Markov processes seems appropriate in such cases (Bennett 2013, Malykh 2014), particularly for hospitalized patients because of the clearly manifested periodic observation as well as physician's decision making.

Precedent-based methods of decision making, including physician's decision support, are well known and described in many papers (Zagorujko 1999, Karpov 2010, Varshavsky 2013). The method requires to introduce a metric or a distance into the model to find the nearest neighbour of a given reference case. In the study of (Malykh 2014) an approach to creating a metric for the model was described, including a metric assessment with parameterized measure. In our case, we consider that the most appropriate metric is the weighed sum of rated distances determined for every characteristic of the state. Presence of weights allows to solve the task of training Physician's Decision Support Systems (PDSS) by optimization of metrics. In this study, optimization was not performed, weightages of one point were given, which means an equal informative value of every characteristic of patient's state. Description of a patient's state was not compressed, and the number of characteristics of a patient's state was not reduced.

Metrics could be calculated not only between pairs of states but also between processes. It is also possible to take into account all the process history in metrics defined for a pair of states (Malykh 2015). The Dynamic Time Warping method (DWT) seems promising for building such metrics.

For solving the task of nearest neighbor search, researchers use data structuring on small world graphs (Malkov 2014). By computing experiments, creation of small world graphs with 8 edges for each graph node was chosen. Experiments showed that for properly solving this problem, it was enough to randomly choose only 1% of the total number of graph nodes as a starting point when searching the nearest neighbor, and then, to traverse the graph and find the shortest path between those nodes and the given reference state. Small world graphs were drawn layered, with each layer representing a day of hospitalization (1st day of hospitalization, 2nd day of hospitalization, etc.). The nearest neighbor of a given reference state in the control sample was defined in the layer corresponding to the day of hospitalization from the reference. In average, for a learning sample with a 1000 DTP capacity, it was enough to randomly choose 10 nodes. This approach allows to quickly find the nearest neighbor and generate real-time recommendations about DTP for a physician. Calculating the structure of a small world graph is rather labor-consuming as the volume of calculations needed is proportional to the square root of the number of graph nodes. Adding a new precedent (a new node) into the DB requires the number of metric calculations equal to the number of nodes.

2.2 Results

Based on existing evidence of finished clinical hospitalization cases, models of seven nosologies were calculated (see Table 1). The models contain 8015 clinical cases including 91040 states which represent an average of approximately 11 days of hospitalizations per case. The number of cases for every nosology exceeded 1000, with the exception of J13 nosology which was specially chosen for debugging algorithms and programs because of smaller number of cases. For every nosology, control sample covering 10% of all the cases of a given nosology and spread throughout the period was chosen from all the cases. Every DTP lasted an integer number of days. Patient's state was fixed once per day. Every state was described by a set of characteristics having certain general dimensional values. A dictionary of characteristics describing the states of modeled DTP was created for every nosology. According to the adopted model, management implying diagnostic and treatment actions prescribed by a physician was defined among the set of characteristics. Management values were integrated according to the adopted model. Integral management characteristics are included into the state of the managed object thus reducing memory effect, which allows to get closer to a process where management becomes a function of state. Management capacity was rather high, ranging between 10^2 and 10^3 depending on the nosology.

Within every particular nosology for all states of the control sample which is not terminal (the latest states in DTP implementations), the closest neighbor states were calculated using the defined metric and data from learning sample structured on small world graphs. If closest neighbor states were not terminal in their DTP implementation, integral characteristics of all managed variables for a state following the closest neighbor were accepted as diagnostic and treatment recommendations. Recommended values of integral characteristics of managed variables were compared with actual integral characteristics of corresponding states from control sample. Depending on the resulting comparison, every generated recommendation was allocated into one of two classes: 1) recommended management level for a given characteristic corresponds to the observed control level; 2) recommended management level for a given characteristic does not correspond to the observed control level. These two classes of management are shown in columns 3 and 4 of Table 1. The third class of management corresponds to the cases in which, in control sample, a physician applied management actions (i.e. diagnostic and treatment actions) which were not included into the list of recommendations. This class of management is shown in column 5 of Table 1. For every class, a total number of cases was calculated and a relative number of cases showing the percentage for every class. The results are shown in Table 1.

Table 2. Evaluation of accuracy of diagnostic and treatment recommendations for seven nosologies

ICD-10 Code / Name	**Power of clinical precedents / Power of control set** Number of States / Number of Controls	The number of accurate recommendations Total / In percent	The number of recommendations not matched by value Total / In percent	Could not recommend Total / In percent
J13 / Pneumonia due to Streptococcus pneumoniae	**166 / 11** 2938 / 118	2865 / 34.4%	3923 / 47.2%	1530 / 18.4%
K80.1 / Calculus of gallbladde with other cholecystitis	**1018 / 128** 12853 / 931	16078 / 35.8%	18390 / 40.9%	10490 / 23.3%
H25.1 / Age-related nuclear cataract	**1205 / 121** 5509 / 293	2983 / 80.4%	539 / 14.5%	189 / 5.1%
H26.2 / Complicated cataract	**1255 / 126** 5778 / 249	2745 / 57.5%	1617 / 33.9%	408 / 8.6%
I67.4 / Hypertensive encephalopathy	**1336 / 134** 23165 / 1431	28115 / 31%	37563 / 41.4%	25060 / 27.6%
I67.9 / Cerebrovascular diseas unspecified	**1403 / 141** 24875 / 1518	26202 / 33.7%	32447 / 41.7%	19117 / 24.6%
N20.1 / Calculus of ureter	**1632 / 164** 15922 / 205	7541 / 25.3%	9948 / 33.4%	12291 / 41.3%

3. CONCLUSION

The results given in Table 1 speak for themselves. In average, for all of seven considered nosologies, the results are clear: 1) in 33% of cases, the method permitted to generate absolutely precise diagnostic and treatment recommendations; 2) in 40% of cases, a physician acts according to generated recommendations but changes their integral level (dosage, etc.); 3) in 27% of cases, a physician prescribes diagnostic and treatment actions that are not included in generated recommendations. Account should also be taken of the fact that there were no system training and, thus, the obtained results can potentially be improved. One of the advantages of precedent-based approach is that there is no need to constantly involve and formalize new knowledge. New knowledge (new treatment methods) 'falls' within the precedents as it is inputted into EMRs when databases of MISs are being updated. With this, we could make quite a general conclusion that supporting an empirical component of medical knowledge in MIS requires less effort then formalizing and supporting a scientific one. This offers encouraging prospects for designing and developing decision support systems for physicians based on empirical components of medical knowledge. This approach also corresponds to existing case-based character of management and decision making in medical practice. So far, the results indicate that precedent-based approach has a high effectiveness and could naturally enhance other approaches to supporting physicians decision making, particularly knowledge-based ones.

The constraints of precedent-based approach include the need for a representative database of verified precedents excluding medical errors. From another perspective, precedents with corrected errors are of particular interest to physicians training and further prevention of such errors. The information about the results of these errors and possible ways of correcting them is also valuable. Thus, precedent-based approach could be widely spread as an educational tool. On the other hand, the precedent-based approach does not imply formalization of medical knowledge, which entails poor cognitive justification of generated recommendations. Consequently, justifications only describe how other patients were treated in similar clinical cases. There are also problems with optimization of provided metrics, compression of state descriptions, and construction of training procedures. These problems are connected with high dimensionality of the space of state characteristics and samples of clinical precedents. However, discussion of these issues and possible ways of addressing them has been left outside of this research.

There are two other factors that draw attention to systems based on precedent approach.

First of them is that there are trends in the development of our civilization which include an explosive development of information technologies (among them M2M, Big Data and IoT), their strong need for formalized knowledge and practical absence of qualified experts who could formalize that knowledge. The chief editor of the Rational Enterprise Management (REM) magazine (Russia) holds regular discussions on a wide range of problems including the above mentioned. Results of the discussions are published in the REM editor's column. The guests of a recent discussion (Vasilyeva E. 2015) included Igor Rudym (Intel), Dmitriy Tameev (PTC), Alexander Belotserkovskiy (Microsoft), Igor Girkin (Cisco), Igor Kulinitchev (IBM). All the participants agreed that, nowadays, the key challenge of IT development is not associated with hardware or software, but it needs breakthrough approaches to data analysis.

As for the second factor, it is obvious that nowadays there are no qualified experts in the field of knowledge even in key branches. The actual situation is even more critical as the experts who are able to solve at least a part of these problems are not able to cope with ever increasing information flow. From this point of view, precedent-based DSSs practically need no experts. Experts may be needed for enhancing or optimizing existing DBs and KBs.

ACKNOWLEDGEMENT

The study is supported by the Ministry of Education and Science of the Russian Federation (project No. RFMEFI60714X0089) and grant No. 13-07-12012 provided by the Russian Foundation for Basic Research.

REFERENCES

Book

Zagorujko N.G., 1999. *Applied methods of data analysis and knowledge (in Russian)*. Novosibirsk: Institut Matematiki im. S.L.Soboleva SO RAN [Sobolev Institute of Mathematics of the Siberian Branch of the Russian Academy of Sciences], Russia .

Journal

Bennett C. and Hauser K., 2013. Artificial Intelligence Framework for Simulating Clinical Decision-Making: A Markov Decision Process Approach. *Artificial Intelligence in Medicine*, No.1 (2013), pp. 9–19.

Butko S.N. and Olshanskiy V.K., 1990. New medical support systems abroad. *Automation and Remote Control* No. 6 (1990), pp. 3-19.

Herrett E. et al, 2015. Data Resource Profile: Clinical Practice Research Datalink (CPRD). *International journal of epidemiology*. 2015, pp. 827–836. doi: 10.1093/ije/dyv098

Karpov L. and Yudin V., 2010. The case-based software system for physician's decision support. *Lecture Notes in Computer Science*, Vol. 6266, pp. 78-85. DOI: 10.1007/978-3-642-15020-3_7

Kohane, I.S., 2009. The twin questions of personalized medicine: who are you and whom do you most resemble? *Genome Med*, Vol. 1, No. 1, Article 4.

Malkov Y. et al, 2014. Approximate nearest neighbor algorithm based on navigable small world graphs. *Information Systems*, Vol. 45, pp. 61-68.

Malykh V.L. and Guliev Y.I., 2014. Mathematical model of clinical processes implemented in the class of stohastic Markov control processes with memory (in Russian). *Informacionnye tehnologii i vychislitel'nye sistemy [Information technologies and computing systems]*, No.2 (2014), pp. 62-72.

Malykh V. and Guliev Y., 2015. Precedent approach to decision making in clinical processes. MEDINFO 2015: eHealth-enabled Health. Studies in Health Technology and Informatics, v.216, pp. 957, 2015, I.N. Sarkar et al. (Eds.) © 2015 IMIA and IOS Press. doi:10.3233/978-1-61499-564-7-957

Varshavsky P. and Alekhin R., 2013. The method of finding solutions in intellectual decision support systems with case-based reasoning (in Russian). *International Journal "Information Models and Analyses"*, No.4 (2013), pp. 385-392.

Vasilyeva E. 2015 Industrial Internet of Things (IoT) (in Russian) *Rational Enterprise Management*, No 4 (2015), pp.8-14;

Conference paper or contributed volume

Arutyunyan G.R. and Dmitriev A.S., 2015. IBM Watson Oncology Solution: Application Experience. *Data Analytics and Management in Data Intensive Domains*, XVII International Conference DAMDID/RCDL'2015/ October 13-16, 2015, Obninsk, Russia. Proceedings of the Conference, pp. 140-142.

Gavrilov A. and A.S.Semenikhin. New Era of Cognitive Computing: IBM Watson. *Data Analytics and Management in Data Intensive Domains*, XVII International Conference DAMDID/RCDL'2015/ October 13-16, 2015, Obninsk, Russia. Proceedings of the Conference, pp. 135-139.

Nazarenko GI et al, 2015. Multifunctional clinical decision support system based on clinical practice guidelines. *Proceedings of 2015 International Conference on Biomedical Engineering and Computational Technologies (SIBIRCON)*, Russia, Technopark of Novosibirsk Akademgorodok, 28-30 October 2015, pp. 72-75. IEEE Conference Publications.

Malykh V.L. and Belyshev D.V., 2015. Case-based Reasoning in Clinical Processes Using Clinical Data Banks. *Proceedings of 2015 International Conference on Biomedical Engineering and Computational Technologies (SIBIRCON)*, Russia, Technopark of Novosibirsk Akademgorodok, 28-30 October 2015, pp. 211-216. IEEE Conference Publications DOI: 10.1109/SIBIRCON.2015.7361885

Palchunov D. et al, 2015. Software system for the diagnosis of the spine diseases using case-based reasoning. *Proceedings of 2015 International Conference on Biomedical Engineering and Computational Technologies (SIBIRCON)*, Russia, Technopark of Novosibirsk Akademgorodok, 28-30 October 2015, pp. 205-210. IEEE Conference Publications.

EVALUATION OF CAREGIVERS' FIRST TIME USE OF A SEAMLESS DEPLOYMENT PLATFORM

Stefan Wagner[1], Femina Hassan Aysha Beevi[1], Christian Fischer Pedersen[1], Jorge Miranda[2], Jorge Cabral[2] and Blaise Ravelo[3]
[1]*Aarhus University, Department of Engineering, Denmark*
[2]*Universidade do Minho, Centro Algoritmi, Guimarães, Portugal*
[3]*IRSEEM, Engineering School ESIGELEC, Saint-Etienne-du-Rouvray Cedex, France*

ABSTRACT

The CareStore project is investigating the feasibility of creating an open and flexible infrastructure for facilitating seamless deployment of healthcare devices and applications. Such seamless deployment is relevant in order to optimize work flows and save resources, but must deliver a calm user experience in order to be effective. This paper describes a study with the aim of investigating the user experience of non-technical care staffs' first time use of a technology platform designed for supporting seamless deployment of telemedicine and ambient assisted living equipment in care facilities. We invited eight caregivers working at an assisted living ward of a North German care facility, two nurses and six otherwise trained caregivers, to go through a realistic deployment and usage scenario. The specific objectives were to: i) evaluate the feasibility of the system during first time use, and ii) gain insights into the caregivers understanding and user experience of the technology. Results indicate that the caregivers could effectively use the touch-to-install deployment system to deploy a new healthcare device and carry through measurements with the device. Also, the nurses found the system effective and easy to use in a first time context, and were optimistic about future possibilities with the system. However, more work is needed to assess the longitudinal effects of this type of seamless deployment technology, including field studies of actual daily usage.

KEYWORDS

Ambient assisted living, pervasive healthcare, telemedicine, seamless deployment, calm technology.

1. INTRODUCTION

The developed countries are experiencing a demographic shift with a rising population of seniors and chronic patient groups, causing significant pressure on national healthcare systems and resulting in fewer staff resources being available to treat more patients than today (Keye and Zitzelberger 2006, Dall et al. 2013).
To mitigate this challenge the pervasive healthcare field aims at optimizing work flows through the use of improved healthcare technologies, including healthcare devices suited for self-measurement efforts (Varshney 2009, Bardram 2008).

Easy deployment of such healthcare devices into nursing homes and other settings of intended use remain to be a challenge, especially in terms of introducing new devices without relying on technical staff (Wagner et al. 2013). The CareStore project is investigating the feasibility of creating an open and flexible infrastructure for facilitating seamless deployment of healthcare and ambient assisted living (AAL) devices and applications on heterogeneous platforms, with a focus on providing a calm technology experience for its end-users (Weiser and Brown 1996).

The aim of this study was to investigate the feasibility of supporting seamless deployment of telemedicine and AAL equipment by non-technical care staff in a care facility setting. Specifically, we wanted to learn how a group of caregivers, nurses and other trained staff performing care tasks at a nursing home, would experience deploying and using a new healthcare device during first time use.

This include evaluating the user experience of caregivers using the seamless installation features, also called "touch to install", of the common recognition and identification platform (CRIP) developed as part of the CareStore project (Wagner et al. 2013). Also, we wanted to learn how the nurses would experience automatic downloading of drivers and subsequent automatic capture and sending of healthcare measurements to relevant care applications. Finally, we wanted to learn whether the caregivers would initially accept using biometrics and radio-based credentials to login to a touch screen care platform designed to be used in the home of the individual residents of the care facility.

2. PRELIMINARIES

In order to understand the study's experimental setup, this section introduces the relevant technical components of the CareStore platform that was used for the evaluation.

2.1 CareStore

CareStore is a platform for enabling seamless deployment of AAL applications and devices in care facilities (Wagner et al. 2013). The platform consists of the following three main components:

i) The Common Ambient Assisted Living Home-Platform (CAALHP), which is an open-source runtime execution platform for reliable and secure execution of services, applications, and healthcare device drivers, as well as a user interface for staff and residents;

ii) The Common Recognition and Identification Platform (CRIP). which is used for identifying staff, residents, and devices using radio-based and biometrics-based recognition and identification technologies (Miranda et al. 2015), and

iii) The CareStore Marketplace, which is an online shared platform for uploading and storing healthcare and AAL drivers and applications, and for allowing CAALHP and CRIP to download drivers and applications in an automatic and seamless fashion.

2.2 User Interface

The CAALHP platform is typically deployed on a Windows Embedded touch screen computer mounted near the entrance at the home of the individual resident. The CRIP is usually mounted below the CAALHP.

The HomeScreen application (see Figure 1) is the starting point of all user interaction with the CAALHP. From here, a range of care and healthcare applications and services can be accessed.

Depending on which user is logged in, e.g. a visiting caregiver, the HomeScreen will display relevant applications for the user to interact with. The available applications are either installed initially when the system is deployed in the home or when a caregiver brings a new healthcare device to the home of the patient.

Installation of new applications and drivers can either occur explicitly when the user selects the Marketplace app and then selects a CareStore application for download or through automatic deployment using the CRIP to identify devices and download CareStore drivers and applications (Wagner et al. 2013).

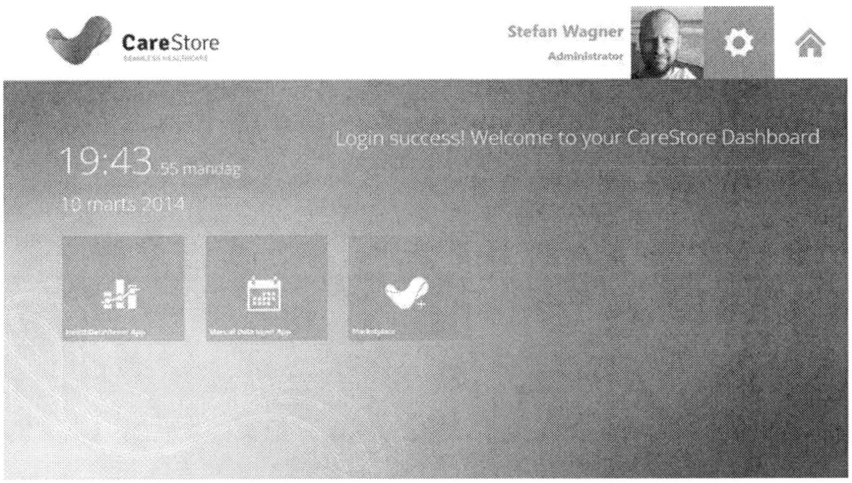

Figure 1. The HomeScreen application showing the installed applications as a collection of clickable icons, each of which will activate an application, such as the Healthcare Data Viewer app

2.3 Seamless Installation with CRIP

If a caregiver wants to deploy a new saturation device, such as the Nonin 9560 blood oxygen saturation device (Nonin Medical Inc. 2014), and associate it with the resident, she can do this by touching the saturation device on the CRIP platform utilizing the built-in near field communication (NFC) reader to read its contact information.

The CRIP in turn identifies the device via the Marketplace web service, and then contacts the CAALHP, which automatically registers that the healthcare device, the caregiver, and the resident, should be joined into a device/user group. Then, the CAALHP automatically starts the download and installation process. After this, all measurements made on the Nonin 9560 device will be associated with the patient and made available to all CareStore applications running on the CAALHP.

The CRIP can automatically detect a range of devices using automatic radio discovery techniques based on the ISO/IEEE 11073 protocol and Bluetooth. However, for most devices, an NFC label is required to make this connection, where a unique identifier is stored on the label and read by the CRIP, after which the CRIP receives relevant MAC addresses and pin codes for pairing with the healthcare device.

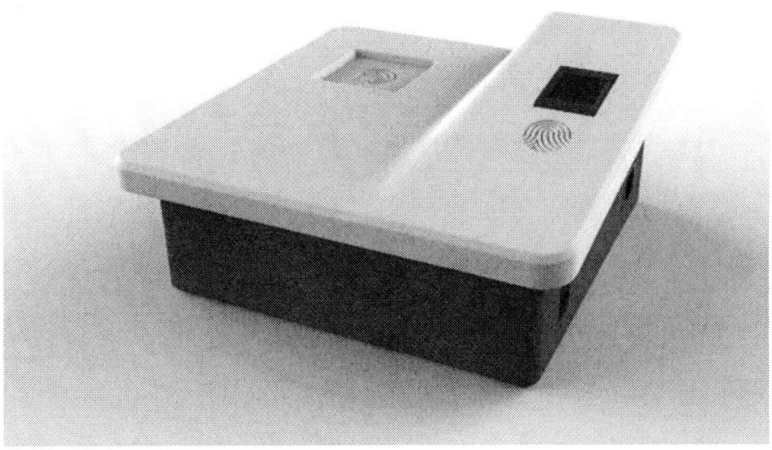

Figure 2. The CRIP unit featuring Bluetooth and NFC radio interfaces for user and device discovery and access control, as well as fingerprint biometrics for additional access control security

2.4 The Healthcare Data Viewer Application

The Healthcare Data Viewer application shown in Figure 3a is an example of an application wishing to subscribe to healthcare data events from new devices being added by the CRIP. This application is automatically coupled to the Nonin 9560 driver following the successful installation of the driver.

Data arriving from healthcare measurement devices attached to the CAALHP causes an instant update of the user interface. Also, a list of historical measurements is fetched from one or more data services upon opening the application. A local healthcare data service exists, while other data services include electronic patient record systems or telemedicine systems, such as the OpenTele open source platform.

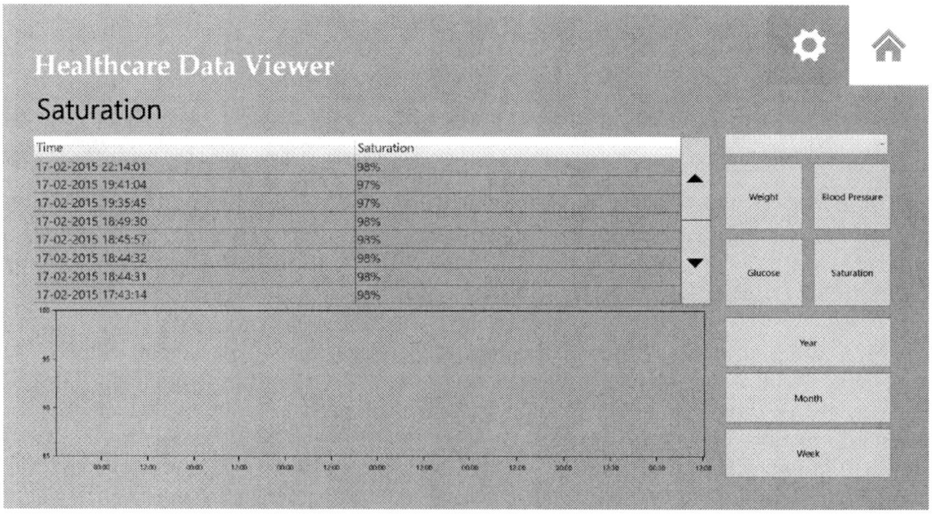

Figure 3a. The Healthcare Data Viewer application showing saturation data of the resident

In our study, only saturation was used for evaluation. However, the preceding demonstration also included a range of other examples, including blood pressure as shown in Figure 3b.

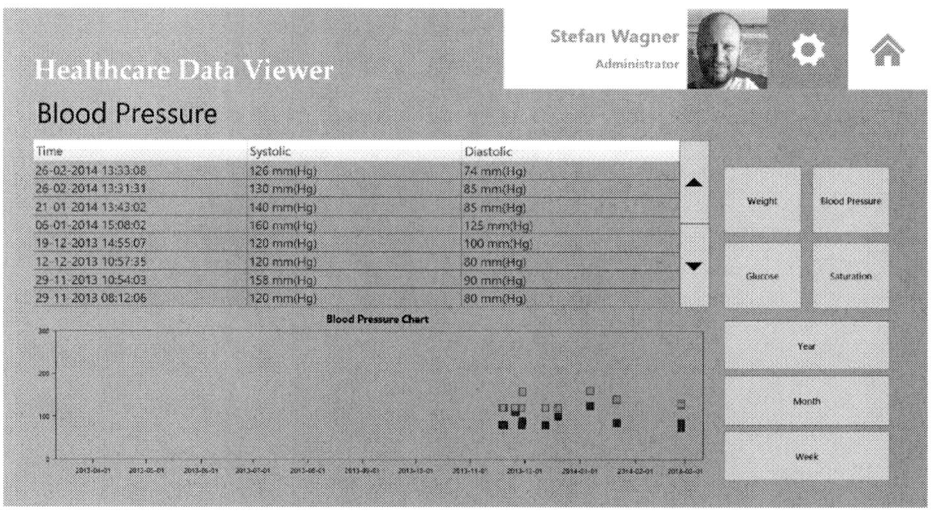

Figure 3b. The Healthcare Data Viewer application showing blood pressure data of the resident

3. MATERIALS AND METHODS

3.1 Evaluation Setting

The evaluation was performed at the Schloßsee Senioren Residenz Glücksburg GmbH & Co KG (SSRG) in Glücksburg, Germany. The SSRG operates 100 assisted living apartments, a patient care service and a nursing station. In addition there is a separate physiotherapy department. Specifically, the evaluation took place at a staff office at SSRG, which is used for project evaluation purposes and interviews.

3.2 Experimental Setup

The experimental setup consisted of a Lenovo Flex 20" touch screen computer running the CAALHP software on Windows 7 Embedded, attached to a CRIP unit. At SSRG one full ward has been equipped with 17 touch screen units, each with a CRIP mounted below it for identification and device registration. The evaluation was done before the care staff was introduced to using healthcare devices for monitoring as part of their daily routine. Thus, the technology was unknown for them in advance.

3.3 Participants and Evaluation Scenarios

We enrolled two nurses and six care staff members from SSRG. All participants were asked to give their informed consent prior to study participation. The study was divided into two parts: i) an observational usability study, and ii) a semi-structured interview on technology comprehension and abilities of the participants. During part i), for each participant, we first provided instructions on how to complete two specific scenarios.

The first scenario revolved around the participant bringing a new healthcare device to the home of a resident, a Nonin 9560 saturation device, and using the "touch to install" feature of the CRIP unit to install the device on the touch screen computer running the CAALHP application.

The second scenario revolved around downloading a health data viewer application from the CareStore Marketplace using the CAALHP Marketplace application, then letting the participant perform a measurement using the healthcare device installed during the first scenario, after having authorized themselves using the CRIP with a combination of fingerprint and NFC identification. Finally, the participants had to start the health data viewer application on the CAALHP and find the self-measured data in the health data viewer application. During part ii), we did a follow-up interview with a structured part, followed by an open-ended part.

3.4 User Experience Evaluation Methods

We asked participants to perform the evaluation of the two scenarios using the think-aloud protocol (Preece, et al. 2015). In this way, we were able to follow the conceptual model of the individual participants. The aim was to time the performance of each participant and identify any major usability problems. All results below 5 minutes were labeled as a successful outcome, as long as the objectives were met without assistance or major breakdowns.

Next, we conducted a semi-structured interview, where we obtained gender, age, formal education, employment period at SSRG, and years of experience with nursing and care work. We also asked a range of closed questions on: i) existing computer skills and competencies as well as on ii) installation of equipment in the home of the residents.

Finally, we invited the participants to provide open comments on the system.

4. RESULTS

A total of eight participants, ages 22 to 59 years with work experience as formal caregivers ranging from 1 to 40 years (average of 15 years) participated in the study.

4.1 Part i) Think-Aloud User Study

All eight caregivers successfully completed the two think-aloud scenarios, spending an average of 82 seconds to complete these. As can be seen from Figure 4, the time to complete the two tasks varied between 56 and 125 seconds.

For 7 out of the 8 caregivers, scenario 2 took longer than scenario 1, and all caregivers were able to solve the scenarios without assistance and within the 5 minutes target time limit.

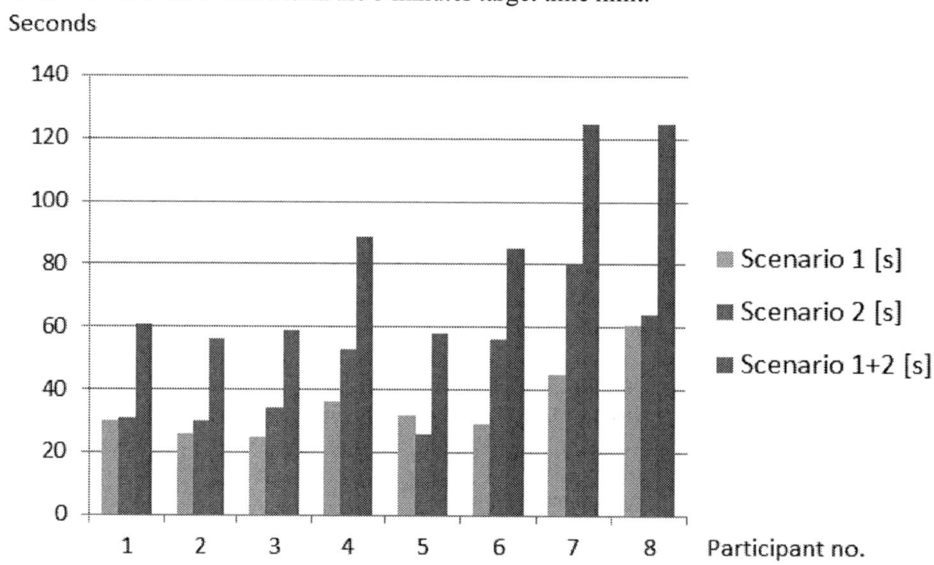

Figure 4. Shows the time the eight participants spent completing each of the two scenarios, as well as the sum of the two scenarios

4.2 Part ii) Semi-Structured Interview

The results of the semi-structured interview are divided into two parts: i) findings from the structured (closed) interview questions, and ii) findings from the open-ended part of the interview.

4.2.1 Findings from Structured Part

Figure 5 illustrates the results of the structured part of the interview session. Data can be seen as the percentage of respondents who either agrees or strongly agrees to the given questions 1.1 to 2.8.

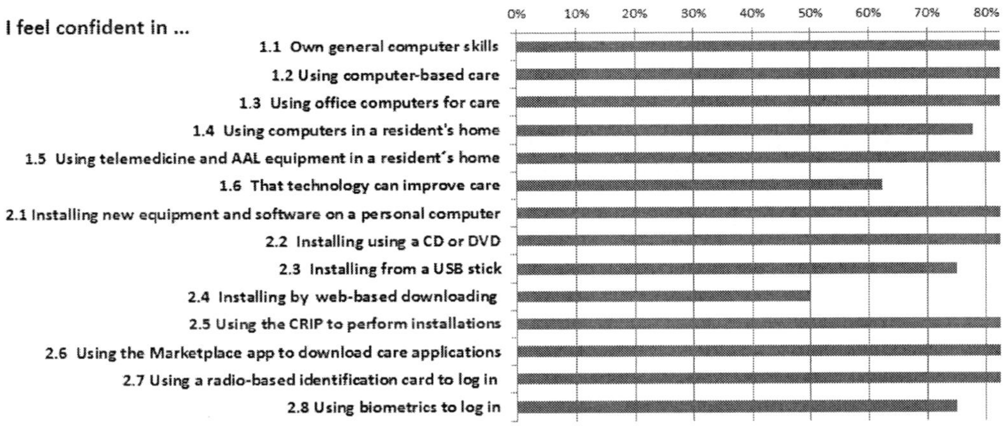

Figure 5. Percentage of interviewed care staff that either agreed or strongly agreed on the interview questions 1.1 to 2.8

4.2.2 Findings from Open-Ended Part

The caregivers provided a range of open-ended answers during the open part of the interview session. Overall, the eight respondents felt that the system was easy to use: "That was easy", Participant 1 (P1), "That's easy" (P2), "Easy!" (P4), "I find it really terrific" (P6), "My confidence in technology is not always too high. But I find this very easy and very smart!" (P8), and "I am really excited to see what happens in the future" (P2).

The caregivers were positive towards using new technology, but there were some doubts on whether the residents would welcome new technology: "Technology is a good thing - it has a very strong transparency - but I don't know whether the residents agree on this" (P1). Some participants reflected about the wider implications of introducing such new technologies: "I find it really good that we can take healthcare measurements in the rooms of the residents. But it could be that we need to bring too many things at a time to do it." (P2), and "For some residents I feel positive. For others, I am not sure." (P3).

Biometrics was less popular with some of the participants "I am not a fan of using biometrics" (P1) and "I don´t feel confident in using fingerprints. It takes too long, should work fast!" (P7), while ID cards raised doubts with others: "I think it is not hygenic to use ID card and you can lose it" (P5).

5. DISCUSSION

The think-aloud evaluation (part i) found that for this group of caregivers, the ability to complete the two given scenarios during first time use was effective (in terms of the ability to solve the given tasks) and efficient (in terms of performance).

All of the participants were able to solve the two test scenarios in around two minutes time or less, without having used the equipment prior to the evaluation and without being given training, besides a short oral instruction.

Also, no usability breakdowns were noted during the think-aloud evaluation. Although the limited number of participants (n=8) poses a limitation to the study, the participants ability to achieve rapid and seamless deployment of new healthcare devices and subsequent succeeding in performing the measurement task and the measurement lookup task, provide a good indication of the potential of the "touch to install" seamless approach.

Results from the second part of the study (part ii) show that most of the participants considered their own technology skills to be overall very high (Figure 2, questions 1.1 to 1.5). Furthermore, 5 out of 8 participants found that technology has the potential to improve care (question 1.6). All of the participants felt confident in using CRIP to install new devices (2.5), as well as using the CareStore Marketplace application to download new care applications (2.6). Furthermore, 7 out of 8 participants felt confident to use radio based login mechanisms (2.7), while only 6 out of 8 participants felt confident using biometrics as a login mechanism.

6. CONCLUSION

The study found that the participating caregivers could effectively and efficiently use the touch-to-install deployment system to deploy a new healthcare device and carry out measurements with the device.

Also, the participants found the system easy to use in a first time context, and were optimistic about possibilities this technology might offer them in the future.

Thus, based on our limited sample size and first time use context, we recommend that seamless deployment technology could be relevant to deploy in nursing homes and similar assisted living facilities in order to provide a more calm technology experience, and that further studies should be carried out.

7. FUTURE WORK

In order to extend our understanding of the mechanisms that influences care staff's user experience during deployment and use under long term conditions, we need to undertake additional detailed interviews with the care staff combined with longitudinal field studies in order to investigate the long time learning effects and challenges. Also, studying usage log files could be relevant.

ACKNOWLEDGMENTS

The CareStore Project is funded by the European Commission as part of the Seventh Framework Program. Grant agreement no. 315158. Thank you to the eight participating caregivers and the management of SSRG, as well as to the other CareStore consortium partners.

REFERENCES

Nonin Medical Inc. - Onyx® II, Model 9560, Finger Pulse Oximeter with Bluetooth® Wireless Technology. Available: http://www.nonin.com/PulseOximetry/Finger/Onyx9560 [9/14/2014, 2014].

Bardram, J.E., 2008. Pervasive healthcare as a scientific discipline. *Methods of information in medicine,* 47(3), pp. 178-185.

Dall, T.M., Gallo, P.D., Chakrabarti, R., West, T., Semilla, A.P. and Storm, M.V., 2013. An aging population and growing disease burden will require a large and specialized health care workforce by 2025. *Health affairs (Project Hope),* 32(11), pp. 2013-2020.

Keye, J. and Zitzelberger, T., 2006. Overview of healthcare, disease, and disability. In: J. Bardram, A. Mihailidis and D. Wan, eds, *Pervasive Computing in Healthcare.* CRC Press, pp. 3-20.

Miranda, J., Cabral, J., Ravelo, B., Wagner, S., Pedersen, C.F., Memon, M. and Mathiesen, M., 2015. Radiated EMC immunity investigation of common recognition identification platform for medical applications. *The European Physical Journal Applied Physics,* 69(1), pp. 11002.

Preece, J., Rogers, Y. and Sharp, H., 2015. *Interaction Design: Beyond Human-Computer Interaction* . 4th edn. John Wiley and Sons.

Varshney, U., 2009. *Pervasive healthcare computing: EMR/EHR, wireless and health monitoring.* Springer.

Wagner, S., Hansen, F.O., Pedersen, C.F., Memon, M., Aysha, F.H., Mathissen, M., Nielsen, C. and Wessby, O.L., 2013. CareStore Platform for Seamless Deployment of Ambient Assisted Living Applications and Devices, *Proceedings of the 7th International Conference onPervasive Computing Technologies for Healthcare, 2013* 2013, pp. 1-4.

Weiser, M. and Brown, J., 1996. *Designing calm technology.* PowerGrid Journal 1.1 (1996): 75-85.

QUALITATIVE EVALUATION OF AN INNOVATIVE 3-D SYSTEM FOR ASSESSMENT AND FEEDBACK OF CHRONIC CONDITIONS: ANKYLOSING SPONDYLITIS AS AN EXEMPLAR CASE

Shijuan Li
Department of Information Management, Peking University, China
No.5, Yiheyuan Road, Haidian District, Beijing, China 100871

ABSTRACT

The assessment of any chronic disease is an integral part of the treatment. In the case of Ankylosing Spondylitis (AS), it is particularly important as it helps to determine whether or not pharmaceutical and/or surgical interventions are required for a patient. There is growing awareness that the current protocols and means of taking measurements are both inaccurate and inconsistent. The objective of this research is to investigate whether utilising 3-D visualisation technology can enhance the current assessment and feedback process by augmenting assessment instruments and by visualizing the function of a subject in ways afforded by a 3-D visualisation tool. From a consideration of the physical therapy process, a 3-D visualisation research tool is prototyped, to include both subjective and objective aspects of current assessment instruments. A qualitative evaluation was undertaken using a pilot study, focus group and individual interview methods to validate the prototype. The results of the evaluation suggested that 3-D visualisation and a prototype for evaluation, as developed in this research, would enhance assessment and feedback from the physiotherapist's perspective. Furthermore, it would be beneficial to the treatment of AS with respect to obtaining subjective and objective measures, and supporting exercise activity.

KEYWORDS

3-D visualization, Chronic conditions, Ankylosing Spondylitis, Virtual Reality, Assessment and Feedback, Evaluation

1. INTRODUCTION

Ankylosing Spondylitis (AS) is a chronic condition which requires clinical assessment to inform the management process. It is a painful progressive rheumatic disease that affects mainly the spine [1]. It can also affect other parts of the body [2]. The intent being to alleviate the symptoms and to improve the health of the suffering person. An integral part of the current assessment is the measurements taken of a number of subjective and objective factors that affect the person's health status. The current measurements taken try to encompass the multidimensional impact of AS. Generally, five core domains were recommended by the "Assessment of SpondyloArthritis International Society" (ASAS), which include pain, spinal stiffness/inflammation, functional ability, patient global assessment of health status and spinal mobility. There is growing awareness that the current protocols and means of taking measurements are both inaccurate and inconsistent. Significant progress in the field has been made since the initial ASAS recommendations, which acknowledged that they could change in the light of new research evidence.

AS's treatment is defined largely visually. This in itself raised the question of whether or not these could be represented and augmented by 3-D technologies. Furthermore, on the innovation side, 3-D visualisation provides the means of integrating data and information that traditionally has been kept separate. AS has multiple symptoms, such as stiffness, pain, fatigue that all have to be measured by different types of instruments that need to be brought together in a coherent way. A further advantage of this approach could stem from the attractiveness of the technology. The AS sufferer is often diagnosed during early adulthood. It is often claimed that the young are more techno-savvy than the older persons. Whether or not this is actually the case, it is true that Virtual Reality (VR) technologies in general and 3-D visualisation in particular are engaging for the users, and this may offer another way of managing. There are more effective ways to manage the assessment process and that this can be done through utilising new innovations in ICT.

2. METHODS AND PROCEDURES

2.1 Development of a 3-D Assessment and Feedback Tool for AS

The 3-D visualisation tool comprises a 3-D virtual person and multi-functional interfaces providing feedback information in both static and dynamic 3-D form. The steps to develop this new form of assessment include: firstly, creation of a 3-D virtual person to represent the subject who needs to be measured and assessed; then, collecting motion data from a healthy volunteer in a laboratory environment using sensors and cameras; thirdly, mapping these motion data to drive the 3-D virtual person and to test the system's limitations. The animated avatar, using these data, was then the centerpiece of the prototype.

A simplified assessment and feedback process is developed and used to understand where a new tool might add value [3]. A 3-D visualisation research tool is prototyped, to include both subjective and objective aspects of current assessment instruments used in the evaluation area as shown in Figure 1. Our initial focus is functional ability and related mobility issues which are strong candidates for the use of visualisation techniques. The former is measured by the Bath Ankylosing Spondylitis Functional Index (BASFI) [4] and the latter by the Bath Ankylosing Spondylitis Metrology Index (BASMI) [5]. The Bath indices are mainly patient reported outcome measures [6], and make heavy use of the visual analogue scale (VAS), a form of assessment that is of particular interest to this study. To overcome the limitations of the 2-D, linear scales used for representing the functional ability of complex entities, a 3-D virtual person is envisaged as being a major component of the tool. To be useful, the virtual person has 3-D motion data mapped to it which required pipe-lining outputs from a set of software tools including Qualisys®, Autodesk® MotionBuilder etc. Methods for developing the prototype were explained in the process of data collection, data process and data output in detail [7]. For the moment this pipeline workflow production needs a healthy adult experiment the selected actions wearing the markers on. The process begins with the possibility of "creation" of a 3-D model, and followed by "characterization" into Motionbuilder®. They have been referred to show the potential of implementation into other environments. BASFI and BASMI were successfully represented in this 3-D visualisation system.

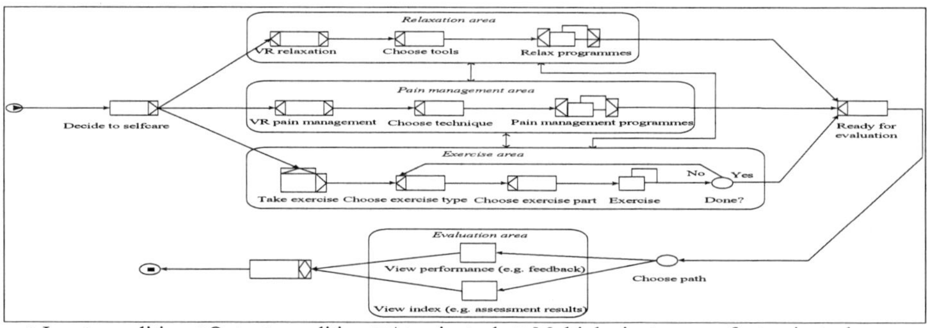

○Condition ◉ Input condition ◉Output condition □Atomic task ⧉ Multiple instances of atomic task xor-split task xor-join task or-join task

Figure 1. A physical therapy process for ankylosing spondylitis

2.2 Functionalities of the 3-D Visulisation Tool

The prototype presents a 3-D virtual person, as shown in Figure 2. Three types of output are available at this stage of the prototype's development: an unadorned 3-D virtual person, called avatar; a 3-D avatar with markers on; and a 3-D video. The 3-D video represents the main output method for the prototype at present. This prototype can provide the following main functionalities at this stage: firstly, it provides 3-D visualisation for a range of activities including assessment activities, physiotherapy exercise activities and objective assessments; secondly, integration capability of several measurements such as integration of subjective assessment and objective assessment and physiotherapy exercises is possible; thirdly, meaningful feedback to clients is available by means of different types of output methods; and it has the potential benefit of integrating other facets, i.e. pain from Bath Ankylosing Spondylitis Disease Activity Index (BASDAI) [8], as shown in Figure 3. Three questions out of BSFI, four kinds of NASS (National AS Society, UK) exercises, two measurements out of BSMI were successfully mapped to the 3-D avatar [9].

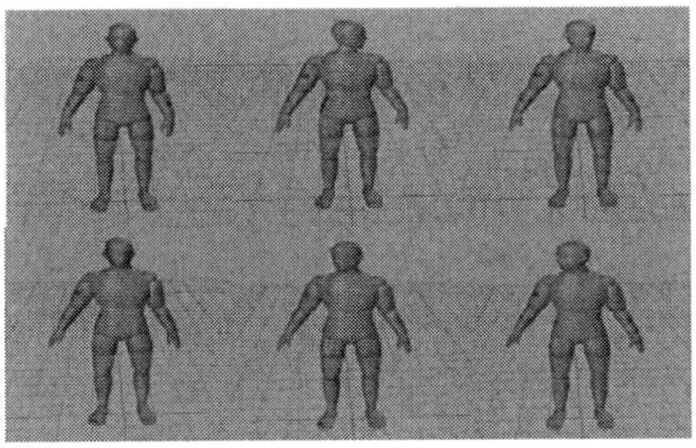

Figure 2. BASFI Q. 8 sequence view of "looking over shoulder"

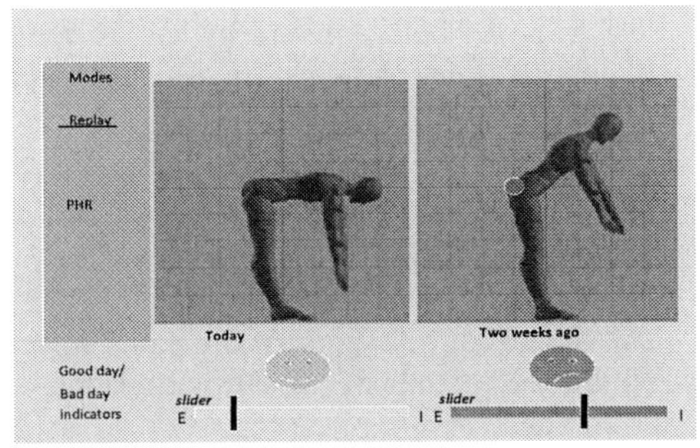

Figure 3. Integration interface

2.3 Qualitative Evaluation Design and Execution

Qualitative evaluation methods were used to evaluate the efficiency and limitations of the prototype developed and the perceptions of healthy professionals towards novel information technology. Focus groups were selected as the method of choice because the group's interaction could produce grounded data on exact issues of this research. Ethical approval was granted prior to starting the study by the Research Governance and Ethics Sub-Committee (RGEC) of the Research Committee, reference number RGEC08/057.

In order to minimize sample bias, participants were purposively recruited from a number of sources, such as different UK NHS trust organizations, rather than from a single source since the goal was to represent a full spectrum of experiences and opinions. The participants were recruited from two major groups of interested parties: academia and practitioners. Individuals were contacted by email or phone with an outline of the nature of the project. Ten out fourteen AS specialists from the Northwest physiotherapist network agreed to participate in the groups; thereafter three groups were held. The participants were all expertise in AS. This homogeneity allowed for more free-flowing conversations among participants within groups and also facilitated analyses that examined differences in perspective between groups. And the groups were mixed rather than segmented by sex, race or age because neither of the factors has been seen affecting the participants' comfort in the discussion or the analyst' ability to make useful comparisons. The first group was also used for a pilot study.

The group meetings were held in a usability laboratory. An experienced moderator moderated the groups and an assistant also attended the whole process. An interview guide was used. The guide was important to ensure consistency between groups however was not adherered to rigidly in order to avoid the tendency to follow a predetermined order of topics in a rigid fashion. As part of the study, formal models of assessment and feedback as developed in [10] were presented and discussed. The device of considering a hypothetical patient at various stages in the assessment and feedback processes was used to design how the prototype could be used and as a framework to elicit comment and create a focus for the discussion. The hypothetical patient (or persona) was selected as being a typical AS stereotype of a young person, who first knew about the condition in her late 20's, early 30s. The scenarios start from the first assessment at the onset of the condition and then the focus moves to later assessments made throughout the course of her life with the disease. The persona and scenarios were approved in the pilot study. A number of logistical issues were raised during the pilot, and these were addressed in order to improve the internal validity of the evaluation. In order to improve comparison and support the analysis process, the participants were asked to go through all the scenarios twice: first time was to collect narratives of procedures in current practice; the second time was for narratives of assessment and feedback with the prototype. The "funnel strategy" was conducted for the evaluation process, which meant each group began with a less structured approach that emphasized free discussion and then moved towards a more structured discussion of specific questions.

All sessions were recorded and transcribed, and the transcriptions have been cross-checked by another researcher prior to the analysis. The data generated from the three sessions were transcribed using Transana® 2.22 and analyzed using Nvivo® 8.0. The interpretation and analysis of the data employed content-based thematic analysis which involved classifying and coding the content into themes. The framework of four scenarios for both assessment and feedback were used to elicit comment and create discussion for the evaluation. These became the general context of the evaluation that provides the environments of generating data and establishing themes.

3. RESULTS

The themes directly address the initial research objectives. Together they provide a comprehensive overview of the interpretations of the participants' perceptions in terms of current practice context and for this prototype.

3.1 The Perceptions of Current Instruments' Limitations

This is a critical issue within the overall research and therefore specific questions were asked regarding what did the physiotherapists feel concerning the limitations of current instruments used for AS.

3.1.1 Limitations of the Bath Indices and VAS

In contrast to the Bath indices', i.e. BASFI, BASMI etc., widely acceptance and usefulness in practice, concerns of less sensitivity and likely bias are still common perceptions from the evaluation sessions. Since the Bath indices were developed in 1990s, a thorough review would seem necessary at this point of time, but that is beyond the scope of the present research. Specifically, regarding question 2 in BASFI, "please indicate your level of ability with bending forward from the waist to pick up a pen from the floor without an aid", one of the physiotherapists has pointed out that it is no longer considered functional or even safe. And they won't suggest it since it may put the patient in danger or pragmatically the patient simply won't obey because they can't do it. The participants reflect again that patients have difficulty in using the VAS utilized in the indices. The patients have problems in conceptualizing the VAS; instead, they express their understanding by writing an essay or drawing two lines etc. Consequently, it can consume more physiotherapy time in explanation and redoing. The patient's response to the VAS is said to vary a great deal.

3.1.2 Inconsistent Measurement of Current Measurements

It has realized that the ways of carrying out objective measurement, i.e. BASMI, are also problematic. They have found that it was inconsistent and varied from individual to individual. For example, they found the way they measured "Schober's flexion" varied between different physiotherapists due to the tools used or the procedure followed. In their networking meetings, measuring in a consistent way was the focus for discussion. There is also a varied time span between each visit due to different requirements and constraints on NHS trusts. In general, the first follow-up visit will be relatively quick, with subsequent visits further apart in their frequency. Given that AS symptoms can flare-up and the "last week" prior to an assessment is not necessarily representative of the whole period between assessments, it can be difficult to understand what the scores mean. Furthermore, the wholly subjective questionnaires are thought to vary in accuracy, according to whether they capture the patient's feelings on a bad day or a good day. This is another difficulty that distorts the object of measuring consistently.

3.2 The Perceptions of the Prototype

The perceptions of the prototype in various ways with respect to assessment, feedback and physiotherapy exercise are discussed.

3.2.1 The Perceived Credibility on Assessment

- Credibility coming from visualizing movements in 3-D
- Potential credibility coming from visualizing pain in 3-D
- Credibility of encouraging multi-disciplinary working

The participants believed that the prototype could be helpful in various ways by visualizing the movements in 3-D. They said it was possible to see how it would improve the assessment of movements with this prototype. The evidence is currently lacking as to what the patient is exactly doing in their daily life in response to the physiotherapist's instructions from an assessment visit. Currently the physiotherapist has to ask the patient to show what exercises they have been doing at their home. If this prototype not only captured the immediate movement but also had a database tracking the movements between visits then it would be possible to replay the patients' movements and therefore provide much more information to the clinician. Furthermore, they felt that the tool made it possible to improve the understanding between the patient and the clinician. No matter how well the exercise is described by the patient, it is agreed that it would be hard to convey the movement in any accurate way without visualizing it. Again, if the patient's performance could be monitored more effectively then the clinician could also assess risks of self-harm through inconvenient or dangerous exercises. By visualizing movements in 3-D, explanation and understanding in an accurate way is more possible. Finally, the comparison of movement between each visit can be performed by the computer, supporting the physiotherapist and the patient with data on improvement or the lack of improvement. Visual evidence can be made available for explanation of clinical terms, providing accurate measurement to statements such as "2cm's difference". Thus, the capability of visualizing the movements in 3-D is deemed potentially useful.

The participants believed that there is a potential from visualizing pain in 3-D. Pain is one of the symptoms that are difficult to measure and describe. The capability of this prototype of integrating multiple measures, such as subjective assessment, objective measurement and pain assessment etc., provides an opportunity to consider the multi-symptoms condition in an integrated way. The participants believe that, since patients like the pain body chart used currently, this prototype is seen to have potential. Visualizing pain in 3-D is an area for further research; the current prototype only illustrates the possibility.

AS usually involves a multi-dimensional discipline (MDT) in the treatment process that oversees different aspects of the disease. The physiotherapist is an important member of the MDT, although the group discussions revealed that they often had different roles and rights. For example, there were different ways to access the case note or patient note if not in a shared record and this depended on different hospital rules. After the first visit, documenting of the different output from the meeting sessions usually depended more on the particular work environment, whether it is an AS clinic or some other environment. There are differences in what the physiotherapist needs and so the physiotherapist will write different things to different people of the team. Another issue mentioned is that there is some repetition of information on the same assessment. The nurses sometimes will re-assess the patient on the next meeting day since they have not got the results yet from the physiotherapist. So it was felt that the prototype might facilitate the communication among the MDT by easily accessing and sharing information within one single environment. It is possible to believe that making such a prototype assist the record keeping and communication for all in the MDT has the potential to save time and resources. However this is beyond the present scope of the research; the current prototype is focused on the assessment and feedback of the physiotherapist-client relationship.

"It would be easier to show them visually what score means, rather than 'cm'. For example, they might score a 2-cm's difference, so you could show them 'last week' or 'last month' you've got, 'You felt you could get this far, and obvious this time, for some reason, you feel you are not going this far.' And that for them 'watch' for sure is easier." (Participant A, Focus Group).

3.2.2 The Perceived Credibility on Feedback

- Credibility related to educating the patient
- Credibility from capturing disease trend and status
- Credibility of feedback information flow changes

Like any chronic arthritic condition, self-care plays a key role in the condition management. Therefore, patient education is vital for the AS sufferers. It is also believed that "education of the physiotherapy way of assessing" may be of more importance than education of the disease for AS sufferers. Feeding back physiotherapy information to the patient with this prototype is believed to be promising in a number of ways. The fact that this prototype helps to show movements in 3-D provides the opportunity to educate the patient about what the physiotherapy requires of them. So the participants thought it might be helpful in feedback by explaining physiotherapy information in 3-D and possibly avoiding misunderstanding during the treatment.

This prototype can represent the disease trend in actually showing the patient's movements in 3-D over a period of time that is more direct than a set of derived numbers, particularly if there are concerns over the meaning of these numbers. A more continuous form of monitoring (say once a day) means that fluctuation and flare-ups can be more easily assessed. Note some of the participants emphasize the use of "markers". It has been envisaged that any future application would minimize the need for such constraints, particularly if it were to be used on a routine basis at home. The capability of store and compare function and disease status is thought to be useful with this prototype, as it can reflect the overall trend of the disease over the years for a particular person.

With this prototype, the feedback process can be modified. The patient now becomes the center actor. And the information flow is not uni-directional any more, i.e. flowing from the patient to the clinician; rather the prototype permits a two-way partnership. The previously clinician-oriented result/scores are a by-product whereas the movement is much less abstract and requires less clinical knowledge for the patient to understand their progress. The new information flow also allows more open interaction during the treatment process between members of the MDT. This model therefore supports the previous requirements for better communications throughout the evaluation process concerning both assessment and feedback.

"... Again, the set is dynamic, isn't it? If you look at a set of 'rotations', it is a very difficult concept for somebody. It is trouble here, because we know too much. You know about your estimation of what you are looking for, which is what you got to try to teach the patient. Because they might be doing the treatment, not moving the bit that you actually ask them to move. That is the fact that would be really helpful." (Participant B, Pilot)

3.2.3 The Perceptions Regarding Physiotherapy Exercises

- Physiotherapist prescribing exercises
- Participants' anticipation of patients' compliance
- Patient transforming exercises

Physiotherapists usually prescribe exercises as a key therapy method to the patient in current practice. There have been previous attempts at trying to standardize exercise prescription through the Physiotherapy Networking Meetings. Currently, personal favorites and choice will continue to play an important role, e.g. breathing exercise is deemed as one of the most important and is therefore being pushed. In contrast, other is happy with the patient initiating the exercise. In those circumstance, guidance works rather than instruction. It seems an enough compromise to provide the patient with the opportunity to initiate the exercise, providing guidance and continuous monitoring. In fact, the physiotherapists try to customize the exercise for the patient in order to increase the possibility of compliance, while it is in conflict with the intention to standardize the exercise prescription. The link between the outcome and the exercise being done is still weak.

It is a complicated issue as to whether the patient is compliant or not with respect to the exercise prescription. It has been noticed in other researcher's work that patients transform the prescribed exercise sometimes, others will not comply at all, and some will obey the exercise prescription strictly. It seems relatively difficult to monitor the patient's exercise activity at home. Although exercise diary provides an opportunity for monitoring, there are doubts from the clinicians in these diaries. Another general perception, either reflecting the practitioners' views or accredited to others, i.e. the researchers, is to understand what exercise the patients are actually doing. Instead of basing it on patients' diary or physiotherapists' anticipation, audit of compliance can be built in this prototype to help the monitoring of the patient, i.e. do they exercise on their own/more than others? The characteristic of engaging in technology brings the promise of improvement of exercise compliance since this prototype could take intervention into consideration. This potential has to be left for further investigation.

Besides compliance, the issues of defining and transforming "exercise" are found to be very complicated. Physiotherapist and patient may well have a different understanding of what "exercise" is. While the communication between the clinician and the patient may be clear, the content that is the "exercise" may be quite different. Nevertheless, the participants have found transforming the prescribed exercise. The complicated issue of transforming exercise may have something to do with the unusual way of thinking with this condition. Unlike the common sense notion that pain is telling you not to do something, it seems essential with this condition to go through the pain barrier and to get the exercises done. The participants thought that if the patient could use this prototype at home and the exercise activity could be captured, then it could be possible to start to build evidence for exercise selection and exercise prescription.

"I think we will guide them. But they had chosen them [exercises], because they [exercises] are similar, you know. There are three or four different exercises for 'lumbar spine rotation'. I think the patient can look at the booklet, and choose which one they want to do. So they would rather do one set on the chair because they can do at work rather than the one they have to go to bed to do it. And you know, by the time they go to bed in the evening to do the exercises, they tend to just fall asleep rather than doing it. So I think if it is a chronic condition, and long term exercise treatment is needed, I am happy the patient has initiated how they are going to perform their exercises." (Participant, Interview)

3.2.4 Discussion and Concerns

Certain design issues were discussed to investigate the prototype's full functionality and interface. The participants liked the neutral appearance of the figure that represents the virtual patient and believed it provided a new platform for data management. From a clinical or care perspective, the participants felt that the neutral appearance of the figure was appropriate because they felt people would rather see an avatar than see themselves, especially when deformation to posture is happening. The avatar was felt to be much more appropriate than simple video.

Although the prototype was clearly experimental and "rough", the participants felt that the current form of the prototype needed to be simplified before it could be taken up and introduced in routine use. It is common to mistake "prototypes" for the "real" thing and expecting it to be closer to product than it actually is. Participants felt that an alarm might be added into the prototype that would be automatically triggered when there was a danger of over assessing the patient; this clearly requires a much more "intelligent" system than the visualisation tool shown. Even so, more sophisticated algorithms for calculation of the score and data processing were desired.

The question of how this prototype can be used with the current instruments has been raised regarding service application. Whether it intends to combine with current instruments or replace current ones completely is an open question. However, the goal of this research was to augment rather than replace. Since the current paper-based instruments for AS were developed back in 1990s, and they have been used in practice for decades of years, it was predictable that there would be resistance from the clinicians with respect to introduction of a new tool. As stated, the purpose of the research includes using this prototype to review current instruments and to seek opinions for the form of this prototype.

4. CONCLUSIONS

The evaluation sessions were carried out and the overall view of the work was positive. First of all, there was agreement from the participants on the limitations of current instruments. The participants perceived that inconsistent measurement existed in the practice due to the varied measurement between individuals; the problems of the patient conceptualising VAS were confirmed according to their experience; and there was a desire from clinicians for computerised documenting of AS treatment process that the prototype would support and surpass. Secondly, the proposed assessment and feedback models were accepted by the participants as the assessment basis, which captured what they did in practice. Thirdly, the participants perceived the prototype positively. The prototype was convincing. It has been perceived as novel and helpful in the visualisation capability of movements and even pain in 3-D figure. This prototype was also perceived as beneficial in educating patient and capturing disease status and trend. And the information flow has been augmented within this prototype by placing the patient as the central actor within the process. Moreover, if this prototype could be used in practice, it may be possibly beneficial to standardise the objective assessment resulting in increased reliability and trust. Now exercise is used as a key physiotherapeutic tool but it is lacking evidence as to which one is better and whether or not the patient is complying. This prototype is thought to be a way to start to build the evidence base but that assumes it can be used directly by the patient in their home. Last but not least, the 3-D visualisation prototype utilises a neutral figure to represent the patient; the participants agreed that the neutral appearance choice is more acceptable than a simple video because people do not like to see themselves especially when the progression of a condition results in a deterioration and/or deformation of their body. Although, there are concerns regarding cost and communication changes, as technology is becoming less expensive over time and people are getting more engaged in technology, it is seen as something that might become readily available for patients to use at home and shared with the professionals.

REFERENCES

1. Calin A., 2006. Ankylosing Spondylitis. *In Medicine*, Vol.34, No. 10, pp 396–400.
2. Walker J., 2006. Ankylosing Spondylitis. *In Nursing Standard*, Vol.20, No. 46, pp 48–52.
3. Li, S. and Kay, S., 2008. Virtual-reality therapies for chronic arthritic conditions. *Proceedings of Healthcare Computing 2008: An invitation to the future*. Harrogate, UK, pp. P21: 1–6.
4. Calin A et al., 1994. A new approach to defining functional ability in ankylosing spondylitis: the Bath Ankylosing Spondylitis Functional Index (BASFI). *In J Rheumatol*, Vol.2, pp 2286-2291.
5. Jenkinson TR et al., 1994. Defining spinal mobility in ankylosing spondylitis (AS): The Bath AS Metrology Index. *In J Rheumatol*, Vol. 21, pp 1694-1698.
6. Prom Group, 2006. Chronic conditions-a structured review. *Report to the Department of Health*. University of Oxford, Oxford, UK.
7. Li, S et al., 2009. A visualisation tool for augmenting assessment of ankylosing spondylitis. *Proceedings of The Medical Informatics Europe (MIE): medical informatics in a united and healthy europe*. Sarajevo, Bosnia and Herzegovina, pp. 821–825.
8. Garrett S et al., 1994. A new approach to defining disease status in ankylosing spondylitis: the Bath Ankylosing Spondylitis Disease Activity Index. *In J Rheumatol*, Vol. 21, pp 2286–2291.
9. Li, S. and Kay, S., 2009. Transforming assessment and feedback for ankylosing spondylitis through the use of 3-dimensional visualisation. *Proceedings of IADIS International Conference: e-Health 2009*. Algarve, Portugal, pp. 175–180.
10. Li, S. et al., 2007. Virtual reality: towards a novel treatment environment for ankylosing spondylitis. *Proceedings of Medical and Care Compunetics 4*. Amsterdam, Netherland, pp. 190–198.

COMPETENCES OF PHYSIOTHERAPISTS REQUIRED FOR THE PRACTICAL USE OF SOFTWARE IN THE THERAPEUTICAL PROCESS

Mag. Karin Messer-Misak[1] and Univ.-Prof. Dr. Rudolf Egger[2]
[1]*FH Joanneum Gesellschaft mbH, Eggenberger Allee 11, 8020 Graz, Austria*
[2]*Karl-Franzens-Universität Graz, Merangasse 70/II, 8010 Graz*

ABSTRACT

The education of health professionals in the German-speaking countries has recently been moved to a tertiary level. With this move, an increased use of new technologies and expanded software solutions is expected in this professional area. Actually the necessary knowledge and competencies to integrate software solutions in the therapeutical process are not dealt with in the educational training. Additional training to enlarge the therapist's professional knowledge after their graduation takes into account either the training of interpersonal skills of therapists or the physiotherapeutic expertise. Only few training units cover the aspects of the usage of special therapeutical equipment. The use of software that is supposed to support the administration and documentation of the therapeutic issues is conveyed on the job. By the example of physiotherapy, the primary objective of this study was to define the specific competencies of therapists that are necessary to integrate the practical use of software in the therapeutical process. These cover the administration, documentation and evaluation of the entire therapy process, including a database with pictures and videos about exercises which can be adapted individually by the therapists. The approach of mixed-methods design was chosen. In the first part a two-stage qualitative study was carried out, followed by a quantitative survey. The results show that for the professional use of software regarding the therapy-related part within the therapeutical process it is not only necessary to expand the therapists' knowledge about the functionality of software. It is also essential to expand some key competencies regarding the social and communicative, the methodological, personal and activity/action competences to embed the use of the advanced functionalities only then it will become possible to integrate software successfully in the therapeutic process.

KEYWORDS

Competence, Skills, Physiotherapy, Software Tools for Therapy

1. INTRODUCTION AND BACKGROUND

The current healthcare system requires more effective management. New media and technology are supposed to support the demands of the current healthcare system (Waegemann 2010: 58-69).

The gap between the available software solutions which can support the entire therapeutical process and their actual use in therapeutic practice is getting bigger (EPatient RSD 2015). The existing and ongoing research predominantly refers to the technology acceptance of clients and/or therapists (Kairy et al. 2013) and shows a number of factors that influence their decision about how and when therapists will use technology in general. On the other hand, there are studies (Hüter-Becker and Dölken 2005) about the competences therapists need in general within the physiotherapeutic process. However, the use of therapy relevant software tools during this process is not considered. This study takes both into account, the aspects of competences that are needed as well as the entire workflow in the therapeutical process.

Software in therapeutic practice is mainly used to streamline administration procedures (Burgsteiner et al. 2015: 182-189). The multifaceted options for interacting with the individual optimization of efficient establishment of findings or the definition of therapeutic targets and strategies as well as the access to a comprehensive database is not yet considered in the therapeutic process (Messer-Misak, Egger 2016).

The aim of this study was to solve a practical problem, namely what kind of competencies are needed to improve the practical use of software that also support therapeutical sequences in a therapeutic process. An increase in knowledge is gained primarily from practice-relevant results. The results were obtained directly in the research field. The objective of this paper is to show which competencies must be met by the therapists to integrate software solutions that support the therapeutic process in their professional field of action to gain the maximum benefit from it. The aim was to refer the competencies to a common model of competences and relate them to the therapeutical process. The physiotherapeutic process follows the logically structured process of treatment as laid down in the International Classification of Functioning, Disability and Health (ICF) as well as in the current model of health issued by the World Health Organization (ÖBIG 2016).

A systematic analysis of existing theoretical and conceptual approaches to competencies leads to the result that there are many different theoretical approaches depending on the perspective and underlying objectives (Weinert 2001). Therefore the fundamental structure, as compiled by the UNESCO (known as the "four pillars of learning") (Zhou 2016) and enlarged by the representatives of the field of competency research particularly active in the research on competencies in health care, were used in this study (Heyse et al. 2012: 154-216, Heyse 2014a: 3-28). The key competencies needed within the medical process were merged and allocated to the four main groups in the model of competence as follows:

- Social and communicative competence (corresponding to the pillar "learning to live together")
- Professional and methodological competence (corresponding to the pillar "learning to know")
- Personal competence (corresponding to the pillar "learning to be")
- Activity and action competence (corresponding to the pillar "learning to do")

2. METHODS

In order to analyse the current state of research and the theoretical background as well as possible, a systematic literature review was initially carried out.

In a first step, specific terms and relevant keywords were determined, and a systematic search within scientific databases (ERIC, SpringerLink, ScienceDirect College Edition, Pubmed, Thieme-connect, Medline), academic publishers and online bookstores was done. The following search criteria were chosen: physioth* and competen*, physioth* and IT, physioth* and software*, physioth* and computer, physioth* and competen* and technic*. The search criteria were restricted to only peer reviewed articles and were limited to the years 2010-2015. In total more than 22,250 hits were recorded, of which 15 studies were selected for their relevance and reviewed.

For the study, the research strategy of mixed-methods (Döring and Bortz 2016: 184-185) was chosen. The work consists of a two-stage qualitative and quantitative study, carried out consecutively and referring to each other.

2.1 Methods Qualitative Research

For the qualitative study, a semi-structured qualitative interview guideline for the expert interviews (n=5) and the interviews with the therapists (n=8) was designed. The interviews were conducted in the period February - April 2014. All interviews were set up as One-on-One Interviews, one hour was allocated. The interview guide contained 18 questions, which were divided into further sub-questions regarding the factors influencing the practical use of software and the competences that are needed to operate with software during the therapeutic process. Only experts were asked the questions which competencies already exist and are well extended through the basic education particularly with regard to the use of software and which competencies will be needed in the future for the professional use of advanced software in the therapeutical process. Interviews were conducted with persons who have relevant qualifications and practical experience in physiotherapy. Various therapy organizations were asked to name interested volunteers (n = 8). It was possible to interview these volunteers at their place of work. The experts (n = 5) were asked in person for an interview because of their specific expertise and their functions in political or management positions.

The interviews were transcribed and a software for qualitative data analysis (MAXQDA) was used. The text material was screened and coded. In a second step, the material was examined for similarities and patterns of action. Codes were created to which the text material was assigned. In a third step the material was selectively encoded and interpreted.

2.2 Methods Quantitative Research

In a subsequent quantitative study in the period March – April 2015 the results of the qualitative part of the study were considered. Using a standardized questionnaire, which previously was subjected to a pre-test, the filtered aspects identified through the qualitative research were tested by a larger sample for their relevance (n = 306, the part of the questionnaire that dealt with the competencies was fully answered by 193 participants). It was sent to approximately 1,000 therapists via a link in an email, with a 30.6 per cent response rate. The mail addresses were exported from the public database of the "Bundesverband der PhysiotherapeutInnen Österreichs". The form was also included in the Facebook page of the Association of PhysioAustria.

The free online survey application "EHR LimeSurvey" was used. The therapists were to complete the questionnaire independently online. Caching the results (after a password was set) and subsequent completion were possible. Upon completion of the survey, the raw data were exported and an analysis was performed by using the statistical software R version 3.1.3 Core Team (2015) and the package psych Revelle (2015). The graphics were created with the package ggplot 2 (Wickham 2009).

The questionnaire comprised 3 topical areas. The first part of the questionnaire included questions about the age, gender, the duration of their professional practice and the medical discipline they are associated with. The second part referred to the competencies needed to use software in connection with the therapeutical process. The specific requirements on the practical use of software in the therapeutical process made up the third part of the questionnaire.

3. RESULTS

3.1 Results of the Qualitative Findings

A total of 68 codings out of five documents and one group of documents (only the interviews with experts had the focus on needed competencies for the use of software within the therapeutical process) were associated with the topic "which competencies related to the practical use of software in the therapeutical process are already well extended through the basic education?" The named competencies were assigned to the four major categories of the expanded competence model for the health care system published by Heyse (Heyse and Giger, 2014), complemented for the special needs for physiotherapists by Becker (Becker 2014: 289-319):

- 27 codings refer to the social and communicative competence
- 25 codings refer to the professional and methodological competence
- 9 codings cover personal competence
- 7 codings relate to the activity and action competence

The results gained from the expert interviews, referring to the competence model, are as follows:

- Social and communicative competences are shown in the key competencies: cooperation skills, consulting capability, relationship management and capacity for teamwork. In addition to these key competences the readiness to understand the clients, communication skills, high adaptability, and dialogue capability are necessary when using software solutions that also cover the therapeutic part.
- Professional and methodological competence are already strongly pronounced: good judgment and interdisciplinary knowledge. Besides this, the experts claim that analytical skills, objectivity, planning behavior and conceptual strength are needed in order to use the therapeutical aspects of the software optimally.
- Personal competence is in particular on balance with normative-ethical responsibility, self-responsibility, self-management, willingness to learn and the competence to reflect on complex relationships.
- Activity and action competence can be applied to the following items: results-driven action and decision-making capacity. These key competences were complemented by the experts, especially by inspiration and initiative as well as readiness to execute the commitments agreed upon for the therapy goals.

The results regarding the question "Which competencies are additionally required/ought to be further developed to successfully integrate the varied applications of software solutions in the therapeutic process to gain maximum targeted support through the software?" led to the following results: 89 codings were recorded regarding the question and were assigned to the four competence groups.

- Social and communicative competence: in addition to the already well anchored competencies there is a need of professional conflict resolution, because technically unexperienced clients are involved in the process and experts expect increased training and consulting efforts to be necessary. For this process the experts require more sense of duty as well.
- Professional and methodological competence: exceptional multidisciplinary knowledge, especially technical understanding, media competence and educational skills should be strengthened. Holistic thinking, conceptual strength and organizational skills are needed because the whole therapeutical process has to be considered. Therefore analytical skills and enhanced judgment are crucial for the assessment of the client´s performance.
- Personal competence especially increased willingness to learn is required and the intention to reflect when and how the use of this advanced software solutions are useful.
- Activity and action competence: Becker (2014: 289-319) claims several competencies that should be extended to be well prepared for the future challenges in the health care sectors. For instance he named innovative spirit and creative drive. The experts´ responses were restrained. Two examples of the expert interviews are as follows:

  "It´s impossible to have a sound grasp of contemporary developments." "The major problem of us older therapists is that we don´t have the competence to use these media. There are fears, we even don´t use email, and actually most of us are still using index and record cards."

In summary it was found, that in all categories there are some competencies that have to be trained more thoroughly powerfully. Especially the activity and action driven competencies should be supported in the future.

3.2 Results of the Quantitative Findings

The sample size does not allow any generalization to all Austrian physiotherapists because rather younger people with little work experience participated in the survey. However it can serve as an indicator of the future use of technologies.

The questionnaire followed the logically structured process of treatment as laid down in the International Classification of Functioning, Disability and Health (ICF) (ÖBIG 2016):

- Problem recognition
- Therapy planning
- Implementation and realisation
- Comprehensive activities

Based on the literature review and the results of the qualitative survey typical plot scenarios within the structured process of treatment emerged in which software solutions are used (see Figure 1):

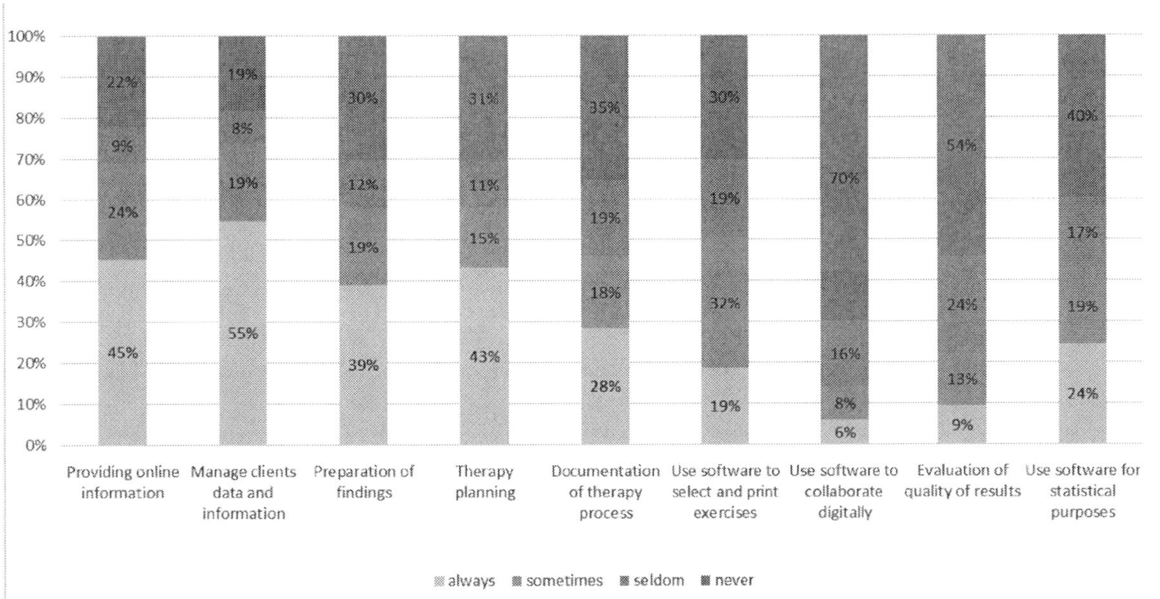

Figure 1. Typical Scenarios – Software Solutions used

The question "For which activities do you use software solutions" (never – seldom – sometimes – always)?" revealed to following results:

The main use of software is attached to the "comprehensive activities" like administrating information (73% often and always) and sharing information (68.9% often and always). Detailed analyses have indicated that there are differences between therapists working as freelancers and therapists working for larger institutions (e.g. hospitals, rehabilitation centers). E.g. therapists working as freelancers more often use software to share information (56.8% always) than therapists working in hospitals (51.6% always) and rehabilitation centers (50% always). But we have to keep in mind that e.g. in bigger hospitals the administrative tasks might not be carried out personally by the therapists. Therapists working in hospitals (35.4% always) or in rehabilitation centers (42.8% always) administrate the clients' record more often digitally than freelancers (23.2% always). The use of software can also be observed with the more challenging activities throughout the diagnostic and therapeutic process. Software is used for generating findings, documentation of the therapy process, therapy planning, selecting individual exercises for the client or evaluating treatment results but just to a certain extent.

In the phase "problem recognition" the medical history concentrates on the health complaints, the social environment, the history and development of the disease and the effects of the disease on the client at the level of "activities" and the level of "participation". The software should provide the possibility of guiding through a standardized diagnostic assessment which can be adapted by the therapists. 57.8% (always or sometimes) of the therapists use these possibilities. Here are also differences in the use between the therapists working in hospitals (45.1% always), rehabilitation centers (42.8% always) and therapists working as a one-man business (29.6% always). This finding is the foundation for the next phase, called "therapy planning". In this phase the therapists determine in cooperation with the client which treatment goals are to be achieved in a realistic time-span and what kind of measures have to be set in order to reach the goal within a planned timeframe. Software solutions offer an optimal way to access databases/videos/images/models by means of which the therapists can explain the effect of therapeutic measures. The recording of the current situation in the form of an image or video (for example, analysis of stance and gait) is desired for the presentation and documentation. 64.5% of the therapists working in a hospital and 50% of them working in a rehabilitation centers use the possibilities software is offering, as opposed to only 31.2% of the therapists working in a one-man business. Based on the findings and the objectives to be achieved, the software should display individually selectable exercise suggestions. The detailed analysis showed that the more complex the requirements are the lower is the usage of software solutions.

Stage of implementation: The selected activities are carried out at this stage. Software solutions are considered to be helpful, if there are possibilities to assemble individually tailored exercises and add individual hints and suggestions for the clients. The individual exercises can be printed for the client or the client has the possibility to obtain an "app". The tools for individual exercise compilation for clients is used by 50.39% (always and more frequently) of respondents. Only few of the respondents (12.9% therapists working in hospitals, 35.7% therapists working in rehabilitation centers and 17.6% free lancers) always take advantage of these possibilities.

Phase Cross-cutting activities: Relief and support in all administrative activities, such as planning appointments and billing as well as documentation and an interdisciplinary exchange of information between the persons involved in the treatment, are vital. Data should be available for future evaluation purposes to contribute to improved data representation in the context of evidence-based therapy, and to evaluate the quality of treatment results. Only 9.6% of the therapists working in hospitals and 6.4% of the therapists working as one-man company (0% of the therapists working in rehabilitation centers) always use the possibility of evaluating the quality of treatment results with the software.

Finally, the therapists were asked which competencies are useless, rather unnecessary, less important and important within the therapeutical process. They also had to answer which of these competencies get more important through the use of software solutions. Remarkable results have been that the therapists realise that the competence that can be categorized into the professional and methodological competence, namely to evaluate the quality of the therapeutic process and the evaluation of the therapeutic results, gain more importance through the use of software solutions. Likewise the competence to advise and guide clients will also gain more importance through the use of software solutions.

Competencies regarding activity and action especially gain more importance in the documentation of client records and the adaption of the therapeutic activities in accordance with the documented records.

Several personal competences are needed when using therapy relevant software. The therapists point out that it is important to become familiar with new methods of therapies. Due to this, the competency "willingness to find new paths in new therapy methods" becomes more important with the use of software.

The results of the questionnaire regarding the social and communicative competence shows (see figure 2) that the therapists expect changes due to the use of software in the therapeutical process especially when it comes to the exchange of information with other health-care professionals. Another point is the need to cooperate in an interdisciplinary team. The therapists expect more competencies to be needed through the use of software solutions.

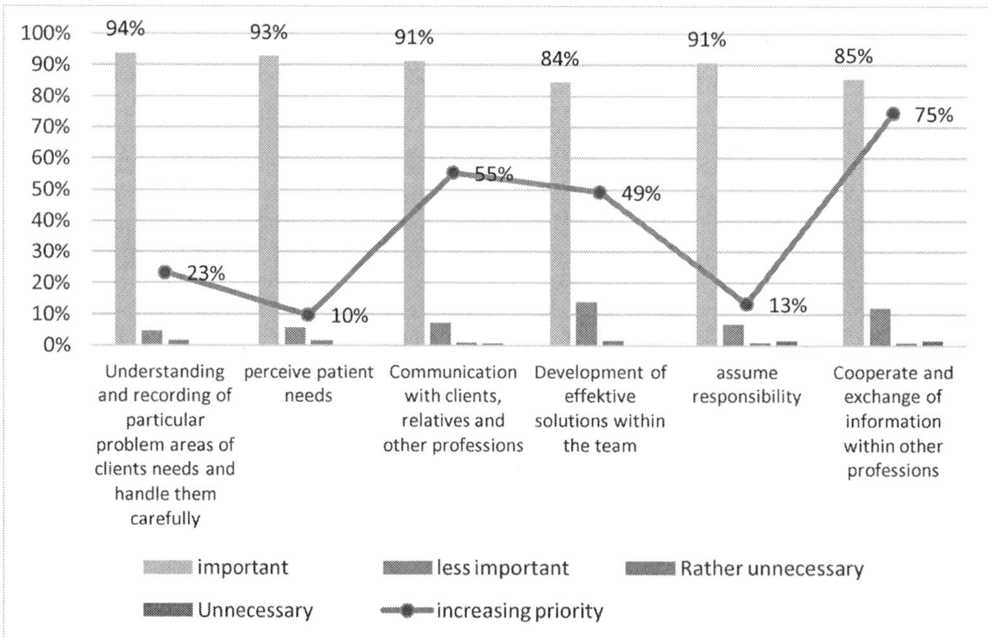

Figure 2. Typical situations – social and communicative competence necessary

4. CONCLUSION AND FUTURE WORK

The digitalisation of entire healthcare systems has triggered complex developments which are not always predictable. Therefore physiotherapists do not only need knowledge about the possibilities of existing software solutions that can support the therapeutical process but also the ability to apply them to specific situations. It is not sufficient to deal selectively with new media, rather a holistic consideration of the physiotherapeutic process is required to enable them to use advanced software for best achievement of the goals set (Rychen et al. 2005).

The detailed evaluations reveal that no holistic, continuous use and integration of software technologies is part of the therapeutic process. The therapists only selectively apply some of the available possibilities and individual functions. So they do not fully benefit from them as support for the entire, continuous therapeutic process.

The adaption of the necessary skills and incorporation of the use of software that opens up new areas of application, or new technologies in general, in the learning contents will need to be considered in the near future. However comprehensive formal knowledge acquisition does not automatically lead to the applicability and implementation of knowledge in practical professional life (Heyse 2014b: 13). Many requirements relate directly to the demands of working and life challenges (Heyse and Schircks 2012: 18-19) and have to be adapted to the actual situations (Heyse 2004: 8, Hof 2002:155). Individuals as well as healthcare institutions have to be responsive, participate actively in opportunities and possibilities in their ongoing development, as work conditions and technologies change rapidly. The application of new work equipment in the physiotherapeutic process requires additional measures in order to use its full potential. The active use of new work equipment within the physiotherapy process depends on many factors, the basis for the necessary skills can be taught during university education.

Openness, interest in innovation, creativity, team spirit as well as interest in interdisciplinary knowledge are required. The organizational and statutory conditions must be guaranteed for technical components to unfold their potential.

REFERENCES

Becker, 2014. *Schlüsselkompetenzen deutscher Physiotherapeuten.* In: Heyse, V., Giger, M.: Erfolgreich in die Zukunft. Medhochzwei Verlag, Heidelberg. pp 289-319.

Burgsteiner, H., et al, 2015. *Development and Evaluation of a web-based Application for Digital Findings and Documentation in Physiotherapy Education*, in: Proceedings of the eHealth 2015. Ammenwerth, Hörbst, Hayn, Schreyer, Wien, pp 182-189.

Döring, N. & Bortz, J., 2016. *Forschungsmethoden und Evaluation in den Sozial- und Humanwissenschaften*, Springer-Lehrbuch, 5., vollst. Überarb., akt. u. erw. Aufl. 2016, Springer Verlag Berlin, pp 184-185.

EPatient RSD GmbH, 2015. *'4. Nationaler ePatient Survey 2015: Die Internetnutzung von Patienten und deren Auswirkungen auf ihre Therapie und den Gesundheitsmarkt.* Die jährliche Befragung unter 50 Mio. Gesundheits-Surfern im deutschsprachigen Internet'. URL: https://medonline.at/2015/e-patient-survey-2015/, last access: 8.2.2016

Heyse, V., 2004. *Kompetenzen erkennen, bilanzieren und entwickeln*, Waxmann Münster, p8.

Heyse, V., 2014(a). *Schlüsselkompetenzen. Erfolgreiche Brücken zu selbstbestimmten Patienten*. In. Heyse, V., Giger, M.,: Erfolgreich in die Zukunft. Medhochzwei Verlag, Heidelberg, pp 3-28.

Heyse, V., 2014(b). *Aufbruch in die Zukunft. Erfolgreiche Entwicklungen von Schlüsselkompetenzen in Schulen und Hochschulen: Aktuelle persönliche Erfahrungen aus Deutschland, Österreich und der Schweiz*, Waxmann Verlag, Münster, p 13.

Heyse, V., Giger, M., 2014. *Erfolgreich in die Zukunft: Schlüsselkompetenzen in Gesundheitsberufen.* Medhochzwei Verlag, Heidelberg.

Heyse, V., Pfister, C., Schircks, A, 2012. *KompetenzAtlas Humanmedizin (Schweiz).* In: Heyse, V., Schircks, A.: Kompetenzprofile in der Humanmedizin. Bd. 8. Waxmann, Münster, pp 154-216.

Heyse, V., Schircks, A.D., 2012. *Kompetenzprofile in der Humanmedizin: Konzepte und Instrumente für die Ausrichtung von Aus- und Weiterbildung auf Schlüsselkompetenzen*, Vol. Bd. 8, Waxmann, Münster, pp 18-19.

Hof, C., 2002. *(Wie) lassen sich soziale Kompetenzen bewerten?*, in: 'Kompetenzentwicklung in der beruflichen Bildung', U. Clement & R. Arnold, Schriften der Deutschen Gesellschaft für Erziehungswissenschaft (DGfE), Leske und Budrich Opladen, p 155.

Hüter-Becker, A. Dölken, M., 2005. *Behandeln in der Physiotherapie*: Physiolehrbuch Basis. Georg Thieme Verlag.

Kairy, D, et.al, 2013. *The Patient's Perspective of in-Home Telerehabilitation Physiotherapie Services Following Total Knee Arthroplasty,* Int. J. Environ. Res. Public Health 2013, 10(9) 3998-4011; doi:10.3390/ijerph10093998.

Messer-Misak, K., Egger, R., 2016. *Specific Requirements of Physiotherapists on the Practical Use of Software in the Therapeutical Process,* accepted for the eHealth 2016 conference in Vienna, to appear in: Informatics meets eHealth, published by IOS Press.

ÖBIG – Österreichisches Bundesinstitut für Gesundheit im Auftrag des Bundesministeriums für soziale Sicherheit und Generationen, 2004. *„Berufsprofil der/des diplomierten Physiotherapeutin/Physiotherapeuten": Projekt MTD Qualtiätssicherung.* URL: https://www.physioaustria.at/system/files/general/berufsprofil.pdf, last access: 8.2.2016.

Rychen, D.S., Salganik, L.G., McLaughlin, M.E., 2005. *Definition and Selection of Competencies: Theoretical and Conceptual Foundations*, Executive Summary, http://www.oecd.org/pisa/35693281.pdf. Version: 2005.

Waegemann, C.P., 2010. *Patienten- und Arztkommunikation mit der Elektronischen Patientenakte*, in: Personalisierte Medizin & Informationstechnologie, Niederlag, Lemke, Rienhoff, Dresden, pp. 58-69.

Weinert, F. E., 2001. *Concept of Competence: a conceptual definition*. In: Rychen, D.S.; Salganik, L.H., eds. Defining and Selecting Key Competencies, WA: Hogrefe &Huber, Seattle, p46.

Zhou Nan-Zhao, 2016. *Four „Pillars of Learning" for the Reorientation and Reorganization of Curriculum: Reflections and Discussions* URL https://www.ibe.unesco.org/corps/Competencies/PillarsLearningZhou.pdf 23.3.2016.

AN ECOSYSTEM BASED ON ONTOLOGIES AND ROBOTIC ASSISTANTS TO SUPPORT THE DEVELOPMENT OF SOCIAL COMMUNICATION SKILLS IN CHILDREN WITH AUTISM

Vladimir Robles-Bykbaev[1], Martín López-Nores[2], Verónica Cevallos León Wong[1], Jorge García-Duque[2], José Juan Pazos-Arias[2], Eulalia Naranjo-Cabrera[1], Diego Lima-Juma[1] and Jhonny Pauta-Pintado[1]

[1]*GI-IATa, Universidad Politécnica Salesiana*
Calle Vieja 12-30, Cuenca, Ecuador
[2]*AtlantTIC Research Center for Information and Communication Technologies*
Department of Telematics Engineering, University of Vigo, Spain

ABSTRACT

The term Autism Spectrum Disorders (ASDs) covers conditions such as autism, childhood disintegrative disorder and Asperger syndrome. In this line, the World Health Organization (WHO) points that core symptoms of ASD are: a mixture of impaired capacity for reciprocal socio-communicative interaction and a restricted, stereotyped repetitive repertoire of interests and activities. In light of the above, it is fundamental for a person with ASD to develop skills to communicate with his/her peers, share ideas, and express feelings and needs. On those grounds, in this paper we present an intelligent ecosystem to support the development of social communication skills in children with ASD. This ecosystem relies on ontologies and robotic assistants, and is able to adapt the intervention plans according to the patient's profile. The robotic assistant has been put to test in a pilot experiment carried out with 32 children of regular schools. The achieved results are positive due to the versatility of the robot and the proposed system.

KEYWORDS

Autism Spectrum Disorder, ontology, robotic assistant, mobile applications.

1. INTRODUCTION

World Health Organization (2016) data show that 1 in 160 children present an Autism Spectrum Disorder, and some other resources (CDC, 2016) point about 1 in 68. Far from being a disease, this condition obeys to a developmental disorder, which implicates impairments in social interaction, social communication disabilities, repetitive behavior and restricted interests (American Psychiatric Association, 2014). New conceptualizations of autism are presented within the latest version of the Diagnostic Statistic Manual (American Psychiatric Association, 2014) to adjust the diagnosis to the particular characteristics of individuals. In this line, a person with Autism is diagnosticated considering the main traits of the spectrum, but also describing his/her performance in different areas such as cognition, language and the seriousness level of the main symptoms.

Neurobehavioral models of autism consider it as a "disorder of complex information processing system" in which skilled motor, complex memory and language, and reasoning domains present impairments, while, in contrast, attention, simple memory and language, and visual–spatial domains show a high level of performance (Minshew and Goldstein, 1998). Considering the fact that visual-spatial domains are a strength in these individuals, most of interventions proposed to enhance their life quality are based on visual supports, addressed to attract the child's attention, anticipate facts and reduce the associated anxiety, understand abstract messages by making them more concrete, as well as helping them to express ideas and feelings (Rao and Cagie, 2006).

These visual supports consist in words, images or perceptible objects which the child can easily recognize and interpret, especially when he/she has to handle abstract or unfamiliar information (Hayes et. al, 2010). Visual tools allow a child with autism to organize better his/her surrounding world, which is perceived as uncertain and consequently generates anxiety and uncomfort. Research has shown that picture based systems reduce the symptoms associated with Autism Spectrum Disorders (Williams et. al., 2006 in Hayes et. al., 2010).

The rest of the paper is organized as follows. Section 2 presents a brief overview of some relevant works related with the use of robotic assistants for the intervention in of patients with ASD. The proposed ecosystem and the psychological basis of this work are described in Section 3. Section 4 presents a pilot experiment carried out with 32 children of regular schools. Finally, in Section 5 are described the conclusions and some ideas of future work.

2. RELATED WORK

Recent studies have shown that children with disabilities can be more receptive towards robotic assistants in comparison to their peers. In this line, (Lee and Hyun, 2015) present a research where the robot iRobiQ was programmed with several activities and exercises to promote the language development in children with communication disorders. This approach was tested with four children with autism and intellectual disability, and as result of the therapies, the children learned to initiate conversations with the robot with the emotional exchange of expressions (Lee and Hyun, 2015).

Likewise, Wainer et. al (2014) present a novel proposal of a triadic collaborative game which involves the participation of two children with Autism Spectrum Disorder (ASD) and KASPAR, a robotic assistant. One of the main goals of this research is to prove that is possible to use a robotic assistant to improve the social communication and social skills of these children in a collaborative environment. The pilot experiment was carried out with 6 children during 78 playing sessions, and all participants have shown an important improvement in their social behavior.

Bugnariu et. al (2013) have developed a robotic assistant that analyzes kinematic data and uses a Dynamic Time Warping algorithm with the aim of evaluating and quantifying the imitation behavior of children with ASD. This robot named Zeno has a human form and has been developed by Hanson Robotics. During the data collection, each child and Zeno are put one in front of the other, and the robot interacts with child through this behaviors: verbal dialog (look at me, and follow me), imitate my facial gestures, imitate my head-eye motion, imitate my arm and hand movement. The results of this experiment show that children with ASD consistently perform worse the imitation behavior than their age-match control pairs (children without ASD). All children have reacted in a positive way to the robot possibly because they perceive that it is a toy (Bugnariu et. al, 2013).

As it can be seen, the most of the developed approaches are focused in the treatment of some specific disorders and areas of psychological intervention. In the same way, the proposed approaches do not consider some important aspects like the following:

- The informatics systems, robotic assistants, and in general, the intelligent ICT tools must be able to adapt itself to the specific requirements of each patient and his/her environment (family, scholar, etc.).
- Even two patients with ASD have the same diagnostic, could require different kind of therapeutic exercises or activities.
- The tools used to provide support in diagnostic and intervention of children with communication disorders and ASD must contain information about the progress of each patient as well as the response to tools and to therapy.

In light of above, in this paper we propose a complete ecosystem able to provide a comprehensive support for diagnostic and intervention of children with ASD and communication disorders. In the same line, in this document we will describe some of the most relevant features of our proposal as well as the preliminary results achieved during a pilot experiment conducted with children of regular schools (as the first stage the incorporation of the ecosystem to specialized centers that work with children with ASD in Cuenca, Ecuador).

3. PROPOSED APPROACH

Our intelligent system relies on two layers and several modules. This aspect allows us to easily incorporate new functionalities in a given module, without affecting the function of the other modules and layers. Below we describe some of the most important aspects of each layer that constitutes up our system (Figure 1):

- The **psychologist** and the **patient** interact with the robotic assistant through a mobile device (tablet or smartphone). As a first step, the psychologist selects the patient's profile from a remote database, and the system automatically loads the ontology that relates the therapy guidelines, activities, and patient's profile. The ecosystem provides a set of **web services** that can be accessed from **desktop, mobile** or **web-based** environments.
- The **robotic assistant** has several costumes that can be easily changed to improve the engagement with each patient (according to his/her preferences and needs). In the same way, some of the most relevant features of the robot are the following:
    - Has two different systems to provide energy (one for Raspberry and other for servo motors).
    - Has four degrees of freedom (head, arms, and can move in a table/ground through the two wheels of his base).
    - Has a low consumption of energy, given that uses a resistive display, can take pictures and make videos through a RaspiCam that is under his jaw.
    - Has six hours of autonomy in the complete functioning mode (using the Raspberry and moving the arms, head, and wheels)
    - Can play videos, songs, and incorporates a Text To Speech System (TTS) and an Automatic Speech Recognition System (ASR).

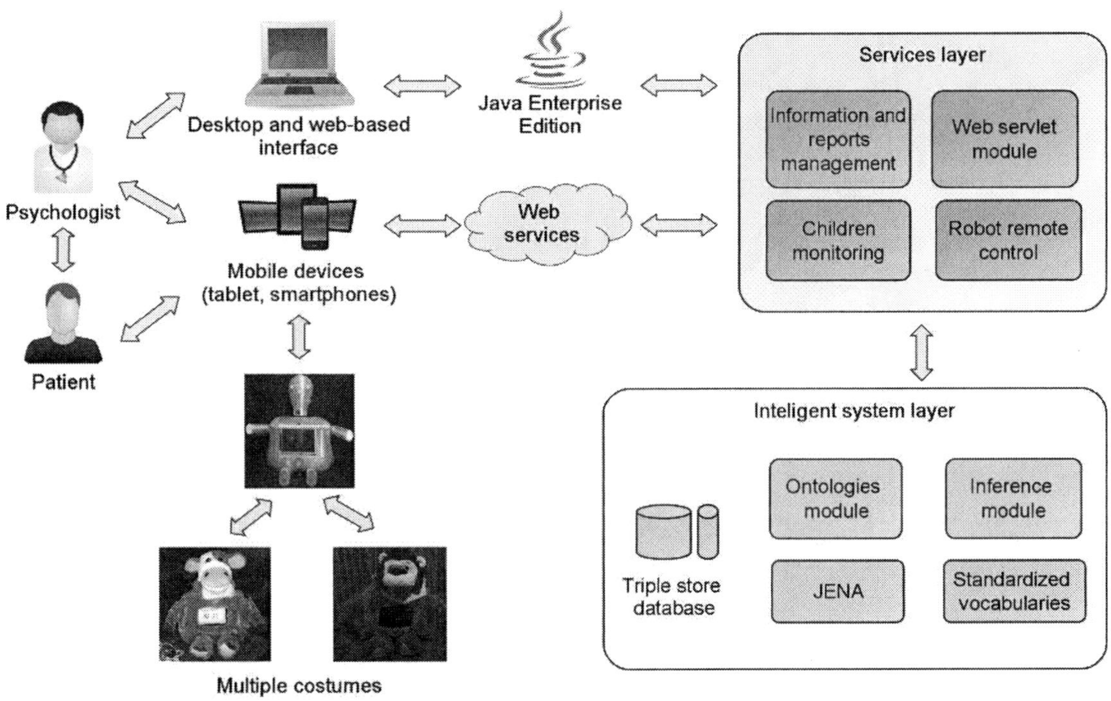

Figure 1. General overview of services, layers and components that make up the proposed ecosystem

- In the **services layer** the system provides several functionalities that allow **monitoring the patient's** progress, registering therapy sessions, **generating reports**, and **control remotely** the robot. Likewise, this layer contains the published ontology, as well as the Resource Description Framework (RDF) instances (**web servlet module**).

- The **intelligent system layer** allows managing all the information of our system as graph structures for semantic queries with nodes. For this purpose we use the **Apache Jena** (*https://jena.apache.org*). With the aim of extending and sharing our proposal through the semantic web, we use **standardized vocabularies** like DSM-5 and ICD-10. In the same way, this layer contains an ontology-based reasoner **(inference module)**, and a **triple database store**.

3.1 Communication and Social Skills Development

Children with ASD present difficulties to adjust their behavior to social situations, which frequently leads them to remain excluded from their peer group. In serious cases of autism, children may show impairments to even recognize simple conventional behavior such as associating objects with specific places or different types of behavior needed to execute simple tasks, such as using silverware correctly. Visual tools have shown to effectively enhance the development of communication and social skills, by helping the child with ASD to understand and integrate conventional information and develop social skills that may allow them to be more tolerated and accepted within the peer group.

Deficits in conventional language and communication are associated with poor social skills in this population; this demands a progressive intervention in which the child can be introduced to basic social patterns and eventually enhance his/her relationship with other significant human being. These interventions are aimed to get the child to be better tolerated and accepted by his/her peer group, by helping him/her to integrate social skills through stories presented in a visual manner.

In addition, children with ASD usually present a preference to interact with objects rather than with humans, which may cause withdraws during the therapeutic work. In this scenario, it is needed the assistance of mediators, which nowadays can be represented by robotic devices. The implementation of artifacts of this kind would reduce the anxiety that children with ASD commonly present towards the interaction with adults or peers, allowing the interventions to be more progressive and respectful with the child's particular characteristics and needs.

3.2 A Progressive Visual Tool to Develop Social Communication Skills

Considering the visual style to process information in individuals with ASD, a series of interfaces containing multiple exercises based on images related to the child's common situations have been developed. In the Figure 2 it is possible to see a screen capture of the screen that contains some options to work with social codes at home, school, and external places.

Figure 2. A screen capture of the menu to select the activities to develop and learn "social codes"

These exercises are aimed to better visualize the situations in which the child may present drawbacks, so he/she can be able to anticipate other's behavior and different situations, to rehearse some conventional responses he/she needs to give in return (Figure 3).

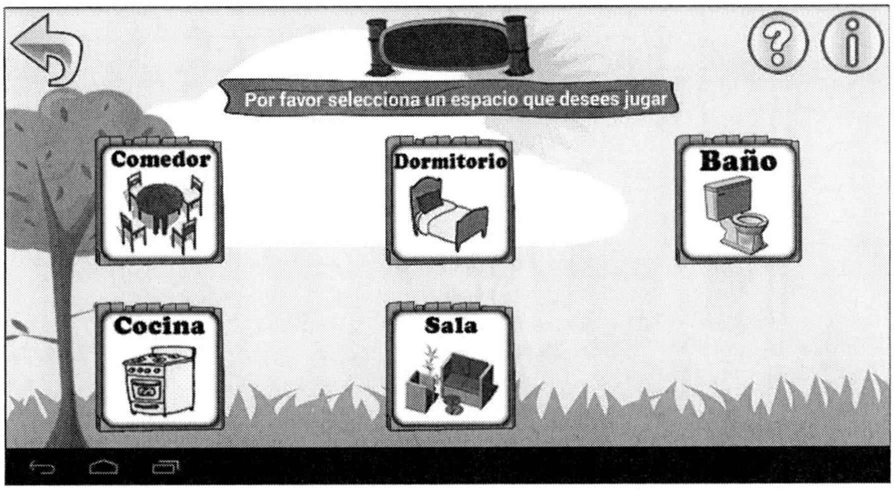

Figure 3. A screen capture of the menu that contains exercises to develop communication skills that are needed at some home's places

The intervention program has a progressive approach, so the child initially works with some conventional information related to daily life behaviors such as using silverware or depositing the garbage in the correct place, and eventually rehearse more complex interactive behavior that usually appear at home or school.

The exercises present a situation in interrogative format, and three different options of answer, to which the child needs to choose the correct response, integrating new alternatives to his/her behavioral spectrum.

Interventions of this kind may offer the child with ASD the structured behavioral and communicational patterns he/she needs to develop satisfactorily within his/her own environment.

3.3 An Ontology to Infer Therapy Exercises and Activities

With the aim of determining which exercises and activities are most suitable for each patient (according to his/her medical and psychological profile), we have designed an ontology that relies on other existing models like Friend of a Friend (FOAF) (Brickley and Miller, 2012) and Autism Spectrum Disorder Phenotype (ASDPTO) (McRay, Trevvett and Frost, 2014).

On this basis, the functionalities and relations of these ontologies have been extended to include new elements like the patient's profile, intervention plans, serious games for supporting therapy, and several activities/exercises to develop social codes. In Figure 4 are depicted some of the most relevant classes that have been included in the ontology to perform the inferences:

- *Action*: this class represents the interactions that occur between user and system.
- *Game_Component*: represents all the components used by the games modules.
- *Metric*: is a measure that is used to represent the most influential factors of the interactions between patients and games.

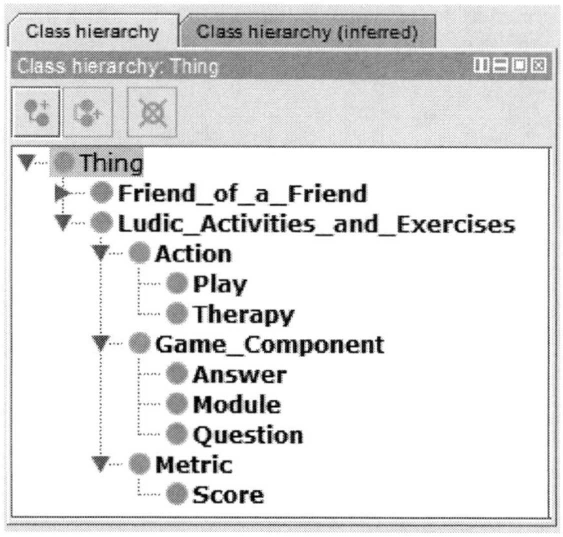

Figure 4. A partial screen capture of the proposed ontology

4. PILOT EXPERIMENT AND PRELIMINARY RESULTS

In order to analyze the real feasibility of our system, we have conducted a pilot experiment consisting with 32 children of regular schools. The children attend to different schools of Cuenca city and are aged between 5 and 14 years (Table 1).

Table 1. Age distribution of children that have participated in the survey

Age (years)	5	7	8	9	10	11	14
Number of participants	3	2	1	3	14	8	1

The aim of this experiment was to analyze the response of children to the robot (aspect and functionalities). The first part of the Table 2 shows the number of children that think that the functionalities of the robot are very good, good, average, bad, and very bad. As we can see, 24 children think that robot is "very good", while 4 of them think that is "good". The second part of the table presents the number of children that prefer a specific costume for the robot. As it shown, the most of the children prefer costumes of dog, cat, and dinosaur.

Table 2. The results of the survey conducted with 32 children of regular schools

What do you think about the robot functionalities?				
Very good	Good	Average	Bad	Very bad
28	4	0	0	0
What animal would you like it to be the robot?				
Dog	Cat	Dinosaur	Rabbit	Other
10	9	7	2	4

Finally, we consider important to know what kind of activities are preferred by children when they come into contact with the robot. In order to obtain this information, we asked children what activities are most interesting to perform with the robot, and the four first choices proposed by children were: playing games, dancing, talking and walking/running (Figure 5).

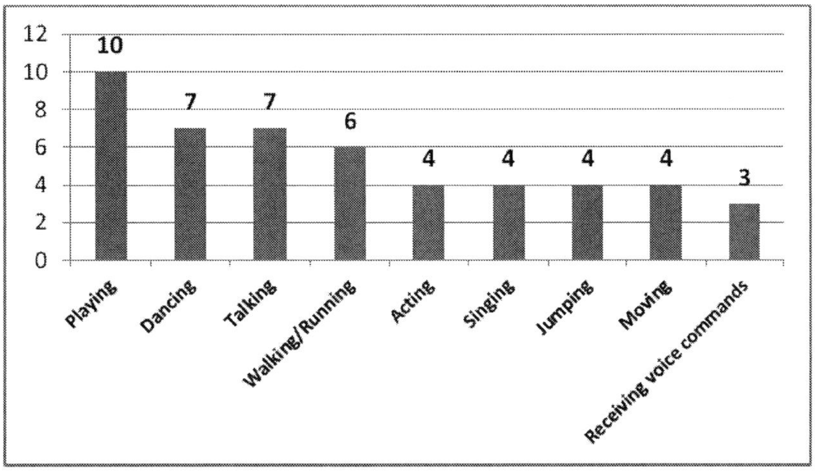

Figure 5. Results of the survey applied to 32 children to answer the question: What activities do you prefer doing with the robot?

5. CONCLUSION

The robotic assistant showed to have a very high level of acceptance among the sample. These aspects allow us establishing master lines with the objective to carrying out the same experiment with children with ASD, and determine the most relevant characteristics of the robot according to age, medical diagnosis, and related disorders.

The sample also exhibited a preference for customs associated with familiar animals like dogs, which can help them feel more comfortable and accompanied in unknown places or situations that commonly causes them to suffer from anxiety.

In the same way, it is fundamental that the system and the robotic assistant have mechanisms to adapt the activities, intervention strategies, and they behavior according to each child's profile, and their needs.

As lines of future work we propose the following ones:

- To develop sensing application for wearables devices, with the aim of determining the physical response of children to therapy and some intervention activities.
- To develop a remote support application for patient relatives, with the aim of gathering information that can provide data about the behavior pattern of children at home.

ACKNOWLEDGEMENT

The authors from the Universidad Politecnica Salesiana have been supported by the "Sistemas Inteligentes de Soporte a la Educación CIDII-010213" research project. The authors from the University of Vigo have been supported by the Ministerio de Educación y Ciencia (Gobierno de España) research project TIN2013-42774-R (partly financed with FEDER funds).

REFERENCES

American Psychiatric Association, 2014. Guía de consulta de los criterios diagnósticos del DSM-5®: Spanish Edition of the Desk Reference to the Diagnostic Criteria From DSM-5®. *American Psychiatric Pub.*

Brickley, D., and Miller, L., 2012. FOAF vocabulary specification 0.98. Namespace document, 9.

Bugnariu, N., Young, C., Rockenbach, K., Patterson, R.M., Garver, C., Ranatunga, I., Beltran, M., Torres-Arenas, N. and Popa, D., 2013. Human-robot interaction as a tool to evaluate and quantify motor imitation behavior in children with Autism Spectrum Disorders. *In 2013 International Conference on Virtual Rehabilitation (ICVR),* pp. 57-62

CDC - Facts about Autism Spectrum Disorders - NCBDDD. 2016. Facts About ASDs. [online] Available at: http://www.cdc.gov/ncbddd/autism/data.html [Accessed 15 Apr. 2016].

Hayes, G. R., Hirano, S., Marcu, G., Monibi, M., Nguyen, D. H., & Yeganyan, M., 2010. Interactive visual supports for children with autism. *Personal and ubiquitous computing*, Vol. 14, No. 7, pp. 663-680.

Lee, H. and Hyun E., 2015. The intelligent robot contents for children with speech-language disorder. *In Journal of Educational Technology & Society*, Vol. 18, No. 3, pp 100-113.

McCray, A. T., Trevvett, P., and Frost, H. R., 2014. Modeling the autism spectrum disorder phenotype. Neuroinformatics, Vol. 12, No. 2, pp. 291-305.

Minshew, N. J. and Goldstein, G., 1998. Autism as a disorder of complex information processing. *Mental Retardation and Developmental Disabilities Research Reviews*, Vol. 4, No. 2, pp. 129-136.

Rao, S. M. and Gagie, B., 2006. Learning through seeing and doing: Visual supports for children with autism. *Teaching Exceptional Children*, Vol. 38, No. 6, p. 26.

Wainer, J., Robins, B., Amirabdollahian, F. and Dautenhahn, K., 2014. Using the Humanoid Robot KASPAR to Autonomously Play Triadic Games and Facilitate Collaborative Play Among Children With Autism. *In IEEE Transactions on Autonomous Mental Development*, Vol. 6, No. 3, pp.183-199.

World Health Organization. 2016. Autism spectrum disorders. [online] Available at: http://www.who.int/mediacentre/factsheets/autism-spectrum-disorders/en/ [Accessed 15 Apr. 2016].

HUMAN-DATA INTERACTION IN HEALTHCARE: ACKNOWLEDGING USE-RELATED CHASMS TO DESIGN FOR A BETTER HEALTH INFORMATION

Federico Cabitza and Angela Locoro
Dipartimento di Informatica, Sistemistica e Comunicazione, Università degli Studi di Milano-Bicocca
Viale Sarca 336 20126, Milano, Italy (Phone: +39-02-64487888)

ABSTRACT

In this paper, we focus on an emerging strand of IT-oriented research, namely Human-Data Interaction (HDI) and on how this can be applied to healthcare. HDI regards both how humans create and use data by means of interactive systems, which can both assist and constrain them, as well as to passively collect and proactively generate data. Healthcare is a challenging arena to test the potential of HDI towards a new, user-centered perspective on how to support and assess data work, especially in current times where data are becoming increasingly big and many tools are available for the lay people, including doctors and nurses, to interact with health-related data. This paper is a contribution in the direction of considering healthcare data through the lens of HDI, and of framing data visualization tools in this strand of research, in order to let the subtler peculiarities among different kind of data and of their use emerge and be addressed accordingly. Our point is that doing so can promote the design of more usable tools that can support data work from a user-centered and data quality perspective.

KEYWORDS

Human-Data Interaction; Primary Use; Tertiary Data

1. INTRODUCTION

Twenty-five years ago, medical informatics was defined as "dealing with the storage, retrieval and optimal use of biomedical data" (Shortliffe et al. 1990). At that time, little emphasis was put on the practices of data production, that is on how medical practice, and single stories of illness, care and recovery are represented, accounted and "datafied" in some objective manner. However, these practices, which include policies, rules, habits, conventions, tools and techniques, have always been intertwined with and affected by the available ITs, as well as by the expectations of the stakeholders on how to make sense and use of health-related data. Different perspectives on these expectations, and on what valuable health data are, lead to manifest chasms between primary use and other uses of health information, as often discussed in the specialist literature (Fitzpatrick and Ellingsen, 2013). To try to cross these chasms, we need to create the suitable language to describe the differences and give some operational definitions.

We distinguish between three different macro-types of data and the related processes in which these data are either produced, processed or consumed: namely, *primary data*, which come from a broad range of sources and are produced *within* a caring process to make its unfolding seamless and smooth (Berg, 1999); *secondary* data that are derived from the primary data for purposes different than care, like accounting and medical billing (Abdelhak, Grostick, & Hanken, 2012); and *tertiary* data, that are produced from the secondary data for any unanticipated need of the potential consumers of health services (see Figure 1).

To illustrate this tripartition, an analogy from the agriculture domain can be drawn (Locoro, 2016): primary data are like the produce of the land, which farmers grow for themselves as well as the external market. Secondary data are the product of a transformation of these primary data, like the one performed in food industry where vegetables are cleansed and chopped. Tertiary data are further transformed from secondary data to make them more easily consumable, that is suitable for and conveyed to a broader population of consumers in terms of information services, like fresh-cut vegetable products can be seen as the service to have vegetables already ready-to-eat.

The definitions mentioned above shed light on the relationship between data and their uses, which cannot be overstated. The tripartition that we propose reflects the different uses and practices in which data are produced and consumed and it calls for a specific area of research focusing on how to support people in interacting with their data of concern and in gaining insight from them: Human-Data Interaction (HDI). Researchers in this field, which lies at the intersection of data science, data visualization and human-computer interaction, *study* how humans record and use data by means of Information Technologies; and *design* interactive systems that allow humans to retrieve and explore complex data sets in order to gain value in their learning, insight in their decisions, and feedback in their action[1]. In what follows we will justify our triparted perspective and propose it as a conceptual framework to focus on the peculiarities and needs of primary and tertiary users, and therefore contribute in the HDI field of research applied to healthcare data.

2. DISTINGUISHING WAYS OF INTERACTING WITH HEALTH DATA

For a straight metonymy, we intend primary data as entangled with the primary use of data, by definition; secondary data with secondary use(s); tertiary data with tertiary uses. The "primary use" of health information is "to use it to directly support patient care", both by aiding medical decision-making and by ensuring continuity of care by all providers, that is both interpretation of medical signs (represented by data) for decision making, and coordination among the actors involved "around" the patient (Berg, 1999). Secondary use regards both other uses of the same data collected for the primary use within the administrative domain, and the generation of derivative data for other aims than care, like billing and reimbursement, performance and care quality evaluation, resource planning and management, service design and public policy making. Tertiary use regards the heterogeneous uses through which tertiary data are put to the test of users' life: this can encompass the publication and dissemination of valuable indications for the citizens and the taxpayers about the available healthcare services, so as to orient them in the choice of the best healthcare provider (or just the closest one to their shelters – cf. Cabitza et al. 2015b); the policy makers and facility or agency manager, to facilitate the monitoring of suitable outcome and performance indicators and to enable the benchmarking and comparison of care facilities; as well as the researchers and epidemiologists to make sense of the great numbers of "similar" patients that are treated and detect trends and patterns of treatment appropriateness and efficacy.

On a more general level, three concepts characterize uses and data in each of these three ambits: primary use is related to what has been often called "*data work*" (Bossen et al. 2016). Secondary uses with *data storing and processing*. Tertiary uses with *data value*. Data storing and processing are self-explanatory concepts in computer science and regard the set-up of efficient data bases and the automation of accurate procedures by which to structure data, run searches and extract complex reports for disparate purposes. As secondary data and secondary processes usually entail little interaction by humans and these are always IT specialists and consultants who have always considered usability a relative concern, we will focus on the primary and secondary domains, and hence on data work and data value.

Data work is quite never defined in the specialist literature; although it is often mentioned, it is a slippery notion. We see it as a modern (but not at all recent) coinage after the word *paperwork*, of which it represents an abstraction, with respect to the medium of data representation; but also an extension, with respect to what people manage as data of their interest (besides accounting and resource management). As such data work is not only "working *on* data", typically producing new data in accounting for and recording a faithful representation of the work done; but also *that* portion of work that can be accomplished only by relying on some accurate data, i.e., "working *by* data". These two kinds of work are usually so deeply intertwined that distinguishing between them is useless and probably wrong: in the healthcare domain, the studies by Berg (Berg, 1999), for instance, clearly show that clinicians record data on the patient record not only to accumulate data for archival reasons (and for many other secondary uses), but also to coordinate with each other, articulate the resources around a medical case, and take informed decision in a written, distributed communication with themselves and the other colleagues taking care of the same patient. In healthcare, data work regards the additional effort paid by caregivers in making the record a "working record" (Fitzpatrick

[1] The reader should mind that *human-data interaction* is not *human data* interaction, that is interaction with human data. We intend HDI as focused not solely on the interaction of humans with human-related data, or with one' own personal data, but *also* on that kind of data and the related dimensions, which have been recently considered by other scholars (e.g., Crabtree and Mortier, 2015).

2004), that is a resource capable of keeping disparate competencies and roles bound together and connected around the same cases over time and space.

On the other hand, tertiary use of tertiary data regards the concept of *data value*, that is how to exploit data to achieve some relevant goal for the user (even just to be informed about something), so that we can say that the user *finds* value in the data (also: she finds data valuable), or she *gains* value to some aim. Whether value is already in the data or rather it is created by an active stance of its consumers looks like an idle question: in a HDI perspective, value is the *result* of interacting with data and being capable to exploit them by tertiary users, that is lay people moved by unexpectable motives and toward unanticipated aims. In other words, on one hand data have got value if they are true and have been made accessible and comprehensible; on the other hand, their value lies in the comprehension itself, in the acquisition of true information, in learning notions, techniques, practices, and in the resulting knowledgeable behaviors, in their turn producing some positive effect, on either the single person or her community[2]. Thus, data value can be defined as the potential of data (for its intrinsic qualities) to enable processes of value creation by someone for anyone (either herself, or the others) in some given context.

3. HDI CONTRIBUTIONS FOR THE QUALITY OF INFORMATION

While this tripartition is general enough to be applied in any domain where many different roles partake to the generation and transformation of data in the same supply chain, so to say, from the practitioner working in the field, to the final consumer of information services, its application to healthcare can be justified for its role in addressing the problem of the still relatively low quality of the data of patient records (Liaw et al. 2013), both in regard to completeness, accuracy and timeliness (Cabitza & Batini, 2016). Indeed it allows, on one hand, to highlight differences, separate concerns, and identifying ambits of use that are irreducible to one another; on the other hand, it allows to see the (existing) "seams" between these ambits (Berg and Goorman 1999) and therefore to address cooperation and interoperability without reckless neglect.

In regard to this latter point, it is time to acknowledge that chasms between primary and other uses exist and they go beyond the usual tension between clinical vs. administrative purposes because it is the concept of "adequate quality" that should be adopted to assess care-oriented data; likewise, crossing the chasm between the secondary and tertiary domains requires more than merely making secondary data more open and accessible, because also the end consumers' readiness to access, comprehend and exploit them, as well as their unanticipatable purposes, are to be considered. Thus, the challenge to bridge the chasms mentioned above can be seen in terms of the problem of reusing primary data, which is "always entangled with the context of its production" (Berg and Goorman, 1999), in different contexts (either secondary or tertiary ones). As Berg and Goorman note, reuse is possible only if data are made "transportable", that is sufficiently disentangled from their context of production; and this can occur only if specialized *additional* work is performed on data. So it is not a matter of improving, automating or changing current data work, but rather to invest on new and different data work, and corresponding new organizational roles (e.g., data nurses).

In regard to the separation of concerns, our distinction between primary, secondary and tertiary data is quite different from other frameworks discussed in the Data Quality literature: for example, the one inspired by the manufacturing domain (Shankaranarayan et al., 2003) which distinguishes between *raw data*, *component data items*, that is semi-processed information and *information products*, which are composed out of these items. Primary data are not necessarily raw (Gitelman, 2013), because they are meaningful to and usable by the primary consumers that generated them. On the other hand, secondary and tertiary data are unfinished information products: the former ones are resources for specialist work (both clerical and managerial) within specific organizational boundaries and processes, while tertiary data result from the enactment of information services conceived for the non specialist and the external consumers, that is for the

[2] Obviously we are aware that the value of something, and hence also data, can be appraised differently than by adopting a consumer-oriented perspective, which regards what in classical political economy is denoted as "use value". Value can be a function of the resources involved or consumed to get the data (mainly by primary users), to verify and cleanse them, and make them fit to interoperable formats typicaly in secondary processes); or, simply, a function of how much someone is going to pay to obtain them (cf. the idea of economic value, that is so common in this age of splendour of online advertising). These different perspectives reflect different conceptions of the idea of value, as it has been formulated in centuries of economic and theoretical speculation (see (Dobb, 1975) for a classic review of those conceptions).

public. The distinction introduced by Shankaranarayan and his colleagues focuses on an incremental definition of the information product, and is often mentioned in regard to how Data Quality (DQ) can be assessed, monitored and continuously improved (e.g., Wang, 1998)

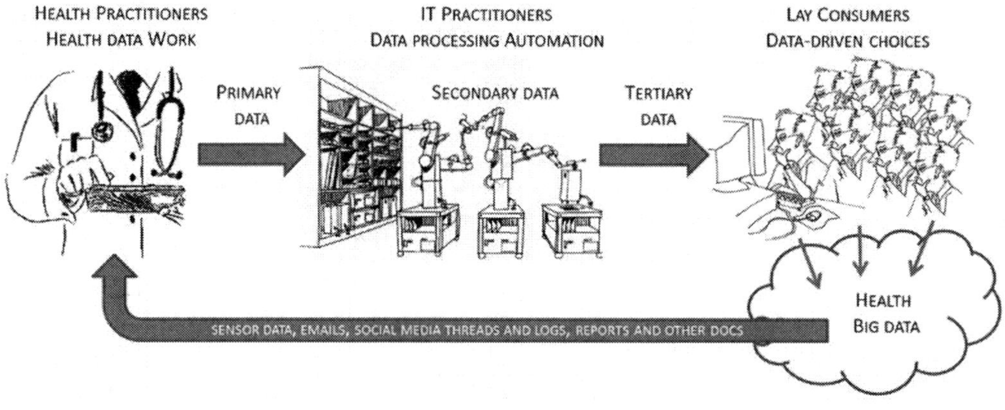

Figure 1. The three kinds of data and ambits proposed for HDI in healthcare

Differently from the Information Product perspective, our triparted perspective focuses on the different roles that produce and consume different information products that can be considered "definitive" only in relation to their context of use. Consequently, a HDI approach to DQ issues grounds on the interpretation of quality of data as their "fitness for use" (Wang & Strong, 1996). Our proposal is aimed at detecting and taking the differences from the various uses in healthcare seriously (Bossen et al. 2016).

According to our metaphor, secondary data and secondary data-centered processes are like industrial processes, that need standardization to achieve regularity and efficiency and that involve relatively few people with specific skills and motives, like the competent and trained operators of assembly lines at the shop floor of plants. This is why HDI regards secondary data and processes only marginally, and to a much greater extent the primary and tertiary contexts. In primary settings, HDI addresses the challenges of fitting "data work" to work, that is of reconciling the articulation of cooperative activities performed by knowledgeable experts with their effort in continuous recording relevant events and information about *what* has been carried out, and often also of *why* and *how*. On the other hand, in the broader ambits of tertiary use, HDI deals with the manifold challenge to design data structures, visualization controls and interaction affordances, for both unintended uses and unanticipated needs, for both the practitioners without particular e-literacy and numeracy, and lay people from the general public.

We would argue that quality of information should not be assessed irrespective of the distinction between primary / secondary / tertiary uses, that is, by adopting the same metrics and methods in a context-independent manner: on the contrary, data used in care processes should be evaluated on the basis of the efficacy they enable appropriate and timely action (fit to use), also on the basis of work conventions and tacit knowledge that are difficult to bring back to the usual dimensions of accuracy, completeness and consistency (Gregory et al., 1995). This is in line with the idea to measure the quality of primary (health) data in terms of appropriateness and as a function of the health outcome they help to achieve. Bad quality should not be related to completeness, timeliness or, even, accuracy, since health practitioners can perfectly make sense of "bad" data and to some extent they even expect them (Bossen et al. 2016); but rather to "information failure" and its impact, that is the number of times data work is either direct or indirect cause of adverse event or near misses (Pipino & Lee, 2011). Despite the authorities that advocated for more research on this concept (Pipino & Lee, 201; Cabitza & Batini 2016), information failure (or error) is still underresearched and a taxonomy of cases to guide analysis and comparison still missing, for instance one that could distinguish between cases in which wrong or late decisions were carried out with accurate and timely data, respectively, and the other way round, a right and timely interpretation has been based on wrong and obsolete data, and so on[3].

[3] Other intersting cases regard the failure to detect relations between either right data or inconsistent ones; the failure to get the right implication from right data or, the other way round, success in detecting right relations and infering right implications from wrong data.

Similarly, further research is needed for the evaluation of the quality of tertiary data: this is highly correlated with the quality of the information services conceived for the general public, which should be evaluated in terms of the extent to which data are "informative" and can be appropriated by consumers and appreciated in their lives (i.e., according to its social value), that is if they affect decisions for the better.

This is where HDI can fit in and be a novel way to address long-standing problems of low quality that neither incentivizing structures nor sanctions seem to solve completely. Indeed, HDI covers three phases: design, development and evaluation of the systems by which to extract information and support knowledge in data-intensive application domains. In particular, HDI regards the user-centered elicitation of better requirements of configuration, adaptation and appropriation of big data analytics cockpits and dashboards to optimize usability and the user experience, i.e., efficiency, effectiveness and satisfaction; the application of End-User Development techniques and tools to allow end-users tweak the tools by which data are extracted and visualized; and user-centered methodology for the assessment and continuous improvement of the quality of the interaction of the humans with their data of interest, so as to reduce both information overloading and information funneling/complacency (Parasuraman and Manzey, 2010), and improve awareness. This also includes the exploration of new techniques toward better interactive visualization environments and above all better data-telling, that is the capability to build and share stories that can explain data and facilitate correct interpretations (e.g., in medical domain by adopting a natural frequency approach - Hoffrage et al., 2002) and to allow, instead of curbing, the social exchange within a community of data-users of multiperspective, sound and viable interpretations around the data that are supplied by the computational systems to inform decision making and knowledgeable action.

Summarizing, HDI can contribute in addressing the reconciliation on needs in health data production and use by providing methods and techniques to investigate the following two research strands:

1) The design of interfaces that could promote mutual awareness between data producers and consumers: on the one hand, by increasing commitment and awareness of consequences in the producers of the primary data; and, at the same time, by raising awareness in the secondary consumers of the contextual and social nature of primary data and hence of their limitations. This can be achieved, for example, by endowing the interfaces by which data are collected and presented with specific affordances that adapt to the context according to specific business rules in order to convey the so called "awareness promoting information" (Cabitza and Simone, 2012); this can be done also by means of simple visual clues like text highlighting or side messages, which do not impose any behavior to the data producers but help their interpretation.
2) The design of interactive visual interfaces that could support the understandability of data analytics. The transformation of data into information services does not necessarily require a massive processing of data but rather the application of state-of-the-art human interaction techniques to develop interactive infographics and highly tailorable dashboards that enable user-friendly online analytical processing and hence the transformation, even by end users (Lieberman, Paternò, Klann, & Wulf, 2006), of secondary data into socially valuable information.

4. EXAMPLES OF USER STUDIES IN HEALTHCARE HDI

In what follows we report two user studies conceived in a HDI perspective. The first study regards primary data in radiology work, and therefore it regards the identification of what data the practitioners should be required to record for their primary aim, and what data they deem less useful to this aim. The second study regards the value of tertiary data about hospital quality and availability, as this is evaluated by citizens that would need to gain information about the hospitals of their catchment area to choose where to go according to their needs and preferences.

In the first study, we asked 15 expert technicians from different radiology facilities of Northern Italy to indicate the usefulness for their work of 31 information items on an ordinal scale, from 1 (negligible) to 6 (essential). The items regarded either the assessment of patients, their management for the execution of

---

It is important to distinguish between all of these cases in understanding the root causes of information failures and understand how to minimize the odds of their occurrence.

Data Item in the main page of the radiographer	Level of priority
Patient ID, Exam Date, Birth Date, Radiographist Name.	Higher
Level of urgency, Name of the patien, Time of execution, Patient Gender, Responsible Radiologist, Access number, Alert, Body section, DICOM ID, Patient fiscal code, Modality Type, Referring facility.	Probably higher
Ordering Radiologist, Exam description, Phone number, Room ID, patient's current address, patient's municipality, how the patient arrived, cost center.	Probably lower
Patient nationality, Patient height.	Lower

Figure 2. The ranking of data item in radiology. Items are enumerated according to their fine-grained ranking within the macro-category of priority level. Levels are established according to statistical testing

digital imaging examinations, or the preliminary evaluation of radiographic media, and they were collected by considering the Italian guidelines for quality assurance in teleradiology (Working Group for Quality Assurance in Diagnostic and Interventional Radiology, 2010).

The aim of the study was to rank those items from the most useful to the least, so that the interface of their information system could be optimized. Here, the concept of optimization should be interpreted as follows: either in terms of having the radiological information system display and require only those data that the radiographers considered certainly useful for their job; or, at least, in terms of a system that would not urge the users to fill in the low priority fields as mandatory data that block the application unless radiographers insert them. The online questionnaire platform displayed the 31 items in different order to minimize order bias and the task was carried out as part of an assignment of a first Level Master degree class on IT management for radiographers.

We then applied an original ranking method, discussed in (Anonymized) by which we classified each item in one of four levels of priorities (see Figure 2). An informal interpretation of the result could consider the first group (see the green block in Figure 2), i.e., Patient ID, patient birth date, exam date, operator name, radiographer name (if these latter are not the same) as those items that cannot be missing in the radiographer screen, and that indeed need to be emphasized at the interface level (e.g. by being rendered at a proper font size) and double checked by the operators involved to avoid potential mistakes and adverse events. The second group of items (including the name of the patient, the ordering radiologist, the access number) are data of lower priority but still very important, that radiographers should consider mandatory to fill in. Conversely, low priority items like patient nationality and patient height (see the red block in Figure 2) could be considered those that should not be imposed to the data work of radiographers, nor displayed irrespective of the specific examination, as these unnecessary items could contribute to clutter the user interface and hide more relevant information.

In our future work, we will perform the same user study in different groups of users of the radiological information system, like the medical radiologists and the administrative clerks, so as to make the interfaces adaptive according to the role logged in, and have the systems display awareness promoting information (Cabitza & Simone, 2012) about the different perception of utility (if any) by different worker categories. For instance, let us assume that the administrative clerks in a given hospital would consider the indication of the cost center and the billing address of the patient essential data items for their paperwork; in this case, radiographers, who conversely attach a low priority to these items (see Figure 2), could be notified that these data are important for someone else in the same facility (e.g., by means of some iconic indication or by a warning message at the end of the work shift) and be made *aware* (see above) of the data-related needs of other professionals that are connected by the same patient case.

In the second study, we collected 330 complete questionnaires from a population of citizens living in an urban area and aged 18 or older. To this aim, we invited personal friends and acquaintances, all of the colleagues of our university department, and the students of two master classes from the same university. Sample representativeness was achieved by weighting the response set by both age and gender according to the latest national census. The respondents had to evaluate the utility of 11 information items in different scenarios of urgency and serious health conditions. The information items had been selected from those that can be conveniently extracted from some of the most important big open healthcare data sets available in the USA, in Canada and in Italy (taken, respectively, from healthdata.gov, www.cihi.ca and dovesalute.gov.it).

The sample exhibited (statistically) significant appreciation for three information items above all the others: the hospital ranking by number of admissions for the pathology of interest; the ranking by average wait time for the pathology of interest; and the ranking of facilities by their reaching time from one's place. The rankings evaluated the least useful were the rankings produced according to either the perceived level of care quality by all the hospitalized patients, or by the percentage of patients that had to be hospitalized again within six months since the first discharge. While the latter information is quite technical, perhaps too much to be appreciated by a sample of lay people, the former indication came unexpected and quite puzzling.

All of the collected indications are useful to design healthcare portals that show their users a minimal set of items with the highest potential to satisfy the users' needs, on default of other information, leaving the other items accessible on demand. Interestingly, we detected significant differences in the perception of information usefulness between different user profiles. For instance, male respondents attached more importance than women to the ranking based on georeferenced information. Not surprisingly, expert users considered the ranking by number of admissions for the pathology of interest more important than people self-declaring non-expert. We also probed the sample preferences for different ways to display the same information, either by simple and essential tables or by richer and more appealing infographics. Also in this case we detected a statistically significant difference between females and males, as the former ones found infographics to be clearer and more intuitive than men; and between young respondents and the elderly as the former ones, quit unexpectedly, found the tabular form more clear than the latter ones. These differences suggest that the usability of a healthcare portal can be maximized even considering basic profiling information and user satisfaction can be improved by asking the user only a few data, like gender and age.

5. CONCLUSIONS

In this paper we have proposed a triparted perspective to health data by which to distinguish between primary data, i.e., the content of health records that clinicians and nurse produce and consume in their decisions and treatments; secondary data, i.e., the content of structured data bases, data warehouses and the reports that automated procedures produce and display on interactive data visualization tools (or dashboards) to support the tasks of clerical and managerial roles (including policy makers); and tertiary data, i.e., particular content that has been selected and optimized in terms of clarity, understandability and appeal, in one word usability, to be offered online for the general public and hence for unanticipated needs and aims. The distinction has been proposed to separate concerns and differentiate agendas. The one-size-fits-all approach to data quality improvement simply does not work (Pipino et al. 2002), nor solve the still present problems of low use of and low satisfaction for the Electronic Health Record (Cabitza et al. 2015a).

Human Data Interaction (HDI), as a research approach, can provide alternative methods to address these challenges, which are becoming more and more urgent especially in our current age of increasing opportunities and threats regarding the datification of medicine and the consequent production of health big data (Murdoch & Detsky 2013). In particular, HDI approaches would focus on evaluating how ICT supports data (producing) work by relieving doctors and nurses from form-filling, coding and other administrative tasks; and on evaluating the usability and the user experience related to interacting with tertiary (health) data, to understand the actual value these convey in decision making and information-intensive tasks.

REFERENCES

Abiteboul, S. et al, 2000. *Data on the Web: From Relations to Semistructured Data and XML*. Morgan Kaufmann Publishers, San Francisco, USA.

Abdelhak, M., Grostick, S., & Hanken, M. A. (Eds.). (2012). Health information (4th ed). Elsevier.

Berg, M. (1999). Accumulating and coordinating: occasions for information technologies in medical work. Computer Supported Cooperative Work (CSCW), 8(4), 373-401.

Berg, M., & Goorman, E. (1999). The Contextual Nature of Medical Information. IJMI, 56, 51–60.

Berg, M., & Toussaint, P. (2002). The mantra of modeling and the forgotten powers of paper: A sociotechnical view on the development of process-oriented ICT in health care. JMI, 69(2-3), 223–34.

Bossen, C., et al. (2016) Data-work in Healthcare: The New Work Ecologies of Healthcare Infrastructures. ACM International Conference on Computer-Supported Cooperative Work and Social Computing, CSCW 2016.

Cabitza, F., & Simone, C. (2012). Affording Mechanisms: An Integrated View of Coordination and Knowledge Management. CSCW, 21(2), 227–260.

Cabitza, F., Simone, C., & De Michelis, G. (2015a). User-driven prioritization of features for a prospective InterPersonal Health Record: perceptions from the Italian context. Comput Biol Med, 59, 202-210.

Cabitza, F., Locoro, A., & Batini, C. (2015b). A User Study to Assess the Situated Social Value of Open Data in Healthcare. Procedia Computer Science, 64, 306-313.

Cabitza, F., & Batini, C. (2016). Information Quality in Healthcare. In Data and Information Quality, 421–438. Springer.

Crabtree, A., & Mortier, R. (2015). Human data interaction: Historical lessons from social studies and CSCW. In ECSCW 2015. Springer.

Fitzpatrick, G. (2004). Integrated care and the working record. Health Informatics Journal, 10(4), 291-302.

Fitzpatrick, G., & Ellingsen, G. (2013). A review of 25 years of CSCW research in healthcare: contributions, challenges and future agendas. CSCW, 22(4-6), 609-665.

Gitelman, L. (2013). Raw data is an oxymoron. MIT Press.

Gregory, J., Mattison, J. E., & Linde, C. (1995). Naming notes: transitions from free text to structured entry. MIM, 34(1-2), 57–67.

Hoffrage, U., Gigerenzer, G., Krauss, S., & Martignon, L. (2002). Representation facilitates reasoning: What natural frequencies are and what they are not. Cognition, 84(3), 343-352.

Liaw, S. T., Rahimi, A., Ray, P., Taggart, J., Dennis, S., de Lusignan, S., ... & Talaei-Khoei, A. (2013). Towards an ontology for data quality in integrated chronic disease management: a realist review of the literature. IJMI, 82(1), 10-24.

Lieberman, H., Patern'{o}, F., Klann, M., & Wulf, V. (2006). End-User Development: An Emerging Paradigm. In EUD, 9, 1–8.

Locoro, A. (2016) A Map is worth a Thousand Data: Requirements in Tertiary Human-Data Interaction to Foster Participation. CEUR Workshop Procs of CoPDA 2015.

Murdoch, T. B., & Detsky, A. S. (2013). The inevitable application of big data to health care. Jama, 309(13), 1351-1352

Parasuraman, R., & Manzey, D. H. (2010). Complacency and bias in human use of automation: An attentional integration. Human Factors: Hum. Factors, 52(3), 381–410.

Pipino, L., Lee, Y. W., & Wang, R. Y. (2002). Data quality assessment. Commun ACM, 45(4), 211-218.

Pipino, L., & Lee, Y. W. (2011). Medical Errors and Information Quality: A Review and Research Agenda. In AMCIS.

Roberts, A., Gaizauskas, R., Hepple, M., & Guo, Y. (2008). Mining clinical relationships from patient narratives. BMC Bioinformatics, 9(11), S3.

Sauleau, E. A., Paumier, J.-P., & Buemi, A. (2005). Medical record linkage in health information systems by approximate string matching and clustering. BMC Medical Informatics and Decision Making, 5(1), 32.

Shankaranarayan, G., Ziad, M., & Wang, R. Y. (2003). Managing data quality in dynamic decision environments: An information product approach. JDM, 14(4), 14–32.

Shortliffe, E.H., Perreault, L.E., Wiederhold, G., and Fagan, L.M. (eds.) (1990). Medical Informatics: Computer Applications in Health Care and Biomedicine. Addison-Wesley.

Smith, P. C., Araya-Guerra, R., Bublitz, C., Parnes, B., Dickinson, L. M., Van Vorst, R., Pace, W. D. (2005). Missing clinical information during primary care visits. Jama, 293(5), 565–571.

Swinglehurst, D., Greenhalgh, T., & Roberts, C. (2012). Computer templates in chronic disease management: ethnographic case study in general practice. BMJ Open, 2(6)

Timmermans, S., & Berg, M. (2003). The gold standard: the challenge of evidence-based medicine and standardization in health care. Temple University Press.

Wang, R. Y., & Strong, D. M. (1996). Beyond accuracy: What data quality means to data consumers. JMIS, 5-33.

Wang, R. Y. (1998). A product perspective on total data quality management. Commun ACM, 41(2), 58-65.

PERSUASIVE E-COACHING FRAMEWORK

Pierre Barralon and Iñigo Dorronsoro Esnal
Tecnalia Research & Innovation
P. Mikeletegi 1-3. Parque Tecnológico
E-20009 Donostia-San Sebastián - Gipuzkoa, Spain

ABSTRACT

Increasingly, persuasive technologies are used to convince, stimulate and motivate users to engage in healthy behaviours. Currently persuasive technologies are based either on social psychology theories (the user compares him/herself with other people), self-determination (human behaviour is a result of a decision that was either taken autonomously or that was rather imposed to a person), goal-setting (process by which people attempt to constrain unwanted urges by gaining control over the impulsive response), or health belief model (individual's perception of the *severity*, *benefit*, *barrier* of a situation or action to perform). However few approaches have tried to integrate user's emotion. In this paper we present a persuasive e-coaching framework which explicitly consider and integrate the emotion in the decision making process.

KEYWORDS

Emotion, e-coach, persuasive technology, physiological parameters

1. INTRODUCTION

Increasingly, persuasive technologies are used to convince, stimulate and motivate users to engage in healthy behaviours. (Fogg2003) defines persuasive technology as "the class of technologies or interactive computing systems that are intentionally designed to change a person's attitude or behaviour." An important element of this definition is that a person being persuaded always has a clear and free choice to change attitude, behaviour or both. In this way, persuasion opposes coercion and deception, in which users are forced or manipulated to change their attitude or behaviour.

The European Artemis "With-Me" project intents to identify and provide complementary elements that will support/persuade the user to follow a diet, engage, maintain or increase physical activity. For that purpose With-Me project has identified and extracted a set of behaviour change techniques, has investigated the effect of conscious and unconscious motivators as well as the role of social interactions. Even though these elements are very important, in this paper we focus on one important additional "ingredient" that is central to our decision-making process: our emotions.

Emotion processing and decision-making are integral aspects of daily life. Indeed any kind of recommendations, suggestions, reminders for a healthier lifestyle (physical activity, nutrition,…) will put the user in a position where he or she will have to decide whether or not to follow the suggested action. At that moment and among other parameters the emotional component will play a role. However, the understanding of the interaction between these constructs is limited. Scientific debate over whether subjective experiences of emotion are functional or maladaptive has been ongoing (Seo2007). Some argue that feelings are a source of unwanted bias (Shiv2005) and thus need to be properly regulated (Gross2003). Others maintain that feelings play an adaptive role in decision making and benefit personal well-being (Loewenstein2003). In both cases, and when aiming at developing e-coaching tools, the ability to detect emotional changes is valuable information that can be either reinforced or mitigated according to the context of the action, the user's profile and goals to be achieved. It is therefore necessary to measure, in real-time, such emotional responses in order to support their management.

In a recent publication described a "recommender system". When using applications with recommender systems the user is constantly receiving stimuli (e.g. visual, auditory etc.) that induce emotional reactions and states. These emotions influence, at least partially (according to the bounded rationality model

(Kahneman2003) the user's decisions on which content to choose. Thus it is important for the recommender system application to detect and make good use of any emotion-related information.

The emotions play a different role depending on the stage of the interaction process. The authors proposed three interaction stages: the entry stage, the consumption stage and the exit stage. This approach is aligned with our coaching approach, however there are situations of functions that are not encountered in Tkalcic approach (e.g. the system is not able to react to an emotional change of the user).

In this paper we are presenting the selected theoretical framework (section 2) from which we have brought some modifications in order to better fit with the context of the With-Me project where users' emotions are integrated into the coaching strategies (section 3). An e-coach should be able to:

- First measure/detect, in real-time, whenever a strong emotion change happens and second react appropriately to this situation by either trying to mitigate or re-enforce this emotional change. The reaction (coach/e-coach) to an emotional stimulus depends obviously on the (emotional) profile of the user and also on the behavioural change technique (BCT) initially selected for this user. Following a strategy of responding to an emotional change, a dimensional model (valance, arousal) of emotions is preferred to a discrete model (e.g. happiness, fear, sadness, hostility, guilt, surprise, and interest). Indeed a dimensional model would be perceived, by the user, less intrusive than a discrete model. This would then facilitate user acceptance. Moreover sensing and detecting discrete emotions are not currently possible with a minimalistic sensor configuration excluding discrete emotional models.
- Deliver to the user some information (reminder, recommendation, warning, information) that can elicitate certain types of emotion with the secondary aim to mitigate or reinforce a recently experienced emotion or simply induce a particular emotion at a particular time.

2. THEORETICAL FRAMEWORK

In 1896 Wundt first proposed a dimensional model of affect as part of his theory concerning mental chemistry, arguing that affect resulted from variations in basic dimensions of pleasure and arousal. Later, based on factor analyses of evaluative language, Osgood developed a dimensional theory of semantic meaning after finding that the most variance in semantic judgments was accounted for by a single factor, hedonic valence, which ranged from unpleasant to pleasant (Osgood1957). The fundamental role of hedonic valence in emotions received further support from studies of language categorization (Ortony1990), which proposed that human knowledge about emotions is hierarchically organized with a superordinate division between positivity and negativity. A second dimension also accounted for substantial variance in evaluative judgments; this was a dimension Osgood et al. labelled arousal. This dimension reflects the activation parameter in affective experience and ranges from an unaroused state (calm, relaxed, sleepy, etc.) to a state of high arousal (excited, stimulated, wide awake, etc.).

A third dimension was termed potency or dominance by Osgood et al. When rating static picture stimuli, dominance ratings are highly correlated with ratings of hedonic valence, with pleasant pictures rated as higher in dominance than unpleasant pictures. The dominance dimension is relatively weak in accounting for variance in evaluative judgments of symbolic stimuli. It is perhaps more potent in social interaction. For instance, dominance usually characterizes differences in unpleasant arousing events, with anger characterized by somewhat higher dominance than fear. Nonetheless, both are rated lower in dominance than highly pleasurable events. For symbolic sensory stimuli (pictures, sounds, words, etc.), there is no actual personal interaction in which social potency can play a role. In this case, dominance covaries highly with hedonic valence. Nonetheless, for more event-like stimuli, dominance will clearly account for some portion of variance in evaluative reports. In (Russell1977), dominance represents the degree of control over the emotion. In the literature there are terms that are commonly used and need a shared understanding. Therefore we present some definitions that are generally accepted about affect, emotion and mood. Then we will describe how emotions are part of the decision-making process.

2.1 Definitions

The constructs of mood and emotion are similar in that both refer to feeling states that can be broadly characterized as pleasant or unpleasant (i.e. positive or negative) and that reflect what is happening with the organism (Parkinson1996, Coan2007). Although the constructs of mood and emotion are closely related, they also reflect a number of fundamental differences including duration, frequency, intensity, and pattern of activation.

Mood is a much broader concept than emotion. First moods are experienced with greater duration than emotion (Gray2001). An emotion episode may last seconds, or minutes at the most. Emotions are intense yet brief. A mood, on the other hand, can last hours or even days. Therefore moods are experienced more consistently and frequently than emotions. True emotion reactions occur rarely. Second, moods and emotions differ in the experiences and events that trigger them (Gray2001).

Emotions are typically linked to a specific, defining moment that causes the response. Because emotions prompt the organism into action or decision, they must be triggered by a particular object. Moods, on the contrary, are not so much a reaction that is triggered by a specific event.

Affect is a broader, more inclusive psychological construct that refers to mental states involving evaluative feelings, that is, states in which a person feels good or bad or likes or dislikes what is happening (Parkinson1996). The second important issue in defining affect concerns the distinction between trait versus state affect. It can be studied either as a temporary, transient state or as a long-term, stable trait. State affects are defined as the experience of moods or emotions that are comparatively short-lived and follow a fluctuating course. They are transient episodes of feeling that are particularly influenced by the current situation (Tellegen1985, Watson2000).Trait affects, on the other hand, are durable dispositions or long-term, stable individual differences that reflect a person's general tendency to experience a particular affective state (Tellegen1985, Watson2000).

2.2 Emotions as Part of the Decision Making Process

During coach-coachee interactions, there is an exchange of information (recommendation, assessment, rewards, ...) which involve jointly cognitive functions and emotional reactions/states. In the context of the With-Me project, we have decided to used physiological data to measure the emotional state of the user (see section 3.3 for the reasons of this choice), therefore we are describing now how the physiological and cognitive components could be interconnected. In (Parker2007) the authors reviewed several theories describing some hypotheses about the origin and nature of emotions.

- The "Commonsense" view of emotion. We perceive and interpret a particular stimulus, movements. "I see a bear, feel fear, experience a flood of physiological reactions, and run because I am afraid";
- The James-Lange Theory. Environmental stimuli triggers physiological responses and bodily movements, and emotion occurs when the individual interprets his or her visceral and these cognitive processes give rise to an emotion that triggers certain physiological reactions and body muscular responses. "I must be afraid because my heart is pounding and I am running like crazy";
- The Cannon-Bard Theory. Emotion is a cognitive event that is enhanced by bodily reactions. Bodily reactions do not cause emotion but rather occur simultaneously with the experience of emotion. "I am afraid because I know bears are dangerous";
- The Schachter-Singer Theory. Emotions depend upon a kind of double cognitive interpretation: We appraise the emotion-causing event while also evaluating what is happening with our bodies. "I am afraid because I know bears are dangerous and because my heart is pounding".

How can emotions and cognitive processes be combined? It has already been identified that emotions could be a behaviour change technique when using narrative persuasion (Moyer-Guse2008). Narrative persuasion refers to the use of narrative messages (narratives) as a persuasion tool. Narratives are messages that present information in a story format as opposed to an abstract format. They are designed to engage the audience, and typically involve a protagonist, and contain emotionally appealing information. For example: (McQueen2011) exposed women to either an informational video about breast cancer, or a narrative video. The videos provided the same information but in a different format. The informational video provided the information in a didactic, expository form, whereas the narrative video provided the information in the form of personal stories of breast cancer survivors. Women who watched the narrative video experienced more positive and negative emotions, reported stronger identification with the message source, and were more engaged with the video than women who watched an informational video. The narrative video was also better liked, enhanced recall, reduced counter arguing, and increased breast cancer discussions with family members. Note that in narrative persuasion emotions are involved together with other determinants such as affective attitude, behavioural intention and beliefs, perceived risk, perceived severity, perceived vulnerability. Based on these observations we are proposing how emotions could be used as a coaching ingredient.

3. EMOTION AS A COACHING INGREDIENT

3.1 Emotions: from Stimuli to Actions

Inspired from the theories presented in section 2.2, we represent how emotions (and emotional profile) are contributing to short and long term actions (Figure 1).

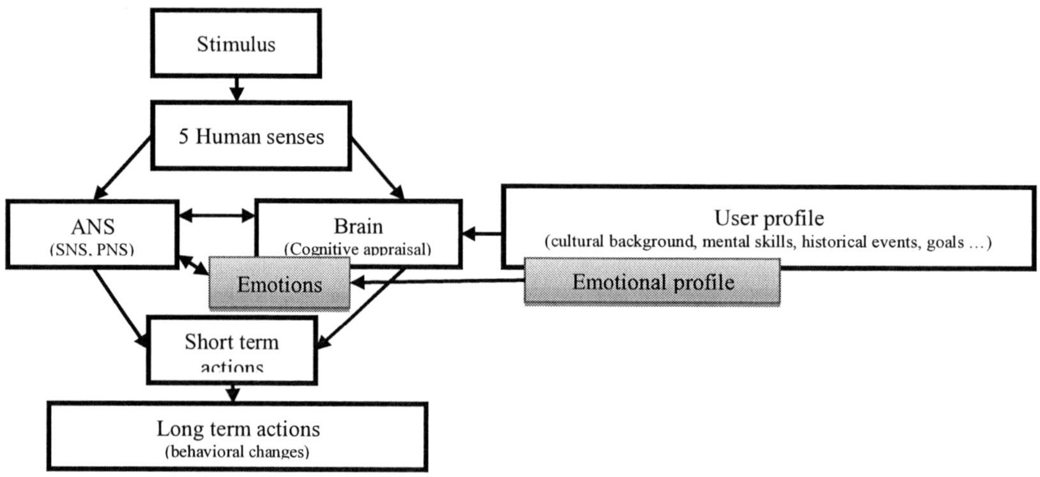

Figure 1. The place of emotion: from stimulus to actions

• Stimulus or (eliciting stimulus). It corresponds to any daily life stimuli: vision or smell of non-healthy food, call from a personal coach to suggest jogging, an internal thought of sitting on a sofa to watch TV, bad news from the doctor about the latest medical tests related to heart conditions, reminder to perform the daily rehabilitation exercise, or perform a stress assessment test.

• Human senses. This step is just to make explicit the fact that stimuli can be in any "dimensions" or different kinds (sight, taste, touch, hearing, smell, proprioception, thought).

• ANS. Autonomic nervous system (ANS) activity is viewed as a major component of the emotion response. See (Kreibig2010) for complete review of the empirical basis for the postulate of emotion-specific ANS activity. Monitoring physiological parameters (e.g. Heart Rate Variability, galavanic skin response, ...) is a way (used in the With-Me project) to quantify the emotion response/state of a person.

• Brain. It refers to theories of emotion which implicate people's personal interpretations of an event in determining their emotional reaction. The most important part of this theory (Cannon1931) is the way we interpret the event (was the event a positive or a negative occurrence) as well as what we think caused the situation.

• User profile. In the field of Human Computer Interaction (HCI), users' individual differences and their tasks are always indicated as the two most important issues addressing usability (Nielsen1993). Users are not homogeneous; they differ in many respects, such as gender, age, physical abilities, educational level, organizational culture, operator skills etc. Beside these standard profile attributes that are relevant for With-Me, it is important to consider the user's "emotional profile". Salovey and Mayer originally coined the term "emotional Intelligence" in 1990. However, Daniel Goleman popularized it in 1995 in the title of his bestselling book, Emotional Intelligence: Why it can Matter More than IQ? (Goleman1995). Goleman defined emotional intelligence as: "understanding one's own feelings, empathy for the feelings of others and the regulation of emotion in a way that enhances living". What made this book so popular were two claims: Firstly, emotional intelligence may be more important for personal success than IQ. Secondly, unlike IQ emotional intelligence can be improved.

• Short term actions. As shown in Figure 1 and explained in (Loewenstein2003), conventional theories of decision making include only "expected emotions" where the decision makers anticipate the upcoming emotions (consequences of his/her choice) and thus take them into account when making decisions. However Loewenstein also puts emphasis on how "immediate emotions" (emotions experienced at the time of the decision making) can influence the decision by altering the decision maker's perceptions of probabilities or outcomes or by altering the quality and quantity of processing of decision-relevant cues. Note that this emotional influence may either serve or not the decision marker. In With-Me one approach is to use emotional changes to trigger a feedback (e.g. via a virtual coach, reminder, real coach reaction) that will help the user to take the right decision toward healthier actions: perform now (short-term action) a recommended physical activity, or eat healthy food for the coming lunch (short-term action). Another approach is to deliver a stimulus (e.g. a message) that specifically designed to elicitate a particular emotion to the user. This might be a positive emotional content to reinforce a particular achievement.

• Long term actions. One of the With-Me objectives is to support behaviour changes that are sustained over a long period of time (months, years). This is why we have added this final block in Figure 2 that corresponds to long-term actions and habits. The user supported by adapted persuasive mechanisms has adopted new behaviours which have become routines. Hypothesis: repetitive feedbacks with positive emotional content sustain better long-term changes.

From the scheme presented we see the importance of two key aspects: being able to (1) elicitate a certain type of emotions and (2) detect and classify human emotional responses when they arise. We are presenting now the different blocks of the figure, focusing on the key element: the coaching module.

3.2 Emotional Coaching Module

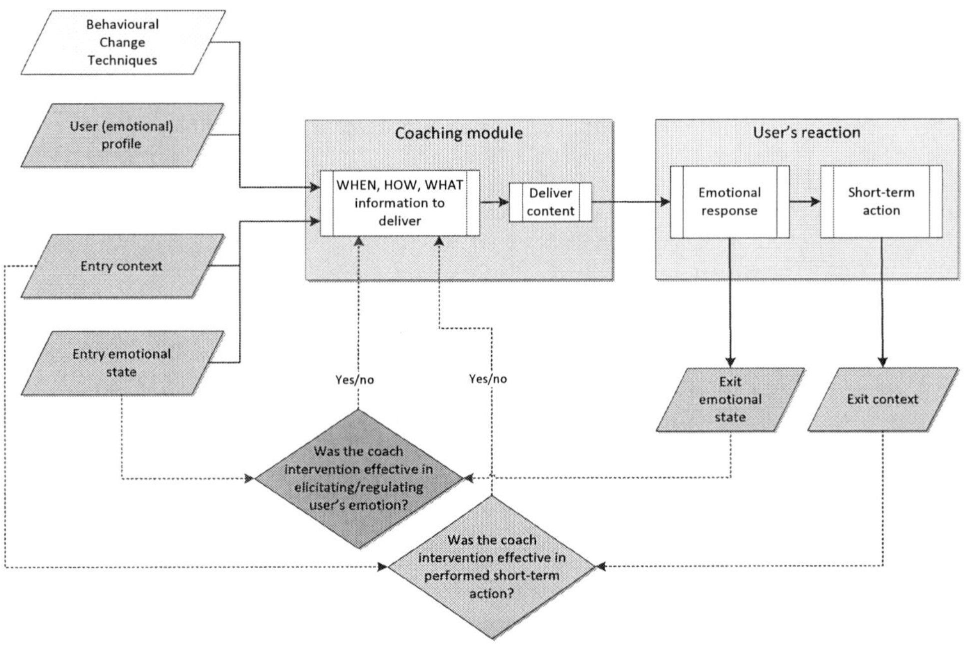

Figure 2. Emotional coaching framework

We are presenting our emotional coaching framework (Figure 2). The coaching framework is composed of a central module (coaching module) which take as inputs (a) the behavioural change technique selected for a particular user, (b) the user emotional profile, (c) the entry context, and (d) the entry emotional state. The coaching module is responsible of deciding WHEN, HOW and WHAT information to deliver/transmit to the coachee.

3.2.1 When

When should the coach intervene? What is the optimal moment? It depends on the combination of all input parameters and their values at a particular moment. As explained in (Loewenstein2003), conventional theories of decision making usually include only the "expected emotions" where the decision makers anticipate the upcoming emotions (consequences of his/her choice) and thus take them into account when making decisions. However Lowenstein (Loewenstein2003) also puts emphasis on how "immediate emotions" (emotions experienced at the time of the decision making) can influence the decision by altering the decision maker's perceptions of probabilities or outcomes or by altering the quality and quantity of processing of decision-relevant cues. Therefore and considering only emotional-related triggering conditions we have identified two relevant intervention moments:

• After an acute emotional change

This could be the situations when a user (with eating disorder) is seeing or smelling of non-healthy food or when a person (recovering from an heart attack) gets bad news from the doctor about his/her last medical tests related to heart conditions, or when a worker is experiencing very high negative emotional change at work after a conversation with a colleague. In these situations the coach would immediately react in order to mitigate the effect of the elicited emotion (e.g. trying to avoid the user to eat the food that has been seen).

• After a specific "emotional response pattern"

In this scenario, the coach is not responding to a specific event in time, but is rather responding to the time history (for example the last hour) of emotional responses/states. An intervention of the coach will be triggered when some conditions are met. For example a recommendation might be sent when the user hasn't experienced strong emotional stimuli for the last hour. This strategy can be selected when it is known that a particular user is strongly receptive to suggestions when being "calm".

3.2.2 How

"How" refers to the modality (format) of the intervention done by the coach or e-coach (text, audio, speech, image or video), but also to the way messages/instructions are transmitted to the user (with or without emotional content):

• Message without emotional content

In this case no particular action is taken to embed any emotional content into the stimulus. This means that the stimulus is only proving neutral (emotional) information. For example, the user is experiencing a strong negative emotion and the message delivered to the person only aims at diverting his/her attention or advice a relaxation technique.

• Message with "on purpose" emotional content

In this situation the message to be delivered is explicitly designed to (1) elicitate a particular and pre-defined emotional response or to (2) mitigate the current user's emotional response. For that purpose one can refer to section 4 for selecting features that have been demonstrated to elicit human emotion. For example, the user is experiencing a strong negative emotion and the message delivered to the person aims at diverting his/her attention and also ties to elicit an "opposite" emotion (e.g. positive).

We can note that a succession of the e-coach intervention with "on purpose" emotional content will contribute or define a "personality of the coach" that should be adapted to the personality of the user, as shown in (Delaborde2013). This relates to the characteristics of an efficient coaching platform able to keep a coachee using the platform because it is pleasant.

3.2.3 What

What should the coaching system suggests or says? WHAT refers to the content of the message that the coaching system delivers to the user. This content will depend on:

• The domain in which the coaching system wants to intervene to coach the user to a healthier lifestyle: nutrition, physical activity or social interaction.

• The type of message to be delivered: positive reinforcement, warning, advices, others.

• The user's personality and what content is more efficient for this kind of user/personality.

• The personality of the coach, for consistency and to reinforce this personality.

3.3 Implementation: Sensing the Emotional State of the User

Since our application aims at using the emotional state/responses as part of an e-coach, it is therefore mandatory to be able to measure/collect quantitative information related to this state. For that purpose a large palette of modalities and sensors has been considered and studied in the last decades: vision-based system, audio/speech system, textual information, touch sensors, physiological and bio signals (HR/HRV, GSR, and EEG). In the context of the With-Me project, we have decided to use physiological signals to detect and quantify the emotional response/state. Advances in medicine, biomedical engineering and computing enable the use of physiological sensing devices that provide high levels of portability, comfort, and reliability offering a descent solution for real-life situations. Besides, physiological measures are difficult to be consciously manipulated, and enable a more reliable representation of inner feelings. Some relevant approaches to physiological emotion detection include the work of Nasoz (Nasoz2004), Kim (Kim2004) or Leon (Leon2007) and the recent review of Kreibig (Kreibig2010). Considering that a real-time estimation of the emotional state of the user is possible, we illustrate three use cases:

- A user, with eating disorder, is seeing or smelling of non-healthy food. In this situation the coach immediately reacts to a strong emotional (compulsive) reaction in order to mitigate the effect of the elicited emotion and tries to avoid the user to eat this food.
- A person recovering from a heart attack gets bad news from the doctor about his/her last medical tests related to heart conditions. The professional who gets real-time information about the user's strong negative reaction to this announcement will adapt/extend his/her explanations.
- A worker is experiencing very high negative emotional change at work after a conversation with a colleague. This person is warned about it and recommendation is proposed to regulate his/her emotional response/state (relaxation exercises).

4. CONCLUSION

An e-coach system aims to support people in preventing diseases but also managing health conditions such as overweight, stress, or heart conditions. However changing its own lifestyle is never easy, it is an every-minute challenge and requires long-term efforts which are usually difficult to handle alone. We therefore proposed to enhance standard e-coach by adding the user's emotional component into coaching strategies (section 3.2) in which we give recommendations about WHEN, HOW, WHAT to transmit to the user by explicitly using emotion-related information. Similar to Tkalcic et al. (Tkalcic2011) the "coaching module" (Figure 2) takes as input the emotion entry state, but also context information, the user emotional profile, and a personalized behavior change technique (BCT). The outputs of the "coaching module" are the exit emotional state and the short term action taken by the user.

We have pointed out that technology is now mature enough to automatically measure and estimate the real-time emotional response/state of a person. The next challenge is to implement artificial intelligence coach able to (a) define the WHEN, HOW, WHAT of the "coaching module" based on the current inputs (see Figure 2) and also (b) adapt, over time, these assignation based on feedback information: "was the coach intervention effective in elicitating/regulating user's emotion" and/or "was the coach intervention effective in generating the appropriate short term user action?".

ACKNOWLEDGEMENT

The research presented in this paper is conducted as part of the With-Me project (European ARTEMIS Call 2012 Project 332885), which is partially funded by the Artemis/ECSEL Joint Undertaking Programme and the National Funding Agencies (in particular from Spain: Ministerio de Industria, Turismo y Comercio).

REFERENCES

Cannon, W. B., 1931. Again the James-Lange and the thalamic theories of emotion, *Psychological Review* 38(4), 281-295.

Coan, J. A. & Allen, J. J. B., 2007. *Handbook of Emotion Elicitation and Assessment*, Oxford University Press, New York.

Delaborde, A., 2013. *User's emotional profile modelling in spoken Human-Machine interactions*, PhD thesis, Université Paris Sud - Paris XI, France.

Fogg, B., 2003. *Persuasive Technology: Using Computers to Change what We Think and Do*, Morgan Kaufmann Publishers, USA.

Goleman, D., 1995. *Emotional Intelligence*, Bantam Books, USA.

Gray, E. K. et al., 2001. *Emotion, mood, and temperament: Similarities, differences–and a synthesis*, Wiley, New York.

Gross, J. J. and John, O. P., 2003. Individual differences in two emotion regulation processes: implications for affect, relationships, and well-being, *J Pers Soc Psychol*, 85(2), 348--362.

Kahneman, D., 2003. A perspective on judgment and choice: mapping bounded rationality, *American psychologist* 58(9), 697.

Kim, K. H. et al, 2004. Emotion recognition system using short-term monitoring of physiological signals, *Med Biol Eng Comput*, 42(3), 419--427.

Kreibig, S. D., 2010. Autonomic nervous system activity in emotion: a review, *Biol Psychol*, 84(3), 394--421.

Leon, E. et al, 2007. A user-independent real-time emotion recognition system for software agents in domestic environments, *Engineering applications of artificial intelligence*, 20(3), 337--345.

Loewenstein, G. and Lerner, J. S., 2003. The role of affect in decision making, *Handbook of affective science* 619(642), 3.

McQueen, A. et al, 2011. Understanding narrative effects: The impact of breast cancer survivor stories on message processing, attitudes, and beliefs among African American women, *Health Psychology*, 30(6), 674--682.

Moyer-Gusé, E., 2008. Toward a Theory of Entertainment Persuasion: Explaining the Persuasive Effects of Entertainment-Education Messages, *Communication Theory*, 18(3), 407--425.

Nasoz, F. et al., 2004. Emotion recognition from physiological signals using wireless sensors for presence technologies, *Cognition, Technology & Work*, 6(1), 4--14.

Nielsen, J., 1993. *Usability Engineering*, Morgan Kaufmann Publishers Inc., San Francisco, CA, USA.

Ortony, A., 1990. *The cognitive structure of emotions*, Cambridge university press, UK.

Osgood, C. E., 1957. *The measurement of meaning*, University of Illinois press, USA.

Parker, M., 2007. Understanding Psychology, R. H. Ettinger, *chapter Emotion and stress*, pp. 325-361.

Parkinson, B., 1996. *Changing Moods: The Psychology of Mood and Mood Regulation*, Longman, UK.

Russell, J. A. and Mehrabian, A., 1977. Evidence for a three-factor theory of emotions, *Journal of research in Personality*, 11(3), 273--294.

Seo, M.-G. and Barrett, L. F., 2007. Being emotional during Decision Making - Good or Bad? An Empirical Investigation, *Acad Manage J*, 50(4), 923--940.

Shiv, B. et al, 2005. Investment behavior and the negative side of emotion, *Psychol Sci* 16(6), 435--439.

Tellegen, A., 1985. *Structures of mood and personality and their relevance to assessing anxiety, with an emphasis on self-report*, *in* A. H. Tuma & J. D. Maser, ed., 'Anxiety and the anxiety disorders', Erlbaum, Hillsdale, NJ, pp. 681-706.

Tkalcic, M.; Kosir, A. & Tasic, J. (2011), Affective recommender systems: the role of emotions in recommender systems, in 'ACM RecSys Workshop on Human Decision Making'.

Watson, D. et al, 2000. Self-other agreement in personality and affectivity: the role of acquaintanceship, trait visibility, and assumed similarity, *J Pers Soc Psychol*, 78(3), 546--558.

EVALUATION OF MOBILE HEALTH EDUCATION APPLICATIONS FOR HEALTH PROFESSIONALS AND PATIENTS

Tareq Aljaber and Neil Gordon
Department of Computer Science, University of Hull, Hull, HU6 7RX, United Kingdom

ABSTRACT

Mobile applications for health education are commonly utilized to support patients and health professionals. A critical evaluation framework is required to ensure the usability and reliability of mobile health education applications in order to facilitate the saving of time and effort for the various user groups; thus, the aim of this paper is to describe a framework for evaluating mobile applications for health education. The intended outcome of this framework is to meet the needs and requirements of the different user categories and to improve the development of mobile health education applications with software engineering approaches, by creating new and more effective techniques to evaluate such software. This paper first highlights the importance of mobile health education apps, then explains the need to establish an evaluation framework for these apps. The paper provides a description of the evaluation framework, along with some specific evaluation metrics: an efficient hybrid of selected heuristic evaluation (HE) and usability evaluation (UE) factors to enable the determination of the usefulness and usability of health education mobile apps. Finally, an explanation of the initial results for the framework was obtained using a Medscape mobile app. The proposed framework - An Evaluation Framework for Mobile Health Education Apps – is a hybrid of five metrics selected from a larger set in heuristic and usability evaluation, filtered based on interviews from patients and health professionals. These five metrics correspond to specific facets of usability identified through a requirements analysis of typical users of mobile health apps. These metrics were decomposed into 21 specific questionnaire questions, which are available on request from first author.

KEYWORDS

Usability Evaluation, Heuristic Evaluation, Metrics, Health Professionals, Evaluation Framework

1. INTRODUCTION

Mobile phones now play a major role in many parts of human life. Enhancements in mobile technology have allowed a wide range of applications to be developed, which can be utilized in various aspects of people's lives (Harrison et al., 2013). One example are the mobile health (mHealth) education applications which have been utilized by several stakeholders, such as patients and health professionals, to improve their knowledge of their own or others' health in certain aspects of their life (Aljaber et al., 2015). Everyday life in general is becoming increasingly complex and there is a need to assist patients in improving their health education to ensure corresponding improvement in their health condition by supporting their well-being via the use of mobile health education applications. Consequently, the accumulative effect of mobile technology has led to significant growth in the number of mobile health mHealth applications and users. Interest from both the health field and software development communities has been shown as a result of this technology revolution.

The inaugural mHealth Summit in 2009, a partnership between the National Institutes of Health, the Foundation for the National Institutes of Health, and the mHealth Alliance, involved 800 people. One year later, the number of people joining the conference was extended to 2,400 (Qiang, 2011). This indicates a significant increase in interest over a short period. Learning and training on health matters are increasingly important issues, particularly as more people are living longer and because of the scale and complexity of support for health and well-being (Shareef, 2006). This type of health education consists of several areas, such as learning how to receive appropriate treatment, how to manage life in an acceptable way and how to manage health requirements without barriers (Glanz et al., 2008). These requirements from the patients' side parallel those of the health professionals who seek to update their medical knowledge and are looking for specific information. According to Istepanian and Lacal (2003), "M-health was defined as wireless telemedicine involving the use of mobile telecommunications and multimedia technologies and their integration with mobile healthcare delivery systems".

As mobile phone devices are small, they have various restrictions, such as battery life and resource storage limitations, which bring a lack of quality to the service and the user experience (Silva et al., 2013). Mobile health apps can create risks of failure equivalent to those of supplementary medical devices, due to mechanical failure, faulty design, poor manufacturing quality, and user error, among other safety issues (NHS Innovations South East, 2014). One consequence of these weaknesses is that most software developers of mobile health education apps do not apply an appropriately specific model/framework in the evaluation of usability for these apps to ensure that they meet certain requirements, such as health education for patients and health professionals. This highlights the need for an evaluation framework to measure and evaluate these apps and make sure they meet the requirements of the stakeholders (Aljaber et al., 2015).

2. THE IMPORTANCE OF MOBILE HEALTH EDUCATION APPS

Recent research has demonstrated that one of the central development areas in computing is mobile phone devices (Ziefle & Bay, 2008). According to Hernandez Munoz and Woolley (2009), mobile phones have become essential and common devices among the general population, which suggests that people depend on mobile phone devices in many aspects of their daily lives, for instance in commerce, transportation and health education. The overall trend is that people are keen to use mobile phones for purposes other than the conventional aspects of using them as flexible replacements for desktop computers. (Traxler, 2009). A wide range of types of information can now be accessed from anywhere at any time to anybody with a mobile phone using wireless technology (Ally, 2009).

In general, there are two types of mobile phone device. The classic type, which contains two parts: a display screen and a keyboard below; and the touch screen type, which is split into either a full touch screen controller or half touch screen controller and a half keyboard under the controller. The first touch screen mobile phone was the Simon Personal Communicator, developed by IBM in 1992, while the first smart mobile phone on the market was the Ericsson R380 released in 2000 (Lobo et al., 2011). Seven years later, the iPhone was released by Apple and was the first smartphone mainly to be controlled by its touch screen (Lobo et al., 2011). Only one year later, the Android operating system was on the market for touch screen mobile phones (Lobo et al., 2011). According to Leonardi et al. (2010), a touch screen set-up is more flexible and easy to use by novice users than the keyboard and mouse set-up, which supports the expectation of increasing reliability on mobile phones.

In addition, a recent estimate has shown that one person in every five in the world owns a smartphone (Heggestuen, 2013). Recent estimates indicate that about 497 million mobile devices were added to the circulation in 2014 and that, of these devices, smart devices represented 26% (Cisco Visual Networking Index [VNI], 2015). According to Hussain et al. (2015), the latest estimates point to the number of mobile devices being enlarged to 1.4 billion in 2017, as reported by the Cisco Global Mobile Data Traffic Forecast 2013 Update (Hussain et al., 2015). This reflects a vast growth in mobile devices, which increases the potential for having more applications installed and used in our mobile phone devices. This brings to the fore the lack of a method for ranking and distinguishing useful apps from the not so useful.

3. USABILITY METRICS

Initially, we proposed a hybrid selection of metrics taken from heuristic evaluation (HE) and usability evaluation (UE) metrics to be filtered by conducting 10 hybrid selected interviews with health professionals and patients. The outcome from analysing the data from the 10 hybrid selected interviews demonstrated that we needed to modify some of the metrics we had selected, as some of the metrics we had proposed and selected did not meet the requirements of the interviewees. The 10 hybrid selected interviews with health professionals and patients reflected that some of the proposed hybrid metrics selected from HE and UE were not essential to their requirements, although some of the metrics needed to be supplementary to our proposed hybrid selection of metrics taken from HE and UE. Moreover, we systematized and categorized our hybrid selection metrics from HE and UE as: A) Memorability, B) Features, C) Attractiveness, D) Simplicity (containing learnability) and E) Accuracy. The definitions of these metrics are:

Memorability: When users have not worked with the interface for a long time, how much of its functionality do they remember and how quickly can they regain their proficiency (Shivade & Sharma, 2014)?

Features: Does the app provide extra features, such as images, video, voice support/help, touch screen facilities or voice guidance (Hussain et al., 2012)?

Attractiveness: The capability of the software product to be attractive to the user, e.g., through the use of colour or graphic design (International Organization for Standardization/International Electrotechnical Commission [ISO/IEC] 9126-1, 2001).

Simplicity: Ease of inputting the data, ease to use output, ease to install and ease to learn (Hussain et al., 2012).

Accuracy: The degree to which it is accurate (Hussain et al., 2012). The capability to provide correct results or effects (ISO/IEC 9126−1, 2001). Figure 1 shows these metrics.

1) Efficiency	A) Memorability
2) Satisfaction	B) Features
	C) Attractiveness
3) Effectiveness	D) Simplicity (containing learnability)
	E) Accuracy

Figure 1. A hybrid of five selected metrics

4. SOFTWARE EVALUATION METRICS AND METHODS

Previous research, such as Shivade and Sharma (2014), has considered that, as a rule, usability is essential for the utility and longevity of an application. If an app is not convenient to operate, people will not be interested in it and they will not use it, which will lead to the failure of the application. Usability metrics can be utilized to help because competition is increasing in mobile app marketing and effective and usable apps are needed. Mobile apps are frequently developed as reasonably small projects, which do not usually support usability testing. Moreover, developers are not always able to test mobile apps.

When we need to evaluate a user interface, we can adopt numerous methods, each method employing different assumptions and reflecting different a priori philosophical approaches and techniques. Broadly, we can decide among the following: first, formal methods, e.g., using a mathematical model; second, automatically, such as a black-box checking algorithm; third, empirically, which is by experiment, using separate cases to study statistical variance; fourth, heuristically, which is by informed opinion; and, finally, socially or ethnographically, by which a systemic approach is preferred and the whole is considered, in a social or cultural context, and empiricism avoided.

In the context of this paper, we need a practical evaluation methodology that we can apply to the present mHealth setting. Formal or automated approaches are not accessible with regard to the present work. Empirical work is prized when the occasion allows, although the quickly advancing and technologically challenging setting of the current work means that other additional pragmatic and heuristic tools are adopted here.

In our paper we conduct a hybrid approach selected from HE and UE.

First, HE is an informed method of usability analysis in which a series of domain experts are presented with an interface solution and mandated to use their specific expertise to comment upon it. The method uses a series of heuristics or rules of thumb about what generally constitutes good practice in interface design (Nielsen & Molich, 1990). There are several advantages to utilizing HE: first, HE is cheap; second, it is easy to motivate people to do it; third, it does not require advanced planning in the sense that an experimental evaluation would do; and, finally, it can be used at every stage of a project e.g., from initial design, to rolling prototype, to final project delivery (Brayshaw et al., 2014).

The most established approach is found in Nielsen (Nielsen & Molich, 1990; Nielsen, 1993), in which he proposes 10 rules concerning good design, which include consistency and standards, error prevention, flexibility and efficiency of use, and aesthetic and minimalist design. Implementing the 10 rules, however, comes at a price and we must be watchful when utilizing a method or technique such as heuristic evaluation. Science in its most empirical structure depends on solid replicable results and a heuristic methodology is not going to convey such exploratory exactitude. We have to be aware of two points: first, some of the time, indicating usability issues without supplying direct suggestions related to solving these issues; and, second, HE is hard for a single individual to do, since one person will never be able to discover all the usability problems in an interface.

Fortunately, experience gained from several different projects has shown that different people find different usability problems. Subsequently, it is possible to enhance the effectiveness of a method significantly by involving multiple evaluators (Nielsen & Molich, 1990). Given the rapidly advancing nature of the mobile application context, HE provides both adaptability and high quality evaluation to advise our research. For health education, it is likely that consistency and the avoidance of mistakes will be key.

Second, UE metrics and methods. Usability is a very broadly held metric for evaluating the learnability and attractiveness of software (Shivade & Sharma, 2014). UE is a critical approach to assessing the function of human-computer interaction (HCI). According to Nielson (1993), usability is defined as follows. Learnability: the user should be able to find the system easy to learn. Efficiency: the user should be able to use the system efficiently and, once learned, the user must have the highest level of productivity potential. Memorability: the user should be able to remember how to use the system easily, so that if the user does not use it for any length of time, he or she should still be able to use the system without learning everything again. Errors: the user should make few errors during the use of the system, so it should have a low error rate so that the user can easily recover from them. Catastrophic errors must not occur. Satisfaction: the user should be satisfied with the system. Software UE is a methodology for evaluating the usability of each kind of software, including mobile applications (Shivade & Sharma, 2014). Usability metrics include response time, accuracy, operability, attractiveness, learnability, satisfaction, productivity, effectiveness and efficiency (Seffah et al., 2006).

5. RATIONALE FOR CONSTRUCTING A FRAMEWORK

There are at least four major reasons for constructing a framework for mobile health education apps. First, in recent decades, several researchers developed UE for systems in general, although the dominant part was for web applications by and large, not for mobile apps or mobile health education apps (Alva et al., 2003; Ivory & Hearst, 2001). Second, recently usability turn into a significant area for smartphones, as it is vital to avoid issues from practicing mobile applications (Baharuddin et al., 2013). Third, according to Constantinos and Kim (2011), the lack of empirical research into the impact of the environment on mobile usability and the significance of user characteristics is a central point. In addition, they specify in their paper that there has been no subjective study on the usability scopes considered in such mobile studies (Constantinos & Kim, 2011). This demonstrates the absence of the construction of a subjective study to enhance, rank and measure the usability of mobile apps in general, and mHealth education apps specifically. Fourth, and most importantly, according to Constantinos and Kim (2011), there is no usability evaluation framework that has yet been created in the context of a mobile computing environment. This provides support for the necessity of our research, which involves building an evaluation framework for mH education apps. Furthermore, according to Smith (2013), smart mobile phone device ownership saw a progression of 10% between 2012 and 2013. The use of mHealth applications began to increase markedly from 2013: over 36% of all mHealth applications in 2014 had been released in the previous year (research2guidance, 2014). According to Dubey et al. (2012), usability inquiries utilize different methods, such as surveys, logging, or interviews. Moreover, UE methods are one of the top approaches in terms of usability to evaluate different aspects, such as systems, applications and prototypes (Bernhaupt, 2009).

6. AN EVALUATION FRAMEWORK FOR MOBILE HEALTH EDUCATION APPS

Initially, we designed our proposed framework to evaluate and measure the usability in mHealth education applications. We had an early design which we then updated but, as time spent on the research passed, we discovered that our framework required a number of amendments to make it more efficient. The early framework design consist: The first step for in early design was based on building and mapping questionnaire questions, based on a hybrid selection from HE and UE. Second, the questionnaire questions had to be used on existing mHealth education apps, such as Medscape. Third, we developed a basic prototype for the purpose of reflecting on basic mHealth education apps. Fourth, we applied the questionnaire to a basic prototype. Fifth, we compared the result between the outcomes from the questionnaire applied to Medscape and the outcome from the questionnaire applied to the basic prototype. Six, we modified the basic prototype depending on the comparison outcome from step 5. Seven, we used the questionnaire in a new basic prototype and, finally, compared the result from this with the results from Medscape in step 2. These steps and the proposed framework are illustrated in Figure 1 (Aljaber et al., 2015).

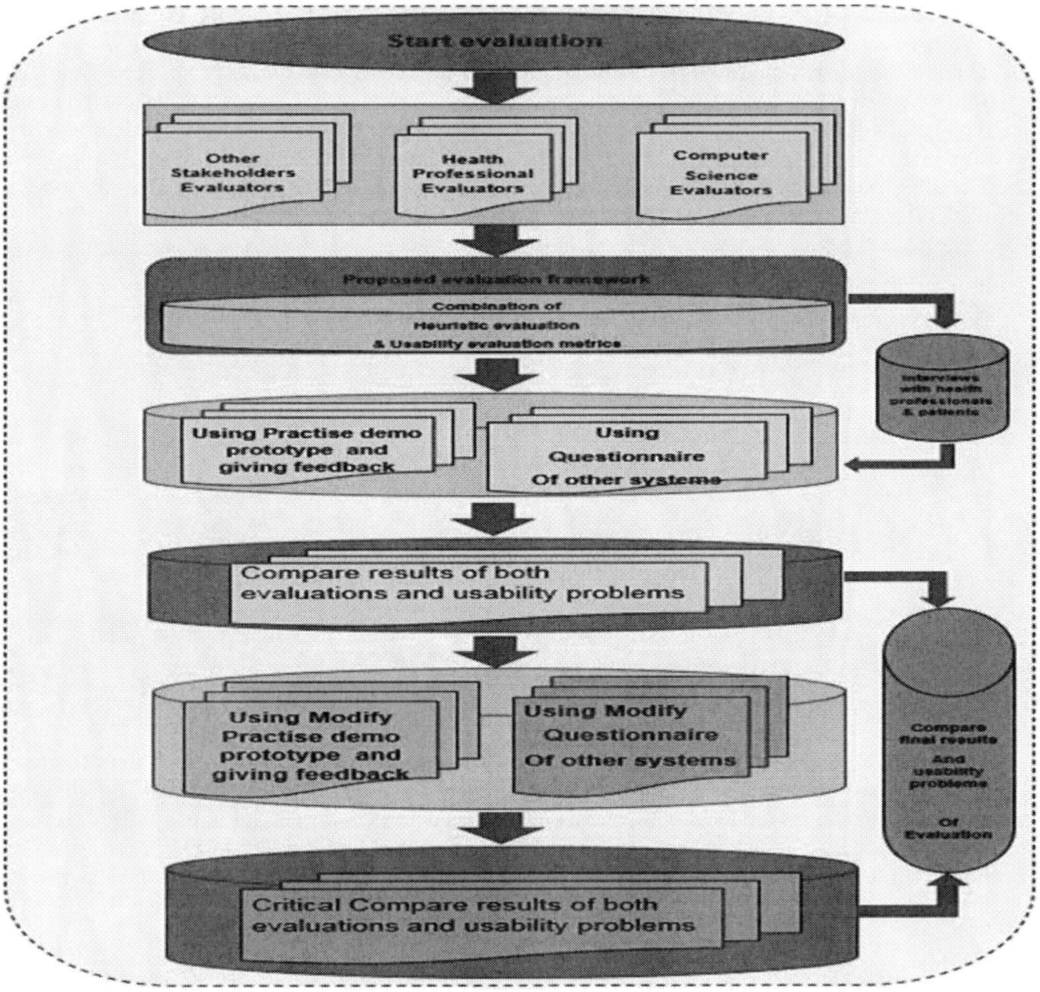

Figure 2. Updated proposed framework design

The updated framework design: The updated framework design depended on the same principle as the early design for the framework, but we modified step 1 of the early design (the building and mapping of questionnaire questions based on a hybrid selection from HE and UE) and mapped it instead with accurate requirements of patients and health professionals, as they are the most frequent users of these apps. In addition to meeting these specific requirements, a hybrid of five selected metrics from a larger set of heuristic and usability evaluation was filtered from the output of 10 interviews with patients and health professionals. According to Nielson (2000), dealing with five test users is sufficient to obtain around 80% of the results. In our research, we dealt with 10 users, which should bring our results to more than 80% of the accurate requirements for health professionals and patients using mHealth education apps (Nielsen, 2000). The five metrics selected correspond to specific facets of usability identified through a requirements analysis of typical users of mobile health apps. These metrics were decomposed into 21 specific questionnaire questions. (The updated design for the framework is shown in Figure 2.)

7. INITIAL RESULTS FOR THE FRAMEWORK

After analysing the feedback data collected from the 10 interviews (five interviews with health professionals and five with patients), we analysed the feedback from both sets of interviewees to address their most important requirements when using mHealth education applications. Our results are shown in Figure 3 and are detailed below.

The most important metric to be measured in meeting patients' and health professionals' requirements is D) Simplicity, shown as having the highest score with an average of 6.9. In second place is B) Features, with an average of 4.6. Third is E) Accuracy, with an average of 3.2. In fourth place is A) Memorability, with an average of 2.4. In last place is C) Attractiveness, which has the lowest score with an average of 0.3. From these results, we can state that the most important metrics for health professionals and patients when using mHealth education applications are D) Simplicity and B) Features.

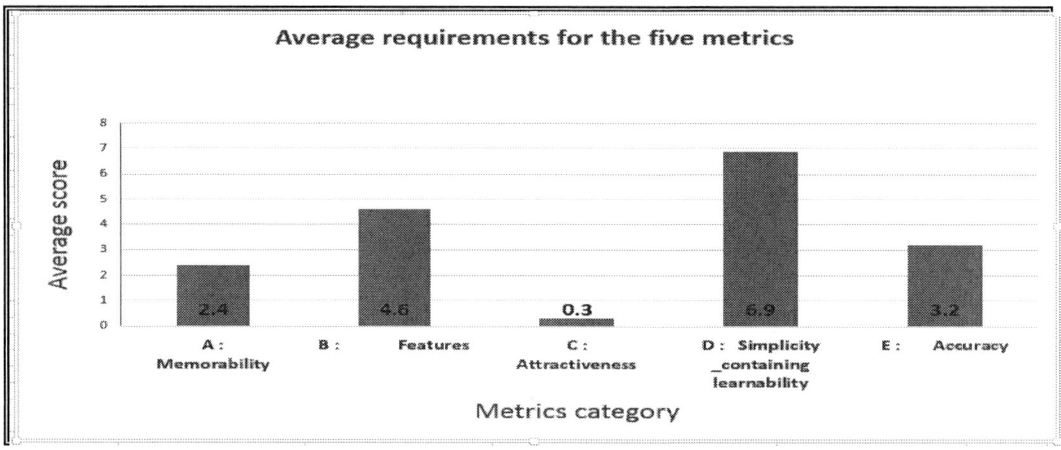

Figure 3. Results from the interviews measuring patients' and health professionals' requirements

To further explain the results we gained from analysing the data collected from the 10 interviews, we have summarized our findings in Table 1. The numbers in the table cells reflect the number of times an interviewee (a health professional or a patient) mentioned a particular metric. For example, interview 1 with a health professional reflects that the interviewee mentioned A) Memorability in his interview five times, B) Features five times, C) Attractiveness once, D) Simplicity 11 times and E) Accuracy 10 times. An overall summary of the results, by taking the aggregated score of each of the five metrics across the 10 interviews, is given below:

Table 1. Scores from the interviews measuring patients' and health professionals' requirements

	A : Memorability	B : Features	C : Attractiveness	D : Simplicity _containing learnability	E : Accuracy
1 : Interview_1_HP	5	5	1	11	10
2 : Interview_10_HP	0	1	0	0	3
3 : Interview_2_HP	3	7	0	11	2
4 : Interview_3_P	0	2	2	3	3
5 : Interview_4_HP	3	6	0	8	2
6 : Interview_5_P	3	5	0	9	3
7 : Interview_6_HP	1	5	0	7	2
8 : Interview_7_P	4	5	0	8	1
9 : Interview_8_P	3	5	0	5	4
10 : Interview_9_P	2	5	0	7	2
Total	24	46	3	69	32

The most important metric to be measured in terms of meeting the health professionals' and patients' requirements is D) Simplicity, showing the highest score (69). In second place is B) Features (46). Third is E) Accuracy (32), in fourth place is A) Memorability (24) and the lowest number was given to C) Attractiveness (3).

From these results, we can state that the most important metrics for health professionals and patients are D) Simplicity and B) Features in meeting their main requirements when using mHealth education applications.

8. CONCLUSION

Health professionals and patients utilize mHealth education applications for the critical reasons of saving time and effort and to simplify their understanding of matters connected with health education. If such a mobile health education application does not offer clients these components, the application will not be utilized effectively as part of the wider goal of satisfying the prerequisites of health professionals, patients, and different partners. What is more, this will not encourage an application to be set up in the first place. An effective way to avoid these issues is by employing HE and UE metrics. This will help to improve the memorability, features, attractiveness, simplicity (including learnability) and accuracy of mHealth education applications, as well as reducing the potential for the failure of these kinds of apps. Moreover, a mHealth education application can be enhanced by evaluating it in the early stages of software design. In this case, our hybrid of five selected usability metrics - Memorability, Features, Attractiveness, Simplicity (containing learnability) and Accuracy - can be improved to increase the chance of these apps being successful and useful. This type of framework could be utilized as an evaluation tool for various sorts of mobile application. Moreover, mHealth education applications are critically important for health professionals and patients, as utilizing low-quality applications could, in extreme cases, lead to the death of patients.

This paper concentrated on highlighting the building of an evaluation framework for mHealth education applications that depends on a hybrid selection of metrics (heuristic evaluation and usability evaluation) and incorporated a hybrid of five selected metrics from a larger set of heuristic and usability evaluations, filtered based on interviews with patients and health professionals. These five metrics correspond to particular aspects of usability identified through a requirements analysis of typical users of mHealth apps. These metrics were decomposed into 21 specific questionnaire questions. The aims of this project are to build a set of tools for the evaluation of mHealth apps to enable them to be ranked, and to assist several of the stakeholders (health professionals, patients and others) in selecting appropriate apps. Our Evaluation Framework for Mobile Health Education Apps is intended to offer guidance on system specifications and different user requirements for software developers of mHealth apps.

REFERENCES

Aljaber, T. et al., 2015. An Evaluation Framework for Mobile Health Education Software. *Proceedings of Science and Information Conference (SAI).* London, UK, 28-30 July, pp. 786-790.

Ally, M., 2009. Introduction. In: M. Ally (ed.) *Mobile Learning: Transforming the Delivery of Education and Training.* Edmonton: AU Press, Athabasca University, Edmonton, Canada, pp. 1-8.

Alva, M. E. O. et al., 2003. Comparison of Methods and Existing Tools for the Measurement of Usability in the Web. *Web Engineering: Proceedings of International Conference, ICWE.* Oviedo, Spain, 14-18 July, pp. 386-389.

Baharuddin, R. et al., 2013. Usability Dimensions for Mobile Applications-A Review. *Research Journal of Applied Sciences, Engineering and Technology*, Vol. 5, No. 6, pp. 2225-2231.

Bernhaupt, R., 2009. Usability and User Experience Evaluation Methods. In: C. Mourlas (ed.) *Mass Customization for Personalized Communication Environments: Integrating Human Factors.* Information Science Reference, New York, USA, pp. 232-233.

Brayshaw, M. et al., 2014. Investigating Heuristic Evaluation as a Methodology for Evaluating Pedagogical Software: An Analysis Employing Three Case Studies. *LCT 2014: Proceedings of 1st International Conference on Learning and Collaboration Technologies.* Crete, Greece, 22-27 June, pp. 25-35.

Cisco Visual Networking Index, 2015. *Cisco Visual Networking Index: Global Mobile Data Traffic Forecast Update 2014-2019 White Paper.* Cisco Systems, Inc., San Jose, USA.

Constantinos, K. C. and Kim, D. J., 2011. A Meta-Analytical Review of Empirical Mobile Usability Studies. *Journal of Usability Studies*, Vol. 6, No. 3, pp. 117-171.

Dubey, S. K. et al., 2012. Analytical Comparison of Usability Measurement Methods. *International Journal of Computer Applications*, Vol. 39, No. 15, pp. 11-18.

Glanz, K. et al., 2008. *Health Behavior and Health Education: Theory, Research, and Practice*, 4th edn. Jossey-Bass Wiley, San Francisco, USA.

Harrison, R. et al., 2013. Usability of Mobile Applications: Literature Review and Rationale for a New Usability Model. *Journal of Interaction Science*, Vol. 1, No. 1, pp. 1-16.

Heggestuen, J., 2013. *One in Every 5 People in the World Own a Smartphone, One in Every 17 Own a Tablet*. [Online] Available at: http://www.businessinsider.com/smartphone-and-tablet-penetration-2013-10?IR=T [Accessed 02/08/2015].

Hernandez Munoz, L. U. and Woolley, S. I., 2009. A User-Centered Mobile Health Device to Manage Life-Threatening Anaphylactic Allergies and Provide Support in Allergic Reactions. *Proceedings of 9th International Conference on Information Technology and Applications in Biomedicine (ITAB)*. Larnaca, Cyprus, 4-7 November.

Hussain, A. et al., 2012. Modeling Subjective Metrics for Mobile Evaluation. *Journal of Information System Research and Innovation*, Vol. 1, pp. 11-20.

Hussain, A. et al., 2015. Existing Trends in Usability Evaluation Method (UEM) for M-learning Apps: A Systematic Review. *Journal of Engineering and Applied Sciences*, Vol. 10, No. 15, pp. 6649-6653.

International Organization for Standardization/International Electrotechnical Commission, 2001. *ISO/IEC 9126-1 Standard*. ISO/IEC, Geneva, Switzerland.

Istepanian, R. S. and Lacal, J., 2003. Emerging Mobile Communication Technologies for Health: Some Imperative Notes on m-Health. *Proceedings of the 25th IEEE Annual International Conference: Engineering in Medicine and Biology Society*, vol 2. Cancun, Mexico, 17-21 September, pp. 1414-1416.

Ivory, M. Y. and Hearst, M. A., 2001. The State of the Art in Automating Usability Evaluation of User Interfaces. *ACM Computing Surveys*, Vol. 33, No. 4, pp. 470-516.

Leonardi, C. et al., 2010. An Exploratory Study of a Touch-based Gestural Interface for Elderly. *Proceedings of the 6th Nordic Conference on Human-Computer Interaction: Extending Boundaries*. Reykjavik, Iceland, 16-20 October, pp. 845-850.

Lobo, D. et al, 2011. Web Usability Guidelines for Smartphones: A Synergic Approach. *International Journal of Information and Electronics Engineering*, Vol. 1, No. 1, pp. 33-37.

NHS Innovations South East, 2014. *App Development: An NHS Guide for Developing Mobile Healthcare Applications*. NHS Innovations South East, Didcot, UK.

Nielsen, J., 1993. *Usability Engineering*. Academic Press, San Diego, USA.

Nielsen, J., 2000. *Why You Only Need to Test with 5 Users*. [Online] Available at: http://www.nngroup.com/articles/why-you-only-need-to-test-with-5-users/ [Accessed 22/04/2015].

Nielsen, J. and Molich, R., 1990. Heuristic Evaluation of User Interfaces. *Proceedings of the SIGCHI Conference on Human Factors in Computing Systems*. Seattle, USA, 1-5 April, pp. 249-256.

Qiang, C. Z., 2011. *Mobile Applications for the Health Sector*. [Online] Available at: http://siteresources.worldbank.org/INFORMATIONANDCOMMUNICATIONANDTECHNOLOGIES/Resources/mHealth_report.pdf [Accessed 12/05/2014].

research2guidance, 2014. *mHealth App Developer Economics 2014*. [Online] Available at: http://research2guidance.com/r2g/research2guidance-mHealth-App-Developer-Economics-2014.pdf [Accessed 20/08/2015].

Seffah, A. et al., 2006. Usability Measurement and Metrics: A Consolidated Model. *Software Quality Journal*, Vol. 14, No. 2, pp. 159-178.

Shareef, A. F., 2006. Introduction to Special Issue on Innovation in Distance Learning Technologies in Developing Countries. *Learning Technology Newsletter*, Vol. 8, No. 3, [Online] Available at: http://lttf.ieee.org/issues/july2006/lt_july2006.pdf [Accessed 15/04/2014].

Shivade, B. and Sharma, M., 2014. Usability Analyzer Tool: A Usability Evaluation Tool for Android Based Mobile Application. *International Journal of Emerging Trends & Technology in Computer Science*, Vol. 3, No. 3, pp. 247-252.

Silva, B. M. C. et al., 2013. *A Novel Cooperation Strategy for Mobile Health Applications*. [Online] Available at: http://ieeexplore.ieee.org/stamp/stamp.jsp?tp=&arnumber=6585879 [Accessed 12/06/2014].

Smith, A., 2013. *Smartphone Ownership – 2013 Update*. [Online] Available at: http://boletines.prisadigital.com/PIP_Smartphone_adoption_2013.pdf [Accessed 30/04/2015].

Traxler, J., 2009. Current State of Mobile Learning. In: M. Ally (ed.) *Mobile Learning: Transforming the Delivery of Education and Training*. Edmonton: AU Press, Athabasca University, Edmonton, Canada, pp. 9-24.

Ziefle, M. and Bay, S., 2008. Transgenerational Designs in Mobile Technology. In: J. Lumsden (ed.) *Handbook of Research on User Interface Design and Evaluation for Mobile Technology*. Information Science Reference, New York, USA, pp. 122-141.

SIMILARITY COEFFICIENTS OF NORMAL DISTRIBUTIONS IN SELECTING THE OPTIMAL TREATMENTS

Elisabeth Rakus-Andersson[1] and Janusz Frey[2]
[1]*Department of Mathematics and Natural Sciences, Blekinge Institute of Technology, 371 79 Karlskrona, Sweden*
[2]*Department of Urology, Örebro University Hospital, 701 85 Örebro, Sweden*

ABSTRACT

In the current research, we aim to define a new form of the similarity coefficient to compare the resemblance grade of two Gaussian density functions. We aim to assess the method utility on a theoretical model. The density functions are stated for a biological marker "*survival length*", observed in three groups of patients, suffering from a hypothetical disease. The first group consists of patients who are not treated, whereas we recommend 2 possible treatment methods for the second and the third group, respectively. All the "*survival length*" assumptions of the model (mean values and standard deviations) are made to exclude the equivocal conclusion, regarding a selection of the better treatment. At the first stage, we apply the measure of similarity to populations: survival among untreated patients contra survival among patients after Treatment 1. Another similarity coefficient estimates a relation between populations: survival among untreated patients versus survival among patients after Treatment 2. The lower value of the coefficient points out the more effective treatment. In order to simplify calculations, proposed in the definition of a similarity coefficient, we approximate the Gaussian curve by a specially designed polynomial, known as the π-function.

KEYWORDS

Gaussian density function, π-function, similarity coefficient, survival length, optimal treatment.

1. INTRODUCTION

In modern times, the progress of technology and biological sciences is an undisputable fact, which gives us the great opportunity to process large amounts of data, contributing to the development of, among others, medical sciences. There are many specific registers of diseases, allowing to pose the most important questions, concerning the diseases, when using the population-based observational methodology of studies. This is a cost-effective and useful compliment to the randomized clinical trials, recognized as a best source for data. Some of the important questions still cannot be answered, even after employing the modern epidemiological/bio-statistical methods.

Our purpose is to develop and to evaluate the method, simplifying normal distribution comparisons, as well as to test its usefulness, efficacy, and reliability. To accomplish this test, we evolve the theoretical model with the "*survival length*" as a biological parameter for the hypothetical disease, which can be managed with two different treatments. We would like to emphasize that the model data only have a theoretical character, and the data values are selected in order to make a differentiation between treatments more difficult. We have decided to build our model only on the basis of the most often measured parameters, such as mean values and standard deviations. We assume that in the biological sciences and medicine, most of data will have the normal (Gaussian) distribution of a density function, especially when a number of observations is high. The standard parameters of the density functions, like the mean value and the standard deviation, are thus estimated on the basis of large samples (Gauss, 1809; Stigler, 1986; Krishnamoorthy, 2006 and many others).

In the medical problem to be solved, we focus on the selection of an efficacious therapy, recommended for patients who are touched by the hypothetical disease, which shortens the humans' lives. We also assume that there exist data, referring to the natural course of the disease without any treatment. Another assumption suggests that the data concerning "*survival length*" of patients, cured with Treatment 1 and 2, are available. The biological parameter "*survival length*" is normally distributed, as stated above. We assume that we, by means of collected data, can estimate the means and the standard deviations for groups of patients examined. The population of untreated patients will reveal the survival lengths constituting a reference set of the model. We confront the reference set of survival lengths with other sets of survival, found in the samples of patients, who have been treated by the therapies recommended.

For the purpose of the model, the following assumptions have been made: the patients with the non-treated disease are marked by the mean survival length of 3.0 years with the standard deviation of 0.9 years. The group of patients, with Treatment 1 proved, has the mean survival length of 4.2 years with the standard deviation of 1.2 years. Lastly, the group, where Treatment 2 has been used, is characterized by the mean survival length of 4.0 years with the standard deviation of 1.3 years. The choice of the superior treatment will be carried out to provide an unequivocal answer.

Generally, to compare how like or dislike the survival lengths are for pairs (reference survival without treatment, survival after Treatment i), where $i = 1,\ldots,n$, is a treatment number, we need a measure of similarity for two normal density functions.

There were some attempts, already made to find this measure. The oldest trial involved the comparison of two confidence intervals, built for two distributions (Neyman, 1937 and many others). Recently, the method has awoken some critical remarks and, in spite of the great popularity, there are authors who avoid it (Hoekstra et al., 2014; Morey et al., 2015).

The Bhattacharyya coefficient is an approximate measurement of the amount of overlap between two statistical samples. The coefficient can be used to determine the relative closeness of the two samples being considered (Bhattacharyya, 1943). In 1989 Inman and Bradley proposed a measure of the agreement between two probability distributions by checking their dissimilarity index. Another proposal appeared in (Mishra et al., 1986), where the confidence interval was constructed for the overlapping coefficient.

In our concept, given two normal density functions, we calculate an area under the common part of the functions to divide it later by the total area under both functions. We produce, as a "*similarity coefficient*" a real ratio value equal to 0 in the case of the total dissimilarity, and a value equal to one, if the densities are identical.

We realize that the less area of the common part provides us with the lesser value of the coefficient. We thus choose the optimal Treatment i, whose "*survival length*" compared to reference survival, generates the least quantity of similarity.

Since Gaussian functions are difficult analytic tools of calculations, then we will try to approximate them by a particular π-function of the second degree, very well adapted to the Gaussian shape.

Section 2 constitutes a general description of the method. In Section 3, we introduce a concept of "*similarity coefficient*". The approximation of a normal density function by a π-function is discussed in Section 4. By comparing similarity coefficients, computed for survivals after different therapies in relation to the reference survival, we finally state a hierarchy of therapies. Some concluding remarks are sampled in Section 6.

2. THE GENERAL MODEL OF A SELECTION OF THE BEST MEDICINE

In the paper prepared, we wish to provide a measure of similarity of two compared normal density functions to decide if two states of the same clinical feature are distributed in like manner or not.

Let us suppose that the clinical attribute "*survival length*" takes values x in space $X = [x_{\min}, x_{\max}]$. The attribute is one of the most important parameters, characteristic of a very invasive disease.

The patients sometimes are not aware of the existence of the disease; therefore there exists a group of patients who are not treated. The sampling of survival lengths for untreated patients creates interval $T \subset X$. We find the mean value μ_T and the standard deviation σ_T for the observations of "*surviving length*", belonging to T. The density function $f_T(x, \mu_T, \sigma_T)$ is normally distributed. This function will be named the "*reference density function*".

To improve some prognoses of "*survival length*", therapies Treatment 1, Treatment 2,...,Treatment n are recommended. We sample the results of the survival in n groups of patients. The values of the survival are allocated in intervals $T_1 \subset X$, $T_2 \subset X, \ldots, T_n \subset X$ after testing the results of Treatment 1, Treatment

2,…,Treatment n, respectively in each group. New density functions $f_{T_i}(x,\mu_{T_i},\sigma_{T_i})$, $i = 1,…,n$, are assigned to the effects of Treatment i. If Treatment i, $i = 1,…,n$, has an expected positive effect for the prolongation of the survival length, then condition $\mu_T < \mu_{T_i}$ should be satisfied.

We aim to extract the most efficacious treatment bringing the most promising survival results, when relating them to survivals of patients who did not get any treatment. Therefore, we pairwise compute values of the similarity coefficients C_i between the reference function $f_T(x,\mu_T,\sigma_T)$ and functions $f_{T_i}(x,\mu_{T_i},\sigma_{T_i})$, $i = 1,…,n$.

Among n coefficients $C_i(f_T(x,\mu_T,\sigma_T), f_{T_i}(x,\mu_{T_i},\sigma_{T_i}))$, $i = 1,…,n$, we select the least value, which points out the most effective treatment in the aspect of prolonging the survival length.

The similarity coefficient is designed by authors as a complement in comparative methods.

3. DENSITY FUNCTIONS AND SIMILARITY COEFFICIENTS

Generally, the probability density function $f(x,\mu,\sigma)$ is given by

$$f(x,\mu,\sigma) = \frac{1}{\sigma\sqrt{2\pi}} e^{\frac{-(x-\mu)^2}{2\sigma^2}} \tag{1}$$

for a mean value μ and a standard deviation σ, computed from the sample.

The domain of $f(x,\mu,\sigma)$, whose graph is known in statistics as the Gaussian curve, is allocated in $X = (-\infty,\infty)$. We suppose in this study that the graphs of $f_T(x,\mu_T,\sigma_T)$ and $f_{T_i}(x,\mu_{T_i},\sigma_{T_i})$ have only one intersection point, due to the assumption $\mu_T < \mu_{T_i}$.

Example 1

We observe the normally distributed attribute "*survival length*" in patients, who suffer from a hypothetical disease (e.g., cancer). Suppose $X = [0, 20]$ (survival provided in years). After sampling the data of survival length in interval $T \subset X$ among patients who did not get any treatment, we estimate the reference density function $f_T(x,3.0,0.9)$, $x \in T \subset X$.

There are two types of recommended therapies, Treatment 1 and Treatment 2, which are expected to improve the length of survival. The survival lengths of patients, cured by Treatment 1, are collected in interval $T_1 \subset X$ and estimated by $f_{T_1}(x,4.2,1.2)$, $x \in T_1 \subset X$. Effects of Treatment 2, seen in interval T_2, are data creating the pattern of $f_{T_2}(x,4.0,1.3)$, $x \in T_2 \subset X$.

The graphs of density functions $f_T(x,3.0,0.9)$, $f_{T_1}(x,4.2,1.2)$ and $f_{T_2}(x,4.0,1.3)$ are sketched in Figure 1.

We note the clear resemblance between $f_{T_1}(x,4.2,1.2)$ and $f_{T_2}(x,4.0,1.3)$. To judge which treatment is more effective, we propose an implementation of our own similarity coefficient for two density functions. Let us thus compare a similarity coefficient between $f_T(x,3.0,0.9)$ and $f_{T_1}(x,4.2,1.2)$ to a similarity coefficient between $f_T(x,3.0,0.9)$ and $f_{T_2}(x,4.2,1.2)$. The lower value of the similarity coefficient will indicate the more efficacious therapy.

We propose a similarity coefficient $C(f_T(x,\mu_T,\sigma_T), f_{T_i}(x,\mu_{T_i},\sigma_{T_i})) = C(f_T(x), f_{T_i}(x))$ stated by the formula

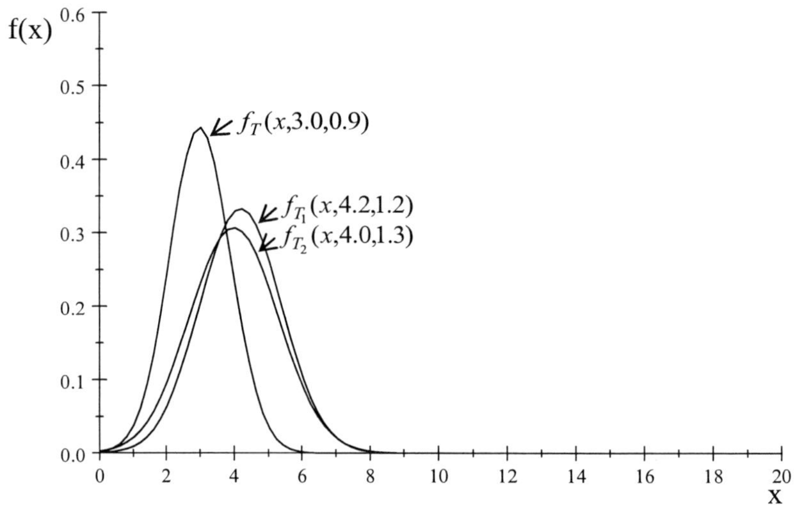

Figure 1. The density functions $f_T(x,3,0.9)$, $f_{T_1}(x,4.2,1.2)$ and $f_{T_2}(x,4.0,1.3)$

$$C(f_T(x), f_{T_i}(x)) = \frac{\text{area under } (f_T(x) \cap f_{T_i}(x))}{\text{area under } (f_T(x)) + \text{area under } (f_{T_i}(x)) - \text{area under } (f_T(x) \cap f_{T_i}(x))}, \tag{2}$$

where

$$f_T(x) \cap f_{T_i}(x) = \{f(x) : f(x) = \min(f_T(x), f_{T_i}(x)), x \in X\}. \tag{3}$$

Let us note that areas under $f_T(x)$ and under $f_{T_i}(x)$ are equal to 1 as sums of all densities.

Example 2

To compute an area of $f_T(x) \cap f_{T_1}(x)$ from Ex. 1, we should at first find the intersection point between $f_T(x)$ and $f_{T_1}(x)$. This generates an equation $\frac{1}{0.9\sqrt{2\pi}} e^{\frac{-(x-3.0)^2}{2(0.9)^2}} = \frac{1}{1.2\sqrt{2\pi}} e^{\frac{-(x-4.2)^2}{2(1.2)^2}}$, which has no exact solution over $X = (-\infty, \infty)$. The domain $X = (-\infty, \infty)$ should be replaced by a continuous but closed interval, and the exponential function should be approximated by a polynomial possessing almost the same shape.

4. APPROXIMATIONS OF *F*-DENSITY FUNCTIONS BY π-FUNCTIONS

To fulfill the suggestions from Ex. 2, we prepare the model of approximation of the Gaussian function $f(x,\mu,\sigma) = f(x)$ by the π-function expanded as an equation

$$f(x) \approx \pi(x, \alpha_1, \beta_1, \gamma_1 = \alpha_2, \beta_2, \gamma_2, \varepsilon) = \begin{cases} 2\varepsilon\left(\frac{x-\alpha_1}{\gamma_1-\alpha_1}\right)^2 & \text{for} \quad \alpha_1 \le x \le \beta_1, \\ \varepsilon\left(1-2\left(\frac{x-\gamma_1}{\gamma_1-\alpha_1}\right)^2\right) & \text{for} \quad \beta_1 \le x \le \gamma_1, \\ \varepsilon\left(1-2\left(\frac{x-\alpha_2}{\gamma_2-\alpha_2}\right)^2\right) & \text{for} \quad \alpha_2 \le x \le \beta_2, \\ 2\varepsilon\left(\frac{x-\gamma_2}{\gamma_2-\alpha_2}\right)^2 & \text{for} \quad \beta_2 \le x \le \gamma_2, \end{cases} \tag{4}$$

in which $\alpha_1, \beta_1, \gamma_1, \alpha_2, \beta_2, \gamma_2$ and ε are parameters and $\beta_1 = \frac{\alpha_1+\gamma_1}{2}, \beta_2 = \frac{\alpha_2+\gamma_2}{2}$.

The graph of the π-function looks like a bell and has intersection points with the x-axis in $(\alpha_1, 0)$ and $(\gamma_2, 0)$. After adjusting parameters of the π-function to the parameters of $f(x,\mu,\sigma)$, the π-function is a very efficient tool of approximating the density function $f(x)$ with a little error.

We suggest that $\alpha_2 = \gamma_1 = \mu$. The value of ε is provided by $\varepsilon = f(\mu) = \frac{1}{\sigma\sqrt{2\pi}} e^{\frac{-(\mu-\mu)^2}{2\sigma^2}} = \frac{1}{\sigma\sqrt{2\pi}}$.

$\pi(x)$ should approximate $f(x)$ with a large precision. Thus, a multiplier k is added to σ. The product $k\sigma$ is a factor of the domain of the π-function, decided as $(\mu - k\sigma, \mu + k\sigma)$. We also calculate $\alpha_1 = \mu - k\sigma$, $\gamma_1 = \alpha_2 = \mu$, $\beta_1 = \frac{(\mu-k\sigma)+\mu}{2} = \frac{2\mu-k\sigma}{2}$, $\gamma_2 = \mu + k\sigma$ and $\beta_2 = \frac{(\mu+k\sigma)+\mu}{2} = \frac{2\mu+k\sigma}{2}$. Equation (4) takes a new shape, which is expressed in parameters of μ, σ and k, as (Rakus-Andersson and Frey, 2014)

$$f(x) \approx \pi(x,\mu,\sigma,k) = \begin{cases} 2\frac{1}{\sigma\sqrt{2\pi}}\left(\frac{x-(\mu-k\sigma)}{k\sigma}\right)^2 & \text{for} \quad \mu - k\sigma \le x \le \frac{2\mu-k\sigma}{2}, \\ \frac{1}{\sigma\sqrt{2\pi}}\left(1-2\left(\frac{x-\mu}{k\sigma}\right)^2\right) & \text{for} \quad \frac{2\mu-k\sigma}{2} \le x \le \mu, \\ \frac{1}{\sigma\sqrt{2\pi}}\left(1-2\left(\frac{x-\mu}{k\sigma}\right)^2\right) & \text{for} \quad \mu \le x \le \frac{2\mu+k\sigma}{2}, \\ 2\frac{1}{\sigma\sqrt{2\pi}}\left(\frac{x-(\mu+k\sigma)}{k\sigma}\right)^2 & \text{for} \quad \frac{2\mu+k\sigma}{2} \le x \le \mu + k\sigma. \end{cases} \tag{5}$$

To find the value of k, we first adopt the property $\int_{-\infty}^{\infty} f(x)dx = 1$ (the sum of all densities in the distribution is equal to 1). This leads to the conclusion that $\int_{-\infty}^{\mu} f(x)dx = 0.5$. Hence, the integral with a half of the π-function as the integrand satisfies the same condition. This means that the equality

$$\int_{\mu-k\sigma}^{\frac{2\mu-k\sigma}{2}} 2\frac{1}{\sigma\sqrt{2\pi}}\left(\frac{x-(\mu-k\sigma)}{k\sigma)}\right)^2 dx + \int_{\frac{2\mu-k\sigma}{2}}^{\mu} \frac{1}{\sigma\sqrt{2\pi}}\left(1-2\left(\frac{x-\mu}{k\sigma}\right)^2\right)dx = 0.5 \tag{6}$$

is satisfied with a minimal error.

Equation (6) has, as an unknown, parameter k. The equation is solved exactly by means of the package Maple 9.

The error of approximating the density function $f(x)$ by $\pi(x)$ is rated as

$$Error(f(x), \pi(x)) = \left|1 - \int_{\mu-k\sigma}^{\mu+k\sigma} \pi(x)dx\right|, \tag{7}$$

since the structure $\int_{-\infty}^{\infty} f(x)dx = 1$ is an estimation of the area under $f(x)$, equal to 1. Expression $\int_{\mu-k\sigma}^{\mu+k\sigma} \pi(x)dx$ stands for the area under $\pi(x)$. The area consists of a sum of four integrals in compliance with Eq. (5).

Example 3

Let us find π-functions fitted for $f_T(x,3.0,0.9)$, $f_{T_1}(x,4.2,1.2)$ and $f_{T_2}(x,4.0,1.3)$. We prepare the following data for $\pi_T(x)$, which fits for $f_T(x)$: $\varepsilon = \frac{1}{0.9\sqrt{2\pi}} = 0.443$, $k = 2.508$, $\alpha_1 = 0.743$, $\gamma_1 = \alpha_2 = 3.0$, $\beta_1 = 1.871$, $\gamma_2 = 5.257$, and $\beta_2 = 4.129$. The domain of $\pi_T(x)$ is interval $[3 - 2.508 \cdot 0.9, 3 + 2.508 \cdot 0.9] = [0.743, 5.257]$.

The equation of $\pi_T(x)$ is provided by

$$\pi_T(x,3.0,0.9,2.508) = \begin{cases} \pi_{T,1}(x) & 0.886\left(\frac{x-0.743}{2.257}\right)^2 & \text{for } 0.743 \le x \le 1.871, \\ \pi_{T,2}(x) & 0.443\left(1-2\left(\frac{x-3.0}{2.257}\right)^2\right) & \text{for } 1.871 \le x \le 3.0, \\ \pi_{T,3}(x) & 0.443\left(1-2\left(\frac{x-3.0}{2.257}\right)^2\right) & \text{for } 3.0 \le x \le 4.129, \\ \pi_{T,4}(x) & 0.886\left(\frac{x-5.257}{2.257}\right)^2 & \text{for } 4.129 \le x \le 5.257. \end{cases} \tag{8}$$

The error of approximation of $f_T(x)$ by $\pi_T(x)$ is evaluated as 0.0000604.
Function $f_{T_1}(x)$ is approximated, due to Eq. (5), by $\pi_{T_1}(x)$ yielded as

$$\pi_{T_1}(x,4.2,1.2,2.502) = \begin{cases} \pi_{T_1,1}(x) & 0.666\left(\frac{x-1.198}{3.002}\right)^2 & \text{for } 1.198 \le x \le 2.669, \\ \pi_{T_1,2}(x) & 0.333\left(1-2\left(\frac{x-4.2}{3.002}\right)^2\right) & \text{for } 2.699 \le x \le 4.2, \\ \pi_{T_1,3}(x) & 0.333\left(1-2\left(\frac{x-4.2}{3.002}\right)^2\right) & \text{for } 4.2 \le x \le 5.701, \\ \pi_{T_1,4}(x) & 0.666\left(\frac{x-7.202}{3.002}\right)^2 & \text{for } 5.701 \le x \le 7.202. \end{cases} \tag{9}$$

The approximation error is computed to be equal to 0.000193.
The third function $\pi_{T_2}(x)$ is fitted for $f_{T_2}(x)$ by equation

$$\pi_{T_2}(x,4.0,1.3,2.506) = \begin{cases} \pi_{T_2,1}(x) & 0.614\left(\frac{x-0.742}{3.258}\right)^2 & \text{for } 0.742 \le x \le 2.371, \\ \pi_{T_2,2}(x) & 0.307\left(1-2\left(\frac{x-4.0}{3.258}\right)^2\right) & \text{for } 2.371 \le x \le 4.0, \\ \pi_{T_2,3}(x) & 0.307\left(1-2\left(\frac{x-4.0}{3.258}\right)^2\right) & \text{for } 4.0 \le x \le 5.629, \\ \pi_{T_2,4}(x) & 0.614\left(\frac{x-7.258}{3.258}\right)^2 & \text{for } 5.629 \le x \le 7.258. \end{cases} \tag{10}$$

The difference error between $\pi_{T_2}(x)$ and $f_{T_2}(x)$ is 0.000144.
Figure 2 contains functions $\pi_T(x)$, $\pi_{T_1}(x)$, $\pi_{T_2}(x)$ (thick lines) and $f_T(x)$, $f_{T_1}(x)$, $f_{T_2}(x)$ (thin lines).

5. THE SELECTION OF THE OPTIMAL MEDICINE

A revised version of the similarity coefficient $C(\pi_T(x), \pi_{T_i}(x))$, provided that $\pi_T(x)$ approximates $f_T(x)$ and $\pi_{T_i}(x)$ fits for $f_{T_i}(x)$, $i = 1,\dots,n$, is proposed as

$$C(\pi_T(x),\pi_{T_i}(x)) = \frac{\text{area under } (\pi_T(x) \cap \pi_{T_i}(x))}{\text{area under } (\pi_T(x)) + \text{area under } (\pi_{T_i}(x)) - \text{area under } (\pi_T(x) \cap \pi_{T_i}(x))}, \tag{11}$$

where

$$\pi_T(x) \cap \pi_{T_i}(x) = \{\pi(x) : \pi(x) = \min(\pi_T(x), \pi_{T_i}(x)), x \in X\}. \tag{12}$$

We note that $C(\pi_T(x),\pi_{T_i}(x)) = 1$, if both density distributions will be identical (the same π-functions), and $C(\pi_T(x),\pi_{T_i}(x)) = 0$ for disjoint domains of $\pi_T(x)$ and $\pi_{T_i}(x)$ (the intersection of domains constitutes an empty set). Thus

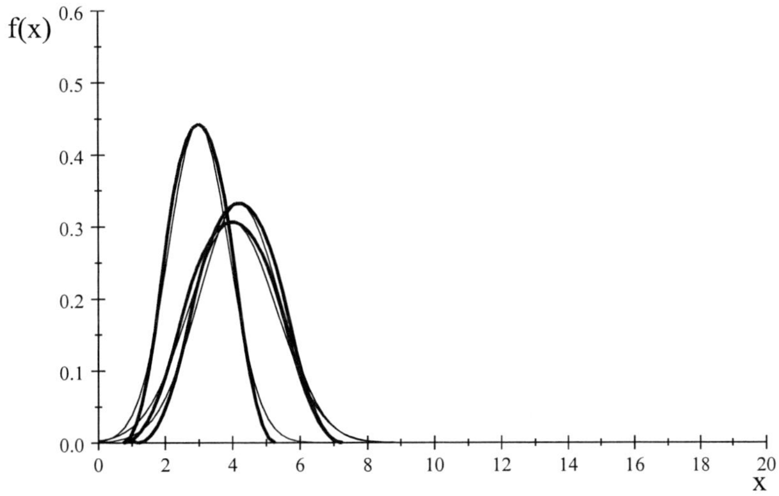

Figure 2. Functions $\pi_T(x)$, $\pi_{T_1}(x)$, $\pi_{T_2}(x)$ (thick lines) and $f_T(x)$, $f_{T_1}(x)$, $f_{T_2}(x)$ (thin lines)

$$0 \le C(\pi_T(x), \pi_{T_i}(x)) \le 1 . \tag{13}$$

It means that the lower value of C is followed by the greater dissimilarity of two compared normal density distributions of different states, assigned to a clinical marker with values in space X.

Example 4

We first evaluate $C(\pi_T(x), \pi_{T_1}(x))$. The process starts with finding an intersection point between $\pi_T(x)$ and $\pi_{T_1}(x)$. Since we have suggested the existence of only one common point, then we will solve the following equations: $\pi_{T,3}(x) = \pi_{T_1,1}(x)$ - no solution in $[3.0, 4.129] \cap [1.198, 2.669] = 0$, $\pi_{T,3}(x) = \pi_{T_1,2}(x)$ - solution is 3.83, $\pi_{T,4}(x) = \pi_{T_1,1}(x)$ - no solution in the common domain, $\pi_{T,4}(x) = \pi_{T_1,2}(x)$ - no solution in the common domain. Hence,

$$\text{area under } (\pi_T(x) \cap \pi_{T_1}(x)) = \int_{1.198}^{2.669} \pi_{T_{1,1}}(x)dx + \int_{2.669}^{3.83} \pi_{T_{1,2}}(x)dx + \int_{3.83}^{4.129} \pi_{T,3}(x)dx + \int_{4.129}^{5.257} \pi_{T,4}(x)dx = 0.543 .$$

The integrands have been prepared by the Matlab program.

We note that $\text{area under } (\pi_T(x)) = 1 - error(f_T(x), \pi_T(x)) = 0.9999396$ and $\text{area under } (\pi_{T_1}(x)) = 0.999807$ due to Eq. (7). Finally, $C(\pi_T(x), \pi_{T_1}(x)) = \frac{0.543}{0.9999396 + 0.999807 - 0.543} = 0.3727$.

By following the same procedure, we obtain $C(\pi_T(x), \pi_{T_2}(x)) = 0.4419$.

Since the lower value of the similarity coefficient assists the more effective treatment, then we will decide that Treatment 1 has the better influence on the survival length in patients who were cured by this therapy.

6. CONCLUSION

The authors' research field is the application of computational intelligence to medicine. Especially, medical decisions have been made in medication, where a hierarchy of therapies, used in a morbid unit, has been stated (Rakus-Andersson, 2008, 2012; Rakus-Andersson and Frey, 2014). The practical examples contained clinical data in these models.

For the first time we have studied the behavior of biological parameter "*survival length*" in the context of its density of appearance in some patient groups. The density function, indicating survival in patients without provided treatment, has been set together experimentally with two other density functions. These have been formed for survivals among patients cured by two alternative treatments.

To judge similarity of two density functions, we have designed "*similarity coefficient*" as a proportion between the area under the common part of functions and the total area under both of them. We have not been able to computationally work with infinite domains of Gaussian functions; therefore we have adjusted π-functions as their thorough approximations with domains, defined in the form of close and continuous intervals. In our experiment, a computer program easily solves all issues, sketched for polynomials. The similarity coefficient is located in interval [0, 1], which agrees with the full dissimilarity of functions for value of 0 and full likelihood for value of 1.

We think that the similarity coefficient, proposed by us, has some advantages. We can create it for an arbitrary amount of pairs, the computations are simple and, most of all, we do not introduce any hypothetical uncertainty. The obtained values of the coefficients are clearly interpretable without being burdened by confidence thresholds.

We believe that the model can help patients and physicians to make a consent and proper therapeutic decision. It could be a valuable tool for patients, e.g., at the same age who will be allowed comparing prognosis of the disease in the current stage with treatment results, achieved in other groups. In this way we may avoid referring to the disease itself, as it is done today. The graphical and computational form of earlier analysed data can also facilitate the reception of treatment effects expected.

In this paper, we have used theoretical data, since our practical experiment contained too little number of patients.

In the future works, we intend to name exactly a disease and treatments after inspecting larger data records. Another task is to suggest a value of the similarity coefficient in order to reconstruct an unknown survival density function after a certain treatment, when comparing it to the reference density function. This allows prognosticating dynamically some demands made for treatments.

REFERENCES

Bhattacharyya, A., 1943. On a Measure of Divergence between Two Statistical Populations Defined by Their Probability Distributions. *In Bulletin of the Calcutta Mathematical Society*, Vol. 35, pp. 99-109.

Davis, C. H, 1857. *Gauss's "Teoria Motus"*. Little, Brown and Company, Boston, USA.

Hoekstra, R. et al., 2014. Robust Misinterpretation of Confidence Intervals. *In Psychonomic Bulletin Review,* in press.

Inman H. and Bradley E., 1989. The Overlapping Coefficient as a Measure of Agreement between Probability Distributions and Point Estimation of the Overlap of Two Normal Densities. *In Communications in Statistics - Theory and Methods* Vol. 18, Issue 10, pp. 3851-3874.

Krishnamoorthy, K., 2006. *Handbook of Statistical Distributions with Applications.* Chapman & Hall/CRC., London, England, ISBN: 1-58488-635-8.

Mishra, S. N. et al., 1986. Overlapping Coefficient: the Generalized t Approach. *In Communications in Statistics - Theory and Methods*, Vol. 15, Issue 1, pp. 123-128.

Morey, R. D. et al., 2015. *The Fallacy of Placing Confidence in Confidence Intervals.* In press.

Neyman, J., 1937. Outline of a Theory of Statistical Estimation Based on the Classical Theory of Probability. *In Philosophical Transactions of the Royal Society*, Vol. 236, pp. 333-380.

Rakus-Andersson, E., 2008. Decision-making Techniques in Ranking of Medicine Effectiveness. *In Advanced Computational Intelligence Paradigms in Healthcare 3,* Springer-verlag, Berlin Heidelberg, pp. 51-73.

Rakus-Andersson, E., 2012. The Parametric *s*-functions and the Perceptron in Gastric Cancer Surgery Decision Making. *In Proceedings of WCCI 2012 World Congress,* Brisbane, Australia, IEEE Computational Intelligence Society, pp. 1852-1859.

Rakus-Andersson, E., 2014. Complex Control Models with Parametric Families of Fuzzy Constrains in Evaluation of Resort Management System. *In Journal of Advanced Computational Intelligence and Intelligent Informatics,* Vol. 18, Issue 3, fujipress, pp. 271-279.

Rakus-Andersson, E. and Frey, J., 2014. The Choquet Integral Applied to Ranking Therapies in Radiation Cystitis. *In Proceedings of IEEE IS 2014,* Warszawa, Poland, Springer, Berlin-Heidelberg, pp. 443-452.

Stigler, S. M., 1986. *The History of Statistics: The Measurement of Uncertainty before 1900.* Harvard University Press, Cambridge, Massachusetts, USA, ISBN: 0-674-40340-1.

AN INTEGRATED SOLUTION FOR IMPROVING EMERGENCY MEDICAL SERVICE SYSTEMS' RESPONSE TIME

Esraa I. Abou El Safa and Ghada A. El Khayat
Information Systems and Computers Department, Faculty of Commerce, Alexandria University
Alexandria, Egypt

ABSTRACT

This paper proposes an integrated system that aims to improve the efficiency of Emergency Medical Services (EMS). EMS systems are responsible for interventions in life-threatening health conditions in which a second counts. A primary measure for EMS performance is its response time; which is the focus of this research. Response time is divided into several components. However, almost all literature tackling its improvement considers the travel time component only. In this research the time taken to report a case and the dispatching time are also considered. The dispatching and relocation policies used affect the system performance. Thus, in this study a modified measure for preparedness and demand forecast is provided. A novel ambulance dispatch algorithm based on these measures is introduced. It lowers the overall system response time, while respecting a predefined threshold of response time for single cases. The contributed integrated system comprises a mobile application for the use of citizens and corresponding applications for the EMS. The citizens' side application allows them to make a faster, easier and more accurate incident reporting. The EMS side applications make use of the proposed policy and rules in order to dispatch ambulance vehicles and relocate them efficiently. In addition, the applications work together to enable the EMS to make better call screening and faster dispatch. Preliminary testing of the citizens' side application was made and results were satisfactory. Data collected by the application will potentially help determine black spots and support decision makers in deciding where to locate new ambulance stations and which roads need maintenance. The proposed solution is expected to lower the overall response time and improve the EMS system's efficiency.

KEYWORDS

Emergency Medical Services (EMS); Response Time; Ambulance Relocation; Ambulance Dispatch; Call Screening.

1. INTRODUCTION

In critical life-threatening situations a second can make the difference between life and death. Therefore the availability of an efficient and effective EMS is crucial. EMS are public services that provide the first line response and pre-hospital care to individuals in an incident of serious illness or accident injury (Chanta et al., 2014), transport them to the appropriate hospital (if necessary), and return back to their base (Iannoni et al., 2009). EMS systems are designed to rapidly provide advance medical care to critical patient cases (Bandara, 2012). Therefore, an EMS system's primary performance measure is set around its response time; the time between an incoming emergency call and the moment of arrival of an ambulance to the reported location. This includes: (1) time spent on the phone obtaining case's information, (2) time spent deciding which ambulance(s) to dispatch, (3) time to contact paramedics, (4) time for paramedics to get prepared and reach the ambulance, and (5) travel time to the case (Ingolfsson et al., 2008). Although, travel time may contribute a longer proportion to the total response time, other components should be given importance as they are uncontrollable and highly variable.

Commonly used EMS performance metrics set around response time are: (1) percentage of the population that can be responded to within a predetermined time threshold T (Nguyen,2015), (2) percentage of calls that can be responded to within a predetermined time threshold T, i.e. coverage level (Bandara, 2012), (3) mean user response time, (4) percentage of calls not serviced within a predetermined threshold T (Iannoni et al., 2009).

EMS systems tasks can be divided into two main categories: serving emergency requests, and transporting patients from and/or to hospital, home and/or medical units. The former task is more complicated than the latter. Efficient management of EMS in case of emergencies is an extremely difficult process due to the randomness and uncertainty associated with its operations. EMS demand, travel times, service time and ambulance availability are never certainly known. On the contrary, the transportation service provided by the EMS system, in non-emergency incidents, is less sophisticated as the demand is received sufficiently in advance so that requests can be scheduled. In this case response time is not of a significant importance unlike the emergency cases (Kergosien et al., 2014). Many EMS systems split their fleet into two sets, managed separately. One set is responsible for serving emergencies, and the other for normal transportation tasks.

EMS systems' operations in case of emergency may vary from one country to another. However, there are a number of operations that are common to most systems. These operations can be classified into: central and external operations. Central operations are related to receiving calls and allocating ambulances to calls while external operations are those related to providing pre-hospital care to patients (Aboueljinane et al., 2013). Central and external operations are described briefly as follows: 1- A call is made to the ambulance organization to request an ambulance (incident detection and reporting), 2- A dispatcher determines the severity of the incident and its degree of urgency (call screening), 3- Ambulance to be dispatched is selected, 4- Crew is notified and gets prepared, 5- The selected vehicle moves to the determined place, 6- First medical aid is provided and 7- The injured is transported to the nearest hospital (if necessary).

This research is concerned with improving the EMS systems efficiency in case of emergencies especially in cases of traffic accidents in order to decrease death rate, prevent potential disability and improve chances for recovery. This is done through providing an integrated system that improves incident reporting and lowers response time. The integrated solution comprises a mobile application for the use of citizens and corresponding applications for the EMS. The EMS applications make use of a set of proposed rules in order to dispatch ambulance vehicles and relocate them efficiently. In addition, the applications work together to allow citizens make a faster, easier and more accurate incident reporting to the EMS and help the EMS make better call screening and faster dispatch.

Moreover, this research aims, on the long term, to make use of the data of reported incidents and the patients' history of disease for analysis by the national instances in order to get insights that can direct public health policies. Furthermore, data about incident locations can be used by government to determine black spots; areas associated with high rate of traffic accidents, in order to decide on future locations for ambulance stations and on roads that need maintenance.

The rest of this paper is structured as follows. In section 2 the related work and the research contribution that fills the literature gap are discussed. In section 3 the contributed solution and the dispatch algorithm are proposed. In 4 the prototype implementation is highlighted. Finally, the conclusion and future work are presented in section 5.

2. RELATED WORK

To improve the efficiency of EMS several problems were tackled in previous research. These include station location, ambulance allocation and relocation, ambulance dispatch, districting, staff scheduling, performance evaluation, demand forecasting…etc. However, the incident reporting and call screening processes did not receive much attention. As clarified previously, response time is divided into several components one of them is the reporting time which is random in nature. Ingolfsson et al. (2008) showed that a model that does not take into account random delays prior to travel time may overestimate the possible service level and underestimate the number of ambulances needed. Despite this, to the extent of the authors' knowledge, no literature has considered reducing reporting time, most literature focused on reducing travel time only. In addition, no literature introduced better means for reporting. A dispatcher uses the gathered information to determine the nature and priority of the call. The nature of the call reflects the type of the incident reported (such as accident, trauma, or breathing difficultly). The priority of a call reflects the extent to which the call is life-threatening. Nature and priority are then used by the dispatcher to make the dispatching decision; the process of deciding on which ambulance(s) should be allocated to a reported incident. Thus dispatching is highly dependent on the screening process. The authors believe that call screening was not thoroughly studied in literature despite its importance. This research introduces a tool for easier, faster and accurate reporting and for better call screening.

EMS response time can be improved through enhancing location and allocation of ambulances. These decisions are strategic in nature and have been extensively discussed in the literature. Another strategic decision that affects response time is how to design pre-determined response areas; this is known as districting. Some operational decisions also need to be considered to enhance response time such as dispatching and relocation of ambulance vehicles. Relocation is considered a type of the ambulance location problem. However, it is dynamic in nature as it is solved in a real-time context in order to relocate idle vehicles among areas of need. To perform these decisions efficiently, measuring the level of preparedness, the ability of the EMS system to cover future demand, and forecasting demand accurately are essential.

Preparedness level and future demand should be considered as inputs for most ambulance optimization problems. Demand forecasting can be very useful for relocation decisions as it may help anticipate shortage before it occurs. The level of preparedness is affected by dispatch and thus may trigger a relocation decision. Cote and Tucker (2001) discussed four common demand forecasting methods: percent adjustment, 12-month moving average, trend line and seasonality. Finarelli and Johnson (2004) introduced a quantitative demand forecasting model for healthcare services, while Myers and Green (2004) proposed an approach for forecasting demand along with capacity needs. Formulae used to calculate forecasted demand and preparedness level will be discussed later.

Few literature contributions considered preparedness as an important factor for dispatch and relocation. Andersson and Varbrand (2007) developed a measure for calculating preparedness level using actual demand rates and made use of this measure in dispatch and relocation decisions. Lee (2011) modified Andersson and Vabrand's measure. He also modified the last step of Andersson and Vabrand's dispatch algorithm to incorporate travel time in the decision making. Dispatch and relocation are dependent on each other. Belanger et al. (2015) showed that considering these two decisions simultaneously rather than separately will enhance the system's efficiency. In addition Gendreau et al. (2001) proposed to dispatch vehicles based on absolute and relative coverage constraints. On the other hand Mayorga et al. (2013) proposed integrated dispatching and districting policies to improve the performance of EMS, they showed that tackling districting and dispatching together had a good effect on response time.

It can be concluded that the literature has either studied factors relating to EMS and EMS decisions separately or through integrating a maximum of three aspects while neglecting the others. Actually, studying, even, a single decision in EMS is a difficult problem due to the random nature of EMS. The more factors and decisions considered in a solution, the higher the problem complexity and the higher the potential improvement as well. This research studies the EMS in a more integrated fashion where reporting, call screening, dispatch, relocation and preparedness are considered.

Several algorithms and rules for dispatch exist. In addition to the previously discussed algorithms, there is a simple strategy known as the nearest vehicle strategy. Schmid (2012) showed that the nearest vehicle strategy may not be the best to adopt when the level of priority is not high; allowing deviation from this rule. While, Toro-Diaz et al. (2013) combined location and dispatching decisions to show that minimization of response time and maximization of coverage can be achieved using the nearest vehicle strategy when dispatching decisions are taken according to fixed preference lists.

Most literature considered either distance or travel time when dispatching. However, traffic patterns are important in determining travel times and should be also considered. Haghani and Yang (2007) used off peak traffic data to predict travel times to decide on which ambulance to dispatch. However, Fleischman et al. (2013) developed a model for predicting ambulance arrival time to hospital and mentioned in their research limitations that using live traffic could have resulted in more accurate results. Meanwhile, as far as the authors know, no literature made use of live traffic data in the dispatch and relocation decisions.

Literature discussed several triggers for relocation. Gendreau et al. (2001) stated that relocation is to be triggered by dispatch. Gendreau et al. (2006) then used the changes in the number of vehicles waiting for call as a trigger for relocation. Andersson et al. (2004) argued that relocation occurs when the ability of the EMS system to cover future demand is decreased. Therefore, Andersson and Varband (2007) triggered relocation by the systems preparedness level. Schmid and Doener (2010) argued that relocation should be triggered as the time period changes. Jagtenberg et al. (2015) introduced a relocation policy that minimizes the expected fraction of accidents that are reached later than a pre-determined threshold T.

From the above review, no literature considered reducing reporting time as most literature focused only on travel time. In addition, most literature related to EMS management dealt with the problem in a fragmented or in a partially integrated way. Moreover, live traffic data were not incorporated into dispatch and relocation decisions.

This research tackles the EMS process in a more integrated fashion where reporting, call screening, dispatch, relocation and preparedness are considered. Dispatch and relocation policies that make use of live traffic data are introduced in order to lower the EMS response time and to raise the system efficiency.

3. THE PROPOSED INTEGRATED SYSTEM

The proposed system is composed of three applications that connect together through the Internet. The first application is the central EMS side application that is used centrally by dispatchers to select appropriate vehicles for dispatch and perform necessary relocations. The second is the ambulance vehicle side application. It is mainly used by en route vehicles that are on their way to a low priority case or returning to their bases and keep self-reporting their positions for urgent cases. The last application is the citizen side application that is used by citizens to report an incident to the EMS system and that keeps track of the citizen's personal data and his/her medical history. Figure 1 provides an overview of the system. Arrows represent sequential processes, dotted arrows represent data flow.

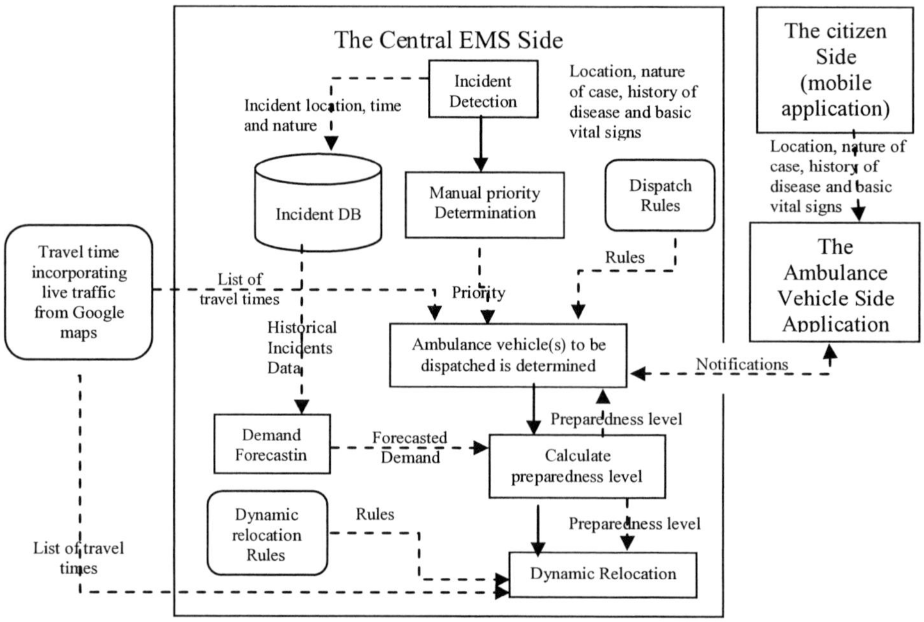

Figure 1. The Detailed Integrated System Framework

The contributed integrated system supports responding to an emergency request when the citizen side application is used to report an incident. When the application is used it captures the injured person's vital signs and determines his/her location. Two scenarios exist. The first is when the vital signs are highly deteriorated. In this case, the application pushes vital signs and location to the nearest en route vehicle if available. In the same time, data are sent to the central EMS side application. If the vehicle receives the data and finds the vital signs received are life threatening, it notifies the EMS application and diverts to the incident. An alternative should be provided in case a vehicle diverts while on its way to a lower priority case.

The second scenario is when the vital signs are not severe or no near en route vehicle is found. In this case, the mobile application sends personal data, location, vital signs, and medical history of the injured to the central EMS side application. The received data are passed to a dispatcher in order to help him/her in assigning priority to the incident and in determining the number of vehicles needed. The dispatcher might contact the reporter to get additional information, if needed, before assigning the priority. Google maps are then used to generate a list of live traffic based travel times between all available ambulances and the reported incident. After this, the algorithm presented in section 3.3 is run to determine the vehicle(s) to be dispatched.

When a dispatch occurs, the system recalculates the preparedness level based on the forecasted demand. If the preparedness level is below a predetermined threshold T, relocation is triggered. A list of live traffic based travel times between all available ambulances and the areas with low preparedness is prepared with the support of Google maps. A decision of which ambulance(s) to be relocated and where to relocate them is then made.

In both scenarios, the data received by the central EMS side application are saved in a database in order to be used for health care analytics and to guide decision makers determine black spots, roads that need repairs or modifications and potential locations for ambulance stations.

3.1 The Citizen Side Application

The authors found that the normal reporting method (phone calls) needs to be replaced by a faster, easier and more accurate method for reporting, in order to easily and accurately determine the exact location of the incident, the injured person's medical history, and the injured person's current vital signs with minimal effort from the reporter and in the shortest possible duration. Current vital signs and medical history of the case should be considered as they may be helpful in the call screening and priority determination process.

The researchers decided to address the previously discussed needs through the development of a mobile application as it is expected to rapidly spread and get accepted among citizens. The application provided is expected to be used by citizens in reporting emergency incidents, especially traffic accidents.

When the application is installed on a smart phone, it asks the user to insert his/her personal data (Name, ID, Address…etc.) and his/her medical history. These data are kept for use in cases of emergency. Figure 2 describes briefly how the citizen side application works. More details about the application are discussed in Abou El Safa and El Khayat (2016).

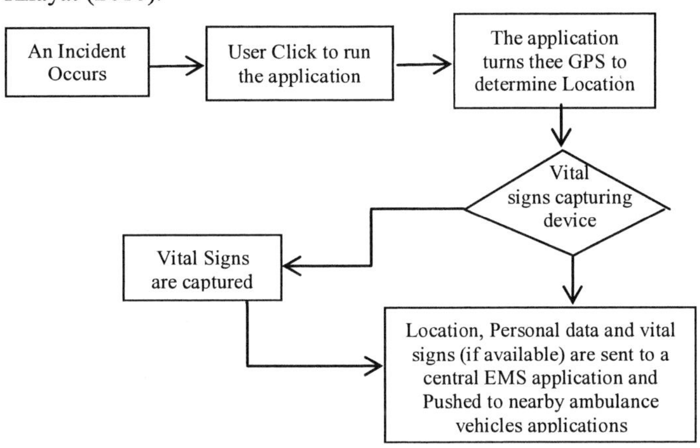

Figure 2. Illustration of How the Citizen Side Application Works

3.2 The Proposed Dispatch Algorithm

Andersson and Varbrand (2007) and Lee (2011) based their dispatch algorithms on the preparedness level, while Gendreau et al. (2001) developed their dispatch algorithm based on a set of covering constraints. The proposed algorithm, illustrated in table 1, uses a combined measure of both preparedness and coverage levels. The preparedness measure used is a modification of Andersson and Varbrand (2007) and Lee (2011) measure. The proposed measure uses forecasted demand rather than actual demand.

Service area is divided into zones Z; the zone preparedness of zone j is mathematically defined as

$$Z_j = \frac{1}{fd_{jp_t}} \sum_{i \in A} \frac{1}{O_i t_{ij}} \quad (1)$$

Where O_i represents the order of ambulance i in set A; the set of all available ambulances ordered in an ascending order based on t_{ij} ; the travel time, based on live traffic, from ambulance i to zone j and fd_{jp_t} is the forecasted demand for zone j at time period t and is calculated mathematically as

$$fd_{jp_t} = d_{jp_{t-1}} + \propto_t \left(d_{jp_{t-1}}\right) \quad (2)$$

Where $d_{jp_{t-1}}$ represents the actual demand of zone j for time period t-1 and $\propto_t$ represents a constant that varies from 0 to 1 representing expected traffic rates at time t.

A zone is said to be either covered or not covered. Therefore C_j, the coverage of zone j will equal to 1 if the closest available ambulance can reach it within a predetermined threshold and 0 otherwise. The complete measure that will be used is

$$D_j = Z_j + \frac{C_j}{t_{ij}} \quad (3)$$

Table 1. The Dispatch Algorithm

1.	Let j be the zone to which an ambulance needs to be dispatched, Let $X[v]$= an array of size V, the number of ambulance(s) that need to be dispatched and let PR= priority level (1,2,or 3) and Dmin=0
2.	Let A[i], an ordered array of available ambulances i that can reach j in less than T_{PR} minutes that is ordered based on t_{ij} ; the travel time from i to j *N.B vehicles assigned to less priority are considered*
3.	If PR= =1
4.	For v=1... V
5.	Set X[v]=A[v-1], i.e. dispatch ambulances in order 1... V
6.	If PR==2 or 3
7.	Check the ambulances in the list, beginning with the closest:
8.	For v=1.....V
9.	For i=1.....I
10.	Remove ambulance i from the array
11.	Recalculate the combined measure of preparedness and coverage Dj for all zones affected by this dispatch
12.	If min(Dj) > Dmin
13.	Dmin = min (Dj)
14.	X[v]= A[i]
15.	For v=1......V
16.	Dispatch A[v]
17.	Update preparedness and coverage levels

The algorithm starts by setting the priority level (1, 2, or 3), it then creates a list with length equal to V; the number of ambulances that needs to be dispatched. It also declares a variable Dmin that is used to compare preparedness, and sets it to zero. For each priority there is a time threshold that should be respected, such that the lower the priority the longer the time threshold. The travel times from all available ambulances to zone j, including ambulances on their way to lower priorities, are checked. If the priority is equal to 1 (high), a list of all ambulances that can reach zone j in less than T_1 minutes (incorporating traffic), the time threshold for priority 1, is declared. The list is ordered in an ascending order based on travel time between vehicle i and zone j. The first V vehicles are selected and dispatched. If the priority is equal to 2 or 3, the algorithm loops over each ambulance in the list, neglects its presence and calculates the forecasted demand for all zones j affected by the dispatch of this ambulance vehicle as illustrated in equation 2. The forecasted demand is then used to recalculate the preparedness of these zones as illustrated in equation 1. Finally, Dj, the combined measure of preparedness and coverage for these zones, is calculated as illustrated in equation 3. When the loop ends, the V ambulances whose dispatch has the minimal effect on Dj are selected and dispatched. New levels of preparedness and coverage are calculated. Relocation may be triggered if preparedness and coverage levels are less than a predetermined threshold.

4. PROTOTYPE IMPLEMENTATION AND TESTING

Android was selected for the implementation of the mobile and the EMS vehicle side applications because of its widespread usage. The two applications can be installed on any smart phone with android 4.4 or above. The authors built the Central EMS side application as an ASP.NET 4.5 web application in order to allow data exchange among the mobile application and EMS vehicle side applications. C# was used as its programming language. Microsoft SQL-Server 2008 was used for creating the database in which incidents data are kept. Language integrated query (LINQ) was used as the data exchange level between the Central EMS side application and the database. It was chosen because it parameterizes input helping reduce standard query language (SQL) injection attacks.

Preliminary tests were made on the citizen side application. Users believed reporting was very quick and could be done with minimal effort. Required data were captured and disseminated accurately. The testing details, the criteria used for evaluating the application and the results can be found in Abou El Safa and El Khayat (2016). The complete testing of the integrated solution will be done after finalizing a number of developments as discussed in the next section.

5. CONCLUSION AND FUTURE WORK

In this paper an integrated solution for improving EMS systems' response time was proposed. Contrary to contributed models in previous literature, this solution considered the time taken to report a case and the dispatching time as well as the travel time. A modified measure for preparedness and demand forecast was provided. A novel ambulance dispatch algorithm was introduced; the algorithm uses a combined measure of both preparedness and coverage rather than a single measure. The dispatch algorithm aims to lower the overall EMS system response time, while respecting a predefined threshold of response time for single cases. The contributed solution used, in its calculations, live traffic based travel times instead of simple travel times. The contributed integrated system comprises a mobile application for the use of citizens and corresponding applications for the EMS. The citizen side application allows them to make a faster, easier and more accurate incident reporting. The EMS side applications make use of the proposed algorithm in order to dispatch ambulance vehicles and relocate them efficiently. In addition, the applications work together to enable the EMS to make better call screening. The results of the preliminary tests of the mobile application showed that reporting through the application was quick and convenient. In addition, the reported incident address, the injured person's history of disease, and vital signs were accurate.

Data collected by the application will be used to help determine black spots and support decision makers in deciding where to locate new ambulance stations and which roads need maintenance. In sum, tackling the problem in this integrated way is very promising and is expected to lead to an observable improvement in response time.

Future research includes the development of a relocation algorithm that aims to improve the response time, the development of a call screening algorithm for automatic priority assignment as well as building a simulation model to test the proposed system and algorithms. The model will be tested against the performance of the authors' country EMS system. The full description of the model and the simulation results will be communicated in a future paper.

REFERENCES

Aboueljinane, L., Sahin, E. and Jemai, Z., 2013. A review on simulation models applied to emergency medical service operations. *Computers & Industrial Engineering*, 66(4), pp.734-750.

Abou El Safa, E.I and El Khayat, G.A., A mobile solution for fast and accurate medical emergency reporting. IEEE 1st International Conference on Connected Health: Applications, Systems and Engineering Technologies: CCH 2016: The First International Workshop on Cloud Connected Health. In-Press.

Andersson, T. and Värbrand, P., 2007. Decision support tools for ambulance dispatch and relocation. *Journal of the Operational Research Society*, 58(2), pp.195-201.

Andersson, T., Petersson, S. and Värbrand, P., 2004. Dynamic ambulance relocation for a higher preparedness. In *Proceedings of 35th Annual Meeting of Decision Sciences Institute,* Boston,

Bandara, D., 2012. Districting and dispatching policies to improve the efficiency of emergency medical service (EMS) systems. Ph.D. thesis. *Clemson University*.

Bélanger, V., Lanzarone, E., Ruiz, A. and Soriano, P., 2015. The Ambulance Relocation and Dispatching Problem. Technical Report. *Interuniversity Research Center on Enterprise Networks, Logistics and Transportation*, CIRRELT-2015-59, 17 pages.

Chanta, S., Mayorga, M.E. and McLay, L.A., 2014. Improving emergency service in rural areas: a bi-objective covering location model for EMS systems. *Annals of Operations Research*, 221(1), pp.133-159.

Cote, M.J. and Tucker, S.L., 2001. Four methodologies to improve healthcare demand forecasting. *Healthcare Financial Management*, 55(5), pp.54-54.

Finarelli Jr, H.J. and Johnson, T., 2004. Effective demand forecasting in 9 steps: shifts in demand for a hospital's services can occur unexpectedly. Demand forecasting can help you prepare for these shifts and avoid strategic missteps. *Healthcare Financial Management*, 58(11), pp.52-58.

Fleischman, R.J., Lundquist, M., Jui, J., Newgard, C.D. and Warden, C., 2013. Predicting ambulance time of arrival to the emergency department using global positioning system and google maps. *Prehospital Emergency Care*, 17(4), pp.458-465.

Gendreau, M., Laporte, G. and Semet, F., 2001. A dynamic model and parallel tabu search heuristic for real-time ambulance relocation. *Parallel computing*, 27(12), pp.1641-1653.

Gendreau, M., Laporte, G. and Semet, F., 2006. The maximal expected coverage relocation problem for emergency vehicles. *Journal of the Operational Research Society*, pp.22-28.

Haghani, A. and Yang, S., 2007. Real-time emergency response fleet deployment: Concepts, systems, simulation & case studies. In *Dynamic fleet management* (pp. 133-162). Springer US.

Iannoni, A.P., Morabito, R. and Saydam, C., 2009. An optimization approach for ambulance location and the districting of the response segments on highways. *European Journal of Operational Research*, 195(2), pp.528-542.

Ingolfsson, A., Budge, S. and Erkut, E., 2008. Optimal ambulance location with random delays and travel times. *Health Care Management Science*,11(3), pp.262-274.

Jagtenberg, C.J., Bhulai, S. and van der Mei, R.D., 2015. An efficient heuristic for real-time ambulance redeployment. *Operations Research for Health Care*, 4, pp.27-35.

Kergosien, Y., Lenté, C. and Billaut, J.C., 2008. A dynamic transportation problem of patients between care units. In *Proceedings of the 34th International Conference on Operational Research Applied to Health Services*.

Lee, S., 2011. The role of preparedness in ambulance dispatching. *Journal of the Operational Research Society*, 62(10), pp.1888-1897.

Mayorga, M.E., Bandara, D. and McLay, L.A., 2013. Districting and dispatching policies for emergency medical service systems to improve patient survival. *IIE Transactions on Healthcare Systems Engineering*, 3(1), pp.39-56.

Myers, C. and Green, T., 2004. Forecasting demand and capacity requirements. *Healthcare Financial Management*, 58(8), pp.34-38.

Schmid, V. and Doerner, K.F., 2010. Ambulance location and relocation problems with time-dependent travel times. *European journal of operational research*, 207(3), pp.1293-1303.

Schmid, V., 2012. Solving the dynamic ambulance relocation and dispatching problem using approximate dynamic programming. *European Journal of Operational Research*, 219(3), pp.611-621.

Thi Nguyen, N.H., 2015. Quantitative Analysis of Ambulance Location-allocation and Ambulance State Prediction.

Toro-Díaz, H., Mayorga, M.E., Chanta, S. and McLay, L.A., 2013. Joint location and dispatching decisions for emergency medical services.*Computers & Industrial Engineering*, 64(4), pp.917-928.

EXPLORING THE INTERPLAY OF DISCOURSE AND INFRASTRUCTURE IN E-HEALTH PROGRAMMES

Egil Øvrelid, Bendik Bygstad and Ole Hanseth
University of Oslo, Department of Informatics, Gaustadalléen 23 B, 0373 Oslo, Norway

ABSTRACT

Research has shown that large IT programmes in e-government and e-health are challenging not only in terms of project failures and high costs, but also that the public and sectorial discourses greatly influences the trajectories and outcomes of mega-programmes. However, few IS studies have investigated this phenomenon in much depth, and the aim of this contribution is to shed more light on the relationship of discourse and mega-programmes. We build loosely on Foucault's discourse concept and on Swanson' term of organising vision, but frame our investigation in information infrastructure theory.

Our empirical evidence is a 15-year study of the growth of the national e-health infrastructure in Norway, which we investigate at three levels, where we analyse the interplay of the eHealth discourse and the various programme initiatives. We identify and discuss a pattern where windows of opportunity open and narrow in waves of discourse and action. We also contribute to the information infrastructure research by highlighting the role of discourse in infrastructure evolution.

KEYWORDS

Discourse, discursive formations, eHealth, infrastructure

1. INTRODUCTION

Our interest in this paper is to understand how large IT initiatives are being influenced and formed by discourses, and – reversely - how the discourses are affected by the practical experiences from these programmes.

Public sector IT mega-projects with political and societal prestige are usually a response to perceived problems, such as poor services or high costs. There is often considerable risks related to this type of projects (Sauer & Willcocks 2007, Currie, 2014), both economical and organizational costs are high, and often these projects will attract considerable attention with negative publicity as a result.

Sauer & Willcocks (2007) describe the turbulent discourse of the failed NHS e-health programme. The main reasons are that these structures are (i) much larger than single organization systems, (ii) technically more heterogeneous and (iii) organizationally more complex because of many actors. Often, no single actor is in control, leading to long processes of power struggles, compromises and complex co-ordination. In addition, they are particularly exposed (as they should be) to public discourse, which Sauer and Willcocks analyse as three "greek choruses"; the public officials who defend the programme, the internal institutions and medical professions that are "sympathetic critics" and the media, consultants and academics that constitute the "professional critics".

Research on discourse is associated to Foucault's archaeological research (1972). The most cited contribution on the role of discourse in IS research is the *organizing vision* (Swanson & Ramiller 1997), which is about how technology is viewed in the light of what it can do for organizations, how discourses about technology is opening up further discourses about societal and organizational implications and possibilities enabled through technological change. Public mega-programs, however, are different from the private company context that Swanson and Ramiller researched, in the sense that the decisions are more complex, involving hundreds of organisations, several technologies, public discourse and political pressures. One way to frame these issues is the information infrastructure research (Hanseth and Lyytinen, 2010), which has successfully investigated the nature and dynamics of large interconnected systems. According to this theory information infrastructures are not designed and implemented, but grow organically and adaptive from an open installed base.

What we know less about is the role of discourse in the evolution of public information infrastructures such as e-health. What are the more specific dynamics of discourse and large IT initiatives? How is discourse translated into action? What is the trajectory from political suggestions, ideas and choices to implementation? Our research questions are:

- Which patterns of interaction between discourse and infrastructure events can we identify?
- How do managers deal with these trajectories and situations?

In his work on scientific programmes and paradigms Lakatos (1969) agued that "programmes" tend to continue in spite of experienced problems, as long as there are no clear alternatives present. This resonates well with documented management practices and the "garbage can" theory (Cohen, March & Olsen, 1972): when alternatives emerge the discourse will change. With this insight we analyse the interplay of discourse and reality of an e-health mega-programme, identify and analyse three strategy shifts. To develop our argument we use Foucault's term *discursive formation*, and we offer a simple, yet powerful model of discourse dynamics.

2. DISCOURSE AND REALITY

Discourse may be simply defined as "talk and texts as parts of social practice" (Potter 2004, p.105). In a practical sense discourse is a shared and fundamental everyday phenomenon. As humans we live our private and professional lives in rhythms of action and discourse; we discuss what to do, then we act, and afterwards we discuss the results of the action, and what to do next.

2.1 Discourses and Space of Possibilities

The French historian Michel Foucault's archaeological research is occupied with discourses that emerge outside the political establishment, sometimes in closed political or organizational forums. Foucault's point of departure is that alternative narratives, speculations and suggestions open up alternative discourses, what he calls "space of interplay" (Foucault 1972). Interplay regards the collaboration between actors that gives new ideas and use established practices to put something into play. Discourses about technological and organizational development are a response to particular problems (Swanson & Ramiller 1997, Ellingsen and Monteiro 2008), and its possible solution exists because there is a collection of participants agreeing that it exists (Swanson & Ramiller 1997). Using Foucault we may say that discursive formations are established through agreement between several stakeholders that a particular way of solving a problem is available. This also means that the different knowledge areas may organize and ally themselves with different "user groups and industry associations to achieve greater political and economic voice" (ibid, 464). When one solution does not fit all, there will be an "application gap", which makes new technology an important driver. The way from organizing vision to organizational complexity in first securing "local" requirements and then seek to obtain alignment, is part of the "continuous shaping and reshaping as the material processes of innovation adoption, implementation and diffusion play out over time" (ibid, 468). The application gap is a space of possibilities open for political or technological movements organized in discursive formations with alternative solutions to the dominant program.

Earlier work using discourse analyses within the IS field has concentrated on the use of narratives and buzzword in management practices (Swanson 2002, Monod et al 2002, Westrup 2002); or on the relation between discourse and practice during implementation of technology or technological routines in organizations (Rose & Kræmmergaard 2002, Oliver & Oliver 2002, Gidlund 2015, Ellingsen & Monteiro 2008). A third group looks at how discourse is used as an existential tool in managing the overwhelming amount of data and information in modern organizations. (Wastell 2002, Edenius 2002)

We use discursive formations as a central tool in understanding shifts in big programs, and we claim that there is a significant relation between the establishment of discursive formations and the opening and closure of an infrastructure. Mega-projects brings with it a coalition of knowledge areas, security issues, management issues, clinical issues etc. "Some parties seek to command the organizing vision, while others trade on its interpretive flexibility" (Swanson & Ramillier 1997, 464). They must be aligned in order to make technological acquisition and organizational improvement possible. This means that the different knowledge areas may organize and ally themselves with different "user groups and industry associations to achieve greater political and economic voice." (ibid)

2.2 Information Infrastructures and Discourse

The literature on Information infrastructures (Ciborra et al. 2000, Hanseth & Lyytinen 2010) describes IIs as open, heterogeneous, performative and emergent. The open and closedness of IIs are often taken as a point of departure, and we lack, within II theory, an understanding of how this openness is obtained and enclosure occurs. IIs literature often focuses on how inscription and translation is obtained (Monteiro 2000). The focus on the pre-existing discourses; the negotiations and arguments which precedes the infrastructure is virtually nonexistent. We believe that discourse matters in the development of IIs and particularly in how IIs opens and encloses through shifting discourses, and that the role of discourses both in establishing and in using spaces of interplay are not well understood in the literature on Information infrastructures.

Using discourse analysis in investigating the establishment of strategies and techniques in political, societal and technological debates, we see that discursive formations are established when a field of association is made available through a flexible understanding of the fields possibilities. They agree on a common object, a common vision. The "vision" is created by attracting a range of people, and the formations of people are again conditioning the vision. This does also explain the shifts in programmes and strategies.

Shifts occur when some understood shortcoming in the program, and when an alternative exists, or when the program fails and alternatives are established through the shift of discursive formations from one program to another. The concept of discursive formations, formations obtained through performative acts of discussions, negotiations and translation, is compatible with Information infrastructures. In addition it extends the understanding of information infrastructure implementation and evolution in that they focus on debates preceding the implementation of the actual infrastructures. In addition they add to the understanding of how infrastructures open and close through extensive discursive formations.

3. CASE AND METHOD

We chose a multilevel case study approach (Greenhalq et al., 2010) in order to investigate the interplay of policies, programs and projects. Norway is a Scandinavian country with 5.2 million inhabitants who enjoy a high standard of living and public health services. The sector is governed by the Ministry of Health and Care, while the Directorate of Health is an implementing agency and health advisory. Primary care is supplied by private GPs and municipal services. Our starting point was the national policies for e-health, which was a continuous public debate issue during these years, and also presented some high-level IT governance and architecture issues. In order to identify the interactions of discourse and materiality our approach was to conduct a systematic analysis of (i) the documented discourses, (ii) the documented events and (iii) the interactions between them.

3.1 Data Collection

We studied the governance and development of a national e-health infrastructure in Norway over a period of fifteen years (2000-2015), at three levels:

- *The national level*: we interviewed top executives and IT managers at the Ministry of Health and Directorate of Health, analysed plans and initiatives, and we analysed the topics of the national e-health conference from 2001-15.
- *Regional level*: we investigated the development of a regional information infrastructure in the South-Eastern Regional Health Authority from 2000 to 2015, through a sequence of programmes.
- *Project level:* we followed two large projects, the Portal project at Rikshospitalet in 2009 to 2011, and; the DIPS project at Oslo University Hospital 2013-2015. We also followed a smaller initiative, the Medicloud project in 2014-15.

3.2 Data Analysis

We conducted the following steps:

Table 1. Data analysis

Step	Description	Output
1	Establishing a chronology 2000-2015	Figure 1
2	Conducting a discourse analysis	Three discursive formations
3	Identifying patterns of interaction between discourse and infrastructure events	Table 2
4	Assessing patterns	Section 6

In step 3 we built on Lakatos' (1969) idea of problem solving: when experiencing a problem within a "programme" work tends to go on, unless there is a clear alternative presented. It is when the power of the discourse is strong enough, that is, when discursive formations are established and gained the sufficient power to present a possible solution to the dominant programme, things changes. We did systematic analyses of the content in the most important annual e-Health conference in Norway, over the past 12 years. We analysed the data chronologically and found three discursive formations. They are shown in the upper part of figure 1.

4. FINDINGS: INTERACTIONS

We first give an overview of the growth of a national and regional infrastructure the past 15 years, and the on-going discourse, in order to establish a chronological baseline. Then we analyse in more detail the interactions between the discourse and the events.

4.1 Growth of National and Regional e-Health Infrastructures 2001-15

Until 2002 all public hospitals were owned by the 19 Norwegian counties. In 2002, the State took over the ownership and organized the hospitals in five health corporations called Regional Health Authorities (RHA) given the names Health North, Mid, West, South and East respectively. In 2007 South and East were merged into the South East Regional Health Authority.

Before the reform in 2002, ICT strategies and decisions were attached to the individual hospital. The motivation behind the reform was a commonly felt need for better coordination and collaboration among the hospitals. And it was a broad consensus about the importance of exchange of information electronically for achieving this. So immediately after the reform, each region established a central organization representing the regional health authorities and an ICT unit as a part of this. In all regions it was decided to focus on standardizing ICT, especially the most important applications, so that all hospitals run the same EPR, PACS, Lab solution etc. All regions also established a governance model based on Gartner's so-called Y-model with organize the overall ICT activities and responsibilities into three roles: control and strategizing (RHF board and management), customer /ordering (hospital) and vendor (regional ICT vendor). All regions transferred most of the ITC staff in the hospitals into a regional organization. The strategy for standardizing applications was based on tendering processes where the RHAs sign a "framework contract" with one vendor for each application running for 5-10years. The hospitals decide when they need a new solution. Then they have to buy the solution from the vendor with which the RHA has signed the "framework contract." In 2003 an organization called "*National* ICT" was established on the Ministry of Health's initiative in order to achieve better coordination of ICT activities and solutions among the RHAs. Specifying a common ICT architecture for all regions based on the SOA model and standardising archetypes for core data elements have been among the highest prioritized activities.

In spite of these, and other, initiatives, the overall ICT portfolio within the hospital sector has remained fragmented and information exchange between applications and organizations problematic. Around 2005 the fragmented eHealth solutions were brought to the attention of national media, through histories of poor patient treatment because of non-integrated IT solutions, and ridiculed the practice of transporting x-ray images by taxi between hospitals. The political pressure on the sector increased, and the answer from the top health executives to the challenge was to establish a new governance regime. One reason for this disappointing situation is the fact that while some applications where standardized within the regions, new ones were continuously introduced to support various new specialist practices, as a part of new digital instruments, etc. And while an increasing number of hospitals were running the same software packages, these packages were configured differently among the hospital making information exchange almost as difficult as if they were using applications from different vendors.

4.2 Patterns of Interaction

The timeline of discourses and growth of the infrastructure is illustrated in Figure 1. It shows, broadly, three phases of discourse and three phases of programmes. The discourses were held at a national level, while we analyse the programmes that were conducted within the largest (South-East) health region. We identified three large *shifts* in the 15-year period. A shift is here understood as a significant change in both discourse and actual programs. The shifts are illustrated in the timeline Table 2.

Historically, each hospital, and even each department, had acquired or developed their own IT systems, reflecting functional specialisation: in addition to shared systems, such as patient record, chart and medication, lab, and imaging, most units had acquired their own systems for births, cancer, and diabetes and so on. In the period 2001-2007 most hospitals in the country conducted their own projects, many of them addressing the increasing need for integration of various "silo" solutions. The discourse during this period included several topics, but after the establishing of the five health regions in 2001, there was an on-going debate on *co-operation* at many levels. Inside each hospital it was increasingly acknowledged that clinicians needed access to medical information produced by other units; likewise, inside each region there was a need to exchange data, particularly between hospitals and primary care GPs.

Finally, at a national level, there was a need to exchange patient information between regions. How should this be done, given the variety of solutions? The answer, according to the Directorate of Health, was to exchange standardized messages, and a long-lasting campaign, the *Message Boost Initiative*, was run for several years.

Progress however was slow, and both the internal and national discourse started to show signs of impatience. We interviewed a key player in policy development, and asked for his assessment of the current situation and the road ahead. He commented:

> "The main problem is the fragmentation of solutions, which has a historical explanation. Each hospital, each clinic – and even each clinician – has had the freedom to choose any solution that was available, during that past 30 years. These choices have often been made arbitrarily, dependent on which vendors were knocking on the door, or other local conditions. The result is hundreds of different solutions, which cannot exchange data, because of the lack of standards, and cannot communicate, because of the lack of integration. Today, this is a hindrance for patient oriented care, and for evidence based medicine. It is also expensive. There is only one solution, which is an overall consolidation to shared systems, and a standardization of data and processes. This requires the courage to establish top-down governance, an integrated architecture and well-financed programmes to implement the strategy."

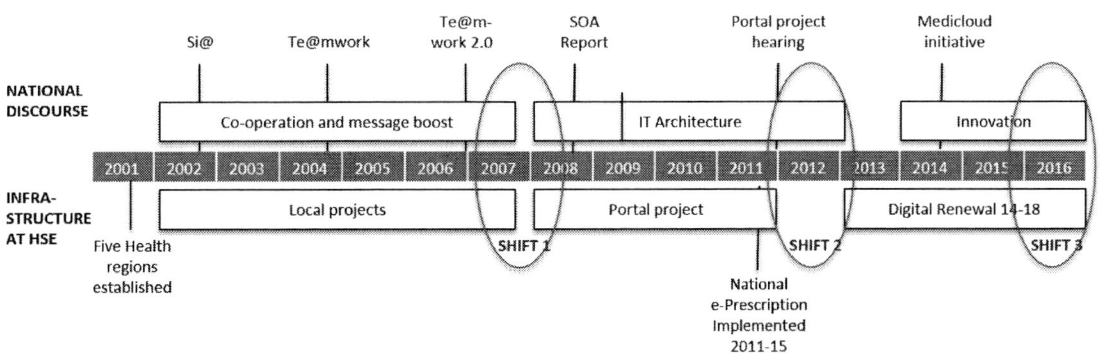

Figure 1. Timeline

4.2.1 Shift 1: From Message-Based Cooperation to SOA

The shift was triggered by the combination of slow progress of the many local projects and the arrival of an alternative, namely service oriented architecture (SOA). SOA had been introduced in the software engineering community in the early 2000s (Erl 2009). In 2004 the national e-health group initiated a project for a national IT architecture, resulting in a large report in 2008. The report recommended a completely new SOA approach for e-health solutions. The aims were quite ambitious, and emphasized patient-centred care, a process perspective (instead of IT silos) and role-based services. A national architecture based on a shared information model and service bus technology was recommended.

In line with these principles a possible pilot was available. At the most prestigious national hospital (Rikshospitalet) the IT department had developed a portal solution, based on SOA. The portal solution was built on the idea of a new layer over the silo applications that gave clinicians role-based access to various services. The solution required some re-engineering of the applications, from GUIs to services, and also dealing with a complex set of security and privacy issues.

A tender for a full solution was won by a New Zealand software company in 2010, and expectations were high, both in the local and national e-health communities. However, the company was not very experienced in e-health solutions, and large problems emerged during implementation.

4.2.2 Shift 2: From SOA to Integration

In 2011 the project was stopped after spending 160 million NOK. The event became a national scandal in the media, and an inquiry was conducted at the Norwegian Parliament. One result of the negative press was that the term *portal* was scandalized. The e-health community embarked on a discourse on *best-of breed* (choosing different applications, and integrating them later) of *suite strategy* (choosing one, integrated solution, such as EPIC). At HSE the decision was taken rather quickly in 2012; because of contractual and procedural issues, "hybrid architecture" was chosen. It consisted of choosing the EPR solution most widely used in Norway (DIPS) as the central application, and to integrate it with other systems with a service bus middleware.

A large program, *Digital Renewal* with a budget of around 1 bn. Euro, was initiated to standardise and implement this architecture for the 70 hospitals in the region. A separate unit, the Integration Factory (specializing on Microsoft BizTalk), was established to program the numerous physical integrations between the central EPS and the other clinical and administrative systems. The key project, implementing DIPS with the needed integrations, was successfully run in 2013-14. The other health regions, with one exception, ran similar projects.

The discourse on e-health architecture continued in the sector, but on a more sober tone; the SOA ideal models were largely put to rest, and the discussion centred on *suite* or *best-of-breed* solutions, actualised by the Copenhagen and Helsinki health authorities' decisions to acquire EPIC. However, a new discourse was entering the field in 2014.

4.2.3 Shift 3: From integration to Innovation

During 2014 it became clear although that the Digital Renewal approach was successful, it was quite expensive, and it provided few new services, since most of the resources went to integration. Many clinicians worried that all other IT initiatives were stopped for the lack of money, which was considered very unfortunate, because of a stream of innovations in the medical field, based on lightweight IT such a sensors, tablets and mobile technology. Also, upstart companies complained that the heavyweight IT communities blocked access to innovations (Bygstad, 2015). This echoed an international discourse in e-health, where a *platform* strategy (Baldwin and Woodard 2008), with ecologies with large vendors and third party innovators, was becoming popular. For instance, Epic and Apple signed a co-operation agreement in 2014. As a response, HSE established a new unit, Medicloud, with a mandate to explore possible solutions to connect heavyweight and lightweight IT. Medicloud, which was part of the IT Service Centre, quickly established relationships with various clinicians and upstart IT companies, and in 2015 a number of pilot projects were initiated. At the annual e-health conference the shift in discourse was evident. Medicloud held a separate event to accommodate lightweight innovations, and the large EPR vendors assured the public that they were quite open to offer APIs to app providers.

5. ANALYSIS: SPACE OF POSSIBILITIES

Analysing our data we identified a repeating pattern of interaction of discourse and growing infrastructure. We analyse this pattern in terms of space of possibilities. The dynamics is illustrated in Table 2 below.

Table 2. Cycle of discourse formation shift

Phase	Step	Description	Effect on space of possibility
De-stabilising	1	A problem is experienced in the information infrastructure	Space opens
	2	An alternative is presented and discussed	Space is open
Stabilising	3	The discourse converges on solution	Space closes
	4	Decision is taken on new approach	Space is closed

The cycle includes two phases and four steps. The first phase describes how established discursive formations is destabilised through two steps; first by a perceived problem at the infrastructure level, and then the discourse on an available solution to the problem. For instance, in the described Shift-1, there was growing dissatisfaction with the “message approach” to connect various IT silo solutions. However, it was the discourse on SOA that really opened the space of possibility. When the space of possibility is open, the discourse often includes many actors (managers, academics, vendors and, in some cases, the general public), and is usually a competitive arena for new initiatives. For example, in the described Shift-2, the dominant EPR vendor worked hard, and succeeded, to solve the portal scandal by a quick decision to choose the DIPS solution. Potential conflicts were mitigated by consensus outside the contractual frame agreements, after 2011-2012 statements about “portal” lacks a common field of association from where discursive formations establishes themselves.

Closing the space of possibility is a process of stabilisation; first the convergence of the discourse through alliances of key actors, and arguments of the benefits of the new approach, followed by formal decisions. In some cases this is done quickly, as in Shift-2, while the process of Shift-3 is likely to take much longer time. However, after the decisions are taken, and the space of possibility is closed, the discourse becomes much more muted, and actors on the losing end will need time to reposition themselves.

6. DISCUSSION

In this section we return to our research questions. We add to the existing literature on the role of discourse in IS in our two contributions:

- We propose a model for understanding the dynamics of discourse and infrastructure
- We extend information infrastructure theory by including discourse as a key factor in infrastructure evolution

Our case illustrates that the e-health field is characterised by shifts of discursive formations. The dynamics of these shifts were theorised in our four-step model, which explains how the discourse and programme events influence the space of possibility. This extends the earlier research on IS discourse dynamics which primarily focus on the formation, but does not deal with the cyclical shifts. Our model, building loosely on Lakatos’ (1969) work on programme shifts, shows that there is a repeating pattern of interaction in these shifts.

We also shed light on how managers relate to this dynamic. When problems occur in a current strategy or project the managerial response will usually be to try to solve them, not to question the whole approach. However, when the discourse on the issue moves from problems (step 1) to alternatives (step 2) some managers may engage in the new discourse, and contribute to open the space of possibility. In the convergent step (3) managers may try to reposition themselves in order to play a role in the new strategy, and in the final decision step (4) managers will aim at closing the discourse, and focusing on “getting to work”.

We also contribute to information infrastructure theory (Hanseth & Lyytinen, 2010), which has not dealt much with discourses, with the exception of Ellingsen & Monteiro (2008). This stream of research has emphasised that infrastructures evolve as a growth of an open installed base, adapting to changes in the environment. Our position is that discourse formations should be included in the installed base, and often play an important role in this evolution. In particular, as our case vividly illustrates, discourse plays a key role in strategy shifts.

While Ellingsen & Monteiro highlighted the flexible character of organising visions (allowing actors with different interests to converge), our findings document a more detailed trajectory of the cyclical nature of infrastructure discourses.

REFERENCES

Baldwin C, Woodward C, 2008, The Architecture of Platforms: A Unified View *Working Paper 09-034*, Harvard Business School

Bygstad B. 2015, The Coming of Lightweight IT *Proceedings of European Conference of Information Systems* (ECIS), Münster.

Ciborra, C. et al 2000, From control to drift – The Dynamics of Corporate Information Infrastructures. Oxford: Oxford University Press

Cohen M., March J.G., Olsen J.P. 1972, A Garbage Can Model of Organizational Choice *Administrative Science Quarterly* 17(1), 1-25

Currie W. 2014, Translating Health IT Policy into Practice in the UK National Health Service *Scandinavian Journal of Information Systems*, Vol. 26, Issue 2, p. 1-24

Edenius M. 2002, Discourse on E-Mail in Use IFIP TC8/WG8.2 Working Conference on Global and Organizational Discourse about Information Technology December 12-14, 2002, Barcelona, Spain

Ellingsen G., Monteiro E. 2008, The organizing vision of integrated health information systems *Health Informatics Journal*, September 2008, vol 14, No 3, 223-236

Erl T. 2009, Service-Oriented Architecture – Concepts, Technology and Design Prentice Hall, Boston

Foucault M. 1972, *The archaeology of Knowledge* Routledge Classics London and New York

Gidlund K. 2015, Three Governments Living Happily Ever After: Discursive Tensions and Conflicting Epistemic Cultures in the Development of Public E-Services *International Journal of Electronic Government Research*, 11 (3), 43-56, July-September 2015

Greenhalgh T et al 2010, The Devil's in the Detail: Final report of the independent evaluation of the Summary Care Record and HealthSpace programmes. London: University College London; 2010.

Hanseth, O., Lyytinen, K. 2010, Design theory for dynamic complexity in information infrastructures: the case of building internet. *Journal of Information Technology*, 25(1), pp.1-19.

Lakatos, I. 1969, Criticism and the Methodology of Scientific Research Programmes *Proceedings of the Aristotelian Society*, New Series, Vol. 69 (1968 - 1969), pp. 149-186.

Monod E, Truex D, Baskerville R, 2002, The Discourse of a Large Organizational Transformation: The reengineering of IBM, 1989-1994" *IFIP TC8/WG8.2 Working Conference on Global and Organizational Discourse about Information Technology* December 12-14, 2002, Barcelona, Spain

Monteiro E. 2000, Actor-Network Theory and Information Infrastructures in Ciborra et al (2000) *From Control to Drift – The Dynamics of Corporate Information Infrastructures*, Oxford: Oxford University Press pp. 71-87

Oliver D., Oliver L. 2002, ERP Adoption: Selling the System" IFIP TC8/WG8.2 Working Conference on Global and Organizational Discourse about Information Technology December 12-14, 2002, Barcelona, Spain

Potter, J. 2005, Representing Reality – Discourse, Rhetoric and Social Construction. London: SAGE.

Rose J, Kræmmergaard P. 2002, Dominant Technological Discourses in Action: Paradigmatic Shift in Sense Making in the Implementation of an ERP System *IFIP TC8/WG8.2 Working Conference on Global and Organizational Discourse about Information Technology* December 12-14, 2002, Barcelona, Spain

Sauer, C. and Willcocks, L.P., 2007, Unreasonable expectations – NHS IT, Greek choruses and the games Institutions play around mega-programmes". *Journal of Information Technology*, 2007. 22: p. 195-201.

Swanson E. B. 2002, Talking the IS Innovation Walk IFIP TC8/WG8.2 Working Conference on Global and Organizational Discourse about Information Technology December 12-14, 2002, Barcelona, Spain

Swanson E., Ramiller N. C. 1997, The organizing vision in Information systems innovation *Organization Science*/ Vol 8, no 5, pp. 458-474

Wastell D. G 2002, Organizational Discourse as a Social Defense: Taming the Tiger of Electronic Government: *IFIP TC8/WG8.2 Working Conference on Global and Organizational Discourse about Information Technology* December 12-14, 2002, Barcelona, Spain

Westrup C. 2002, Discourse, Management Fashions, and ERP Systems *IFIP TC8/WG8.2 Working Conference on Global and Organizational Discourse about Information Technology* December 12-14, 2002, Barcelona, Spain.

THE RISE OF PATIENT 3.0: THE IMPACT OF SOCIAL MEDIA

Dr. Jane Clemensen, PhD[1], Dr. Dorthe Boe Danbjørg, PhD[2], Mette Damkjær Syse[3] and Dr. Ian Robert Coxon PhD[4]

[1]*Associate Professor and Head of Research,*

[2]*Assistant Professor,*

Centre of Innovative Medical Technology, University of Southern Denmark/Odense University Hospital

[3]*Head of Co-Lab, Odense, Centre of Innovative Medical Technology, Odense University Hospital*

[4]*Associate Professor of Experience-based Designing, Institute of Technology and Innovation, University of Southern Denmark*

ABSTRACT

Social media, particularly Facebook (FB), is a new way for patients to find peer patients. By investigating communications within a patient generated FB group over a one year period we looked for evidence of the existence of *Patient 3.0*[1]. The indicators for these kinds of activities were where patients gained a higher level of reflection through sharing and combining knowledge, and were able to use the knowledge gained for taking responsibility and action towards their own health situation[2].

The aim of the research described in this paper was to investigate the communication between patients in a Facebook group looking for signs of *patient 3.0* in the context of *health 3.0* (Gagnon, 2012). Qualitative content analysis was conducted on transcripts of postings on the Facebook group site from the period January 2010 to January 2011. These were categorized into themes based on the intentions indicated in patient updates and comments on the group site. Additionally, a focus group interview consisting of 10 participants was conducted, transcribed and analyzed.

Patients in the FB group made the most of each other's previous experiences by sharing their knowledge. This sharing empowered patients to address issues they had with both their health care professionals and the health system in general. The material indicates that this group of patients exhibit signs of being what has been described as "Patient 3.0" (Ibid). Their experiences also showcase collisions with the health care system and their expectations of it. Patients did not feel that they were being dealt with from a holistic point of view, and that the FB group was able to compensate for their unmet needs.

Being an 'expert', will in the future, mean a lot more than knowing about the disease and treating symptoms. It will require a different approach containing the ability to work with health issues from a holistic point of view. Therefore the roles between patients and the health care providers will change. How these changes will come forward we do not know, but this paper foreshadows what might be expected.

KEYWORDS

Health 3.0; Patient 3.0; Facebook; social media; hip dysplasia, patient empowerment.

ABBREVIATIONS

CDH Congenital Dislocation of the Hip
DDH Developmental Dysplasia of the Hip
FB: Facebook
GP: General Practitioner
HD: Hip dysplasia

---

[1] We take Health 3.0 and by default Patient 3.0, to refer to a more holistic approach to health issues where the body and mind are understood and treated, as deeply connected. In health 3.0 we consider mental wellbeing as important to improvement in physical well-being (Steele, 2014).

[2] This references Winterton (2006) second and third order learning which takes the patients evolution beyond just the level of sharing for social purposes into a form of 'informed' action.

1. INTRODUCTION

First there was health or health 1.0, which basically focused on the body and how to get it back in function, and in more general terms productivity was the main focus. Then Health 2.0 came along, in which the physical focus grew. Patients added technologies into their lives that helped them to monitor, quantify and report health regarded issues. Wristbands, back bands, sleep monitors, calorie counters became a part of daily life in an attempt to quantify and alter behavior patterns (Steele, 2014).

Health 3.0 refers to a more holistic approach where the body and mind are understood and treated, as deeply connected (Gagnon, 2012). In Health 3.0, mental wellbeing is as important as physical well-being. Reducing stress and seeking a life of balance is core to the next wave of health care. Health 3.0 is expected to foster the creation and maintenance of supportive virtual communities within which individuals can help one another understand, cope with, and manage common health-related issues (Nash, 2008). Health 3.0 is aimed at promoting a better quality of life, and to enhance relations between people with similar health issues.

The Internet began as a place for seeking information and has become a place where people interact, share information, gain social support and exchange advice (Antheunis, et al. 2013). Along with the development of these new possibilities, the role of patients in terms of their own care is changing rapidly. Their new, more active role has been pushed along by a healthcare sector that, through new technologies, is able to treat and diagnose more patients. Accelerated courses of treatment are increasing within all specialties (Moffatt et al, 2014) At the same time, due to broader economic pressures brought on by changing demographics, there is a general need to minimize the expense of treating individual patients (Weiner, & Schwartz, 2015). All of these conditions combine within the dawning of Health 3.0 and showcase the value of using a patient's own resources as a way of facing the challenges described above. In parallel to these developments, many patients' usage of the internet, as information seeking platform, and as a broader social media seems to serve three distinctly different purposes: The first is to get facts about their disease which will assuage their need for information. The second is to create communities where they are able to share experiences about their everyday life with an illness or disease, in order to be a more active and empowered participant in the process of their treatment. And thirdly, patients try to help each other to translate biomedical information into practical and useful knowledge (Pols, 2013; Dhillion, et al. 2011; van Uden-Kraan, et al. 2009). People meeting these needs in this way can be described as Patient 3.0, namely an empowered patient who is recreating the health care system in a more empowering way once they become a part of it. In this case empowerment may be referred to as enabling individuals to take control of their own health, wellbeing and disease management, and participate in decisions affecting their health and care (Faber, 2015)

In this paper we will refer to Patient 3.0, as those empowered patients who consider themselves to be equal partners with the health care professionals they are dealing with.

The following three points further characterize what we refer to as 'Patient 3.0'.

1. Gaining a higher level of reflection on their life condition through sharing knowledge using social media.

2. Combining information provided from the health care system with knowledge shared, or experienced by the patients themselves and using this new information to make their own decisions as they negotiate with the health care system.

3. Feel a natural sense of their own responsibility and therefore take concrete actions (ownership) regarding their individual health issues and broader needs for mental as well as physical well-being (Gagnon, 2012).

This aspect of 'patient responsibility' or ownership suggests a paradigm shift in the healthcare sector Gagnon, 2012). There is a general movement towards patients taking more responsibility in the treatment of their own conditions thus shifting a degree of responsibility from the healthcare sector to the patient. Patients have in many instances taken on this added responsibility enthusiastically and are finding new ways to source solutions to concrete problems (Aitken, et al. 2014). This can readily be seen in the use of social media and supportive technologies, which provide patients with a new platform for exchanging experiences and knowledge (Ibid). The use of social networking sites has grown from 8 % of all adults online in 2005, to close to 80% in 2014 (Ibid).

It has been argued that reasons for the growing success in patient use of social media for health matters correlate to the following value adding benefits, (a) Usefulness; (b) the enjoyment of sharing links, photos, films etc. and (c) the possibility of extending their own network through friends (Kuan, 2011). These fast growing patient networks have both negative as well as positive consequences. There is an associated risk that inaccurate or problematic misinformation can be spread rapidly throughout a poorly informed or otherwise unaware group. On the other hand, a large number of people can also have access to reliable, up-to-date, and understandable information. This has the added advantage of informing and empowering large groups who can influence political agendas within which it would otherwise be difficult to effect change. (Hawn, 2009). Facebook has been reported as the fourth most popular source of health information in the U.K. (Aitken, et al. 2014).

In Denmark (5.5 million Inhabitants) there are more than 3.5 million users of Facebook (Sociale Medier, 2016). Of course, FB is used for many different reasons, the most common being friends sharing updates about their everyday life but this can extend to groups with similar interests who want to share knowledge and information with friends from within and outside the group. Facebook is a 24/7 solution, and can be reached from all internet platforms. Which means, it is present everywhere and at all times; making information and relevant groups instantly and endlessly accessible (Klastrup, 2013). In terms of Health 3.0, Facebook has evolved into a readily available avenue for patients to meet with peers, as increasingly it is assumed, that most people already have an account. Local and international groups with an interest in sharing information related to illnesses such as diabetes, breast cancer and chronically obstructive pulmonary disease have emerged and are now well established. Many studies have identified that patients in these information sharing groups experience an increased feeling of empowerment (Wentzer, 2013; Zhang, et al. 2013; Bender, 2011; Greene, 2011; Madan, 2011; Broom, 2005) Based on this rising interest in Health 3.0; in this paper we report the outcomes of our investigation into the impacts on health and wellbeing arising from communications between people exhibiting 'Patient 3.0' characteristics in a Danish-based Facebook group.

2. METHODS

A Danish Facebook group called *We with Ganz* was started in 2009 by one patient, inspired by a leading surgeon in the field. At first the group was open to the public, meaning that everyone on Facebook was able to read what the group members had written. As the group grew larger, some members felt uncomfortable writing very personal things in the public forum; hence a 'private' group was created. Subsequently, as participation increased, more private groups arose. Some of these sub-groups met privately. The FB group included 109 members, of which 50 members were considered active in the sense, that they posted updates and/or comments on the wall. Qualitative content analysis was conducted on the transcripts of patient comments posted on the chosen Facebook group from the period January 2010 to January 2011. This was categorized into themes drawn from the participants implied intentions within their updates and comments on the general group website. Furthermore a focus group was held with 10 patients from the same private access group page in August 2013. The focus group lasted 140 minutes, was recorded and fully transcribed providing approximately 100 pages of text cleaned to exclude photos and any personal identifiers. It is important to remember that the FB group was in general used for sharing information, exchanging experiences, finding emotional support, recognition and understanding.

The research group gained access to this group, as one of the authors had earlier experienced a Ganz condition and therefore had knowledge of the group from previously being an active member.

One of the authors applied for membership, and told the group in a post, about the interest in this kind of community, the subjects discussed and questions asked. The author did not take part in any of the discussions; the group members were informed about the purpose of the research as well as being informed that all material collected would be treated as confidential. The authors acknowledge that this intervention could have influenced subsequent posts and discussions because group members knew that there was a researcher observing their interactions. No significant change in such behaviour was observed. Data was anonymised and informed consent was gained for all member quotations used.

The study was submitted to the Danish Scientific Ethics Committee. This committee determined that additional approval from an internal ethics committee was unnecessary according to the national legislation in Denmark. The Danish Data Protection Agency subsequently registered and approved the study.

3. DEFINITIONS

Hip Dysplasia: This happens when the bones of the hip joint are not aligned correctly. It affects thousands of children and adults each year and is known by many different names: Developmental Dysplasia of the Hip (DDH), Hip Dislocation, Congenital Dislocation of the Hip (CDH) and also as Loose Hips. Hip Dysplasia prevents the hip joint from functioning properly and the joint wears out much faster than normal. It is also a "silent" condition that means pain is not normally felt until much later stages, making it harder to detect. (For more on Hip Dysplasia see http://www.hipdysplasia.org/#sthash.i0z4IIDL.dpuf)

3.1 Incidence in the Danish Population

Prevalence

- App. 1 out of 1000 newborn children have a dislocated hip and app. 10 out of 1000 children might have subluxation
- 1 % of all children are treated for the condition in Scandinavian countries

Sex

- Relation girls/boys is 6/1

Location of Hip

- Joint: 20 % Left Hip Joint: 60 %
- Right Hip Joint: 20 %
- Both

(Source: sundhed.dk)

3.2 Nature of a Ganz Operation

Ganz osteotomy is a surgical treatment for acetabular dysplasia that preserves and enhances the patient's own hip joint rather than replacing it with an artificial part. The goal is to alleviate the patient's pain, restore function, and maximize the functional life of their dysplastic hip. A Ganz operation used to be considered an intrusive operation requiring cuts through tendons and muscles. It is now however, conducted using a minimally invasive technique, and performed through a seven centimeter cut in the skin, without cutting through tendons and muscles. (Source: http://www.soballe.com)

3.3 Data Analysis

The data was analyzed using a method of qualitative content analysis after the style of Lindseth and Norberg (2004). Firstly, we captured a 'naïve' overall impression of the data and formulated a preliminary set of data groupings. Secondly, the data was sub-grouped into themes, relevant to the focus of the study i.e. signs of patient 3.0. Thirdly, the most meaningful meta-themes were condensed and axial coding was applied (Straus and Corbin, 1990) to form sub-themes. Finally, these sub-themes were synthesized inductively, involving a shift upwards from condensation to descriptions (stories) in the style of Van Manen (2006) and Bate (2004). In order to improve validity, two different researchers were involved in the analysis process and collaborated on meaning development. Findings were subsequently discussed in relation to established theoretical procedures within the field of qualitative analysis (Lindseth and Norberg, 2004)

3.4 Results

The analysis of FB comments within the private group described earlier resulted in a set of key categories (see below). The data shows that about 70% of the active members posted at least five updates or comments, and the updates generally prompted responsive comments from other members of the group. In the following we refer to users of the FB group as *patients* and the person who starts a communication thread is referred to as an *updater*.

The key categories that emerged from the data analysis were:

- *Achieving a higher level of reflection when sharing knowledge using social media*
- *Combining knowledge provided from the health care system and from peers*
- *Feeling a natural sense of responsibility leading to actions.*

3.4.1 Achieving a Higher Level of Reflection When Sharing Knowledge Using Social Media

In the following sections we present examples that illustrate how information was shared:

An updater: "How long did it take for your wound to heal and does it look nice"?

Comment: "Was operated three months ago and my scar is quite nice, still red but rather thin and can be covered by my pants, so I don't see it. Remember to massage it, it improves the healing. I have used a scar sucker which I have used very often and it is super-efficient - you can see it in the picture I have added".

Comment: "I'm going to Turkey next week – to cover my scar is impossible in a bikini– therefore I will use sun factor 50.

This update was followed by more discussion and advice about scar care, clothes to wear, protection from sun, and how to get life back on track and still look good. The final comment from the updater was, *"Looking forward to seeing my scar nice and healed".* This communication shows how the patients add to each other's advice or comments, and how the *updater* finally draws some kind of conclusion to bring into her actual situation. Furthermore the other patient can choose to be either active or passive, making use of the whole communication or maybe just a selective part of it.

Updater: Were any of you informed about potential consequences of a Ganz operation? Seven months after the operation I have different problems such as pain, problems with acetabular labrum, longer leg length and a changed position of my foot. With a Ganz operation your muscles, bones and joint change position...it sounded so good with the 97% chance of getting better from the Ganz, but I'm still waiting".

Comment: "remember that you can ask your municipality for help - for example a car with hand- and speeder operated breaks and special designed shoes are some of the things I got".

Comment: "In my case I belong to the 97% - I have no pain in my hips and knees anymore".

This update was followed by a discussion about consequences that they had been informed about and other consequences that were unexpected but learned about through experience. This communication provides new information to patients with similar problems, and the communication shows that they use the new information to act or to reflect on their own situation, and how things affect both of them. The advice is often very concrete referring to specific things they can do for example, how to get their municipality to help; who is the best physiotherapist, and so on.

Sharing knowledge was not altogether without problems in the public version of the Facebook group. Some members of the group determined that it was essential for the group to move into a private space and as a result they reported experiencing communications that became more free and intimate. One stated: *"If I have some kind of sexual problems, or problems in my marriage connected to the HD, then I don't want all my FB friends to see it".*

3.4.2 Combining Knowledge Provided From the Health Care System and Peers

Updater: Going to the surgeon next week to plan for an operation regarding acetabular labrum. Suddenly I'm in doubt, if I should do it? Some of you have opinions/experiences about it. If I should or should not go for it? I'm in two minds. I do a lot of sport, but increasingly I have pain doing it".

Comment: "I can't tell you what to do, but I did it, and now I have run seven half marathons. It is your decision but when you make it – believe in it"

Comment: If you don't get the operation you have a high risk of needing a new hip .....by doing sports with your problem you might worsen it day by day. The same was an issue for me – do it! You can read a lot more about it on this page (followed by a link). Good luck!"

Comment: (updater). Never thought that it could get worse, I can see the consequences, and it scares me"

Comment: Remember to ask the surgeon for consequences next week, it is a big decision". Did you ask the physiotherapist?

Comment: (updater): "I will surely put it on the agenda". My physiotherapist is totally blank and has no sensible suggestions".

Comment (updater, after the consultation with the doctor): "Operation is planned and I could decide the time for it myself – what service! The surgeon just bent my legs and I screamed in pain, and then he asked, "When do you want the operation? Many thanks for all your help. This is a fantastic forum"

The example above illustrates that the patient has doubts regarding the operation. It shows that she has more options, and that the choice is her own, but that she still hasn't received adequate guidance that helps her to make a decision. The impact from the FB communication was that she decided to "put it on the agenda" when she met with the surgeon. The input from the other patients provided her with perspectives she did not know of and they shed a new light on an old problem. The patient became more competent and better at navigating between both professionals and peer-patients.

This situation was again supported in the focus group. One patient said about her experience on the FB forum, *"It is very easy, you send out a question, get a meaningful answer, and you bring the answer to the doctor, who accepts it together with you – it is like taking a detour to find your own diagnosis".*

One of the patients was operated on 10 years ago, and commented on her experience comparing it to where she was now with the support of FB. "*Before I felt completely lost, as if I were alone in the world. The doctor and the physiotherapist dealt with the technical stuff, and you just had to say "okay". Another focus group patient added, "I agree, what is going on? The doctors can't answer the questions we have because they haven't tried this damn operation themselves. Here you get another kind of help than can be offered by the doctors – they are superficial".*

3.4.3 Feeling a Natural Sense of Responsibility Leading to Actions

Sport is a big issue for many of the patients. They worry about what level they will retain after their operation(s). For them, sport is equivalent to well-being, and for many it becomes an important goal. Others have given up sport due to their hip problems and must find well-being in other activities. The following example illustrates how patients take responsibility for regaining the feeling of well-being by acting in concrete ways.

Updater: "I have been an active handball player for many years, but had to stop due to pain – with all kinds of physical activity, even sex! But, after my operation everything seems brighter - what are your experiences – have any of you returned to sport etc.?"

Comment: "I have done a half marathon since my Ganz".

Comment: I have started riding again – my horse has always been my break in a stressful daily life, so it means the world to me to be able to pick it up again"

Comment: Start slowly or find a less challenging sport for your body. I went from handball and football to theatre – it really bothers me!"

Comment: "Right now I exercise with my physiotherapist, so I can avoid another operation. The surgeon told me two years ago that I would be pain free – so not everything he says is true!"

The communications above indicate that for the involved patient, well-being is important and they consider themselves responsible for re-gaining some kind of activity in their lives that contributes to their wellbeing even if it is not what it once was.

An updater:" I'm going to have an arthroscopy, but am waiting for the insurance to find out if they will pay for it. I would like to go to a private hospital, but the insurance is not sure whether they want to pay more for an operation on this condition. I am so irritated about just getting to know this now. If I had known this before I would have worked out how to get an operation at a public hospital – but the waiting lists are so long"

Comment: A Ganz and an acetabular labrum operation are NOT the same condition - you can tell this to the insurance people. About 20% of Ganz patients will need an acetabular labrum operation afterwards. Tell them this and hope they will listen – have a nice day!

Comment: The insurance has to pay – they have to finish the job – they did it for me, good luck!

It can be seen in the exchanges above how patients add to both information and reassurance of the *updater* – so she feels more supported in dealing with the insurance company. Also, the others ask important questions, so that the *updater* is prompted to reflect in a more nuanced way regarding her diagnosis and potential treatment.

During the focus group one patient said: *"at the hospital you get neglected – you have to fit in to the statistics - they don't consider the person behind, it is a craft, it is economy, it is politics! And you get fixed! And, if you don't fit in, then where do you go? To this group, and it is super and relevant".*

These FB patients see the value of connecting. They can discuss how they have managed their illness with each other since the system seems to care only for the 'technical' part of their treatment. As one patient put it, *"at the hospital, I was expected to go home after two days, the surgeon came by, and said it all went well, and then I did not see him again. I couldn't do anything, not even lift my leg from the ground, and I had two small children at home, how should I manage? I cried the night before discharge, but the nurse just said that many before me had felt the same way, but it didn't have anything to do with me – it was bad nursing and care"*

4. DISCUSSION

4.1 Sharing Knowledge

The results indicate that patients gain in many ways from sharing knowledge in a Facebook group. They use the knowledge to make more qualified decisions when engaging with the health care system. The combination of professional knowledge, knowledge gained from the internet and experiences from peer patients enables them to achieve a more complete and holistic view of their situation. Most of the patients in this study exhibit behavior consistent with those described in the introduction as "Patient 3.0" and they desire but do not receive holistic care from the healthcare system. Many doctors still practice in a largely Health 1.0 way (Hawn et.al, 2009), and are not used to taking care of their patient's needs for happiness and wellbeing (Hawn, 2009; Santana, 2011), this is why patients seek support from peer to peer groups such as are appearing on Facebook.

It has been said that many specialists view internet-informed patients as a challenge, and might even take a negative view of Patient 3.0 (Broom, 2009; Hardey, 1999: Randeree, 2009). Some studies have shown that obstacles to self-management arise when health professionals are not comfortable with patients who have an increased level of knowledge (gained through focused education), and that in the worse cases, patients were denied access to self-management resources (*Snow et al, 2013; McMullan, 2006*). Some healthcare professionals have made their way into social media because they have realized that patients are more often seeking medical advice through online mediums such as YouTube. (French, 2012)

In this study, we saw that patients gained knowledge and support through each other leading to increased empowerment, whereas the participants in Snow's (2013) study gained knowledge and empowerment though focused education. Our results correlate with Zhang's (2013) study where they investigated communications in a similar but internationally based FB group for patients with diabetes. Their study found that interaction between FB users provided emotional, informational and appraisal support as well as influencing social behavior whilst providing a sense of companionship (Zhang et.al. 2013)

An important point to remember concerns the activity of posting personal health questions and information on any public media such as FB. The Ganz group changed from public access to private when issues of privacy began to surface. As Hader and Brown argue, the issue of privacy goes both ways. The difference between being a member of a private group instead of a public group is important to the group, and as we found, it was an important theme in our focus group discussions. Patients experienced that when the group was public, the members did not talk frankly since their friends could see updates and comments. Professional providers of all social media should likewise consider how to use their media in a manner that will not have them crossing any boundaries that would otherwise protect patients' confidentiality (Hader and Brown, 2010).

4.2 Health 3.0 and Patient 3.0

Today, highly specialized hospitals are being built with capacity based on future accelerated courses of treatment (P@H, 2015). This will lead to shorter contact times between patients and healthcare professionals.

When patients of the future (Patient 3.0) do not obtain what they request or need from their hospital staff, they will gather together either physically or in more likelihood it will be a virtual meeting place. They will do this in order to fulfil their fundamental need to be taken care of, as a person with both body and mind. As we have seen in this and other studies, Facebook is a place where patients derive great benefit from the company of similar others and gain confidence in managing their own health (Broom, 2005; Zhang, 2013). At the same time Health 3.0 will continue to grow through the same need for supportive partnerships between health care professionals (who also need support as people caught up in the system) and the patients they wish to help. Health 3.0 faces additional challenges with an aging population as well as traditional health care professionals who are not used to communicating through social media. Even if a Health 3.0 type of relationship could be fostered between healthcare professionals and their patients another dilemma arises. The institutionalized health system as a whole must somehow find a way to provide the holistic, personal-experience-based-care that a FB group like Ganz can provide. Healthcare professionals will find it difficult if not impossible to offer the experience-based knowledge that peers can. This suggests that there could be room for collaboration between some social media in collaboration with the established healthcare system that will support new notions of patient empowerment. (van Uden-Kraan et al, 2009; Bartlett, 2011).

To summarize; we have compared the results of this study to literature about Health 2.0 and 3.0 and Patient 2.0 and 3.0. This study shows that patients, at least in many areas, are moving towards being 3.0 patients. They appear to be promoting a more patient-centered health care system through information provision, developing ownership in knowledge and actions through peer-to-peer group togetherness. This will also change the communication patterns of health professionals more broadly (Chou, 2009)

4.3 How can the Health Care System Gain from this New Knowledge?

Based on the findings in this study, it would appear that patients themselves will lead the majority of change within the healthcare system, finding and creating new ways to rethink the delivery of health care services. The role of social media will continue to provide higher levels of empowerment to patients and health care professionals will need to find new ways of understanding patient behaviors and values in line with their expressed needs. This new way of considering healthcare could use social media as a new platform. As Deloitte concluded in their report *Healthcare 3.0. Healthcare for the new normal*: "The task of having increased patient engagement extends beyond social media – it also involves a greater understanding of a redesigned paradigm for understanding and influencing patient behavior" (2013, p.13). Facebook is pushing healthcare professionals into conversations that they should have been having all along according to French (2015), who suggests that Facebook might even help healthcare professionals to be better healthcare providers by facilitating more genuine conversations with patients and their families.

5. CONCLUSION

Across the western world, in Healthcare there is a general movement towards patients taking more responsibility in the treatment of their own conditions. Based on this contemporary phenomenon and increasing international interest in *Health 3.0*; we investigated the impacts on health and wellbeing arising when people exhibiting *Patient 3.0* characteristics went online in a Danish Facebook group to communicate with each other. We found that they obtained three key forms of benefit from their participation.

Firstly, through interactions with their peers they were able to improve the quality of their personal reflection and subsequent decision making. By engaging in online discussion they shared important and useful knowledge that enabled better and more experience-based decisions to be made especially during their interactions with health professionals.

Secondly, they were enabled to combine the various knowledge sources from peers, from systemic sources and from clinicians about their condition. This achieved more personally relevant information that offered a clearer perspective on the decisions they needed to make in terms of the best courses of action for them. They were empowered in their actions through the knowledge they gained.

And lastly, they were also encouraged and given guidance about concrete actions they could take to improve their general wellbeing through practical activities and realistic goal-setting in sport and other life-enhancing projects during and after treatments.

The results of this and other studies support the conclusion that the changing face of health in the form of Health 3.0 and the more active roles of Patient 3.0 are shaping the provision of future healthcare towards a more personally responsible and individual form of healthcare and living well. The role of social media is beginning to make its presence felt in this arena and show signs of being not only a shaping force but a medium for more productive outcomes. The changing relationships dynamics between healthcare professionals and patients can already be seen in the patient 3.0 examples described in this research. They have shown to be beneficial and useful in managing and to some degree, improving the treatment of the conditions studied in this research. A key question remains; in a health system under economic pressure to increase throughput while maintaining treatment quality, and at the same time consider individualized patient needs - how can we achieve these dual but seemingly conflicting needs? The answer will lie not in applying more technologies or necessarily developing new technologies as a solution, but in getting more out of current technologies such as social media thus implementing solutions that are more relevant to Patient 3.0.

REFERENCES

Aitken, M., Altmann, T. and Rosen, D., 2014. *Engaging patients through social media*. IMS Institute for healthcare informatics, Tech. Rep

Antheunis, Marjolijn L.; Tates, Kiek; Nieboer, Theodoor E. 2013. *Patients' and health professionals' use of social media in health care: Motives, barriers and expectations*, Patient Education and Counselling, 92, 3, (426–431) URL: http://www.pec-journal.com/article/S0738-3991(13)00265-6/pdf

Bartlett Y.K and Coulson N.S. 2011 *An investigation into the empowerment effects of using online support groups and how this affects health professional/patient communication*. Patient Education Counselling.83(1):113-119

Bender, J. L., Jimenez-Marroquin, M. C. and Jadad, A.R., 2011. *Seeking support on Facebook: a content analysis of breast cancer groups*. Journal of Medical Internet Research, 13(1).

Broom, A. 2005. *Virtually he@ lthy: the impact of internet use on disease experience and the doctor-patient relationship*. Qualitative health research,15(3), pp.325-345

Chou, W.Y.S., Hunt, Y.M., Beckjord, E.B., Moser, R.P. and Hesse, B.W., 2009. *Social media use in the United States: implications for health communication*. Journal of Medical Internet Research, 11(4).

Deloitte, 2013. Healthcare 3.0: *Healthcare for the new normal*. Accessed 20-01-2015, URL: https://www2.deloitte.com/content/dam/Deloitte/sg/Documents/risk/risk-sea-healthcare-3.0-healthcare-for-the-new-normal.pdf

Dhillion, J.S., Lutteroth, C., Wünche B., 2011. *Leveraging Web 2.0 and Consumer Devices for Improving Elderlies' Health*. HIKM '11 Proceedings of the Fourth Australasian Workshop on Health Informatics and Knowledge Management - V.120 (17-24)

Faber, Nicholai Holm. 2015 *Why don't we just tell patients what to do? – How to reach the potential of Patient 3.0*, Careum [Online Blog] URL: http://blog.careum.ch/blog/author/nicolajholmfaber/

French, S. 2015 *Building relationships on YouTube*. In, Bringing the Social Media revolution to Health Care. Torres, J. (Ed). Mayo Clinic Center for Social Media

Gagnon, S. and Chartier, L. (2012) Health 3.0—The patient-clinician "arabic spring" in healthcare. *Health*, **4**, 39-45. doi: 10.4236/health.2012.42008.

Greene, J.A., Choudhry, N.K., Kilabuk, E. and Shrank, W.H., 2011. *Online social networking by patients with diabetes: a qualitative evaluation of communication with Facebook*. Journal of General Internal Medicine, 26(3), pp.287-292.

Hader A., Brown ED. 2010 *Patient Privacy and Social Medias*. Legal Briefs, AANA Journal.78, 4.

Hardey, M. 1999. Doctor in the house: *The internet as a source of health and knowledge and a challenge to expertise*. Sociology of Health and Illness, 21(6), 820-835.

Hawn C. 2009. *Report From The Field. Take Two Aspirin and Tweet me in the morning: How Twitter, Facebook and Other Social Media Are reshaping Health Care*. Health Affairs. 28, 2. (361-368)

Klastrup, L., 2013. *Danskerne på Facebook: et globalt eller nationalt kulturfællesskab?*. In Facebook. Samfundslitteratur.

Kuan.Yu Lin, HSI-Peng Lu. 2011 *Why people use social networking sites: An empirical study integrating network externalities and motivation theory*. Computers in Human Behaviour, 27, (1152-1161)

Lindseth, A. & Norberg, A. 2004. *A phenomenological hermeneutical method for researching lived experience*. Scandinavian Journal of Caring Science, 18, 145-153

Madan, G., Stadler, M.E., Uhrich, K., Reilly, C. and Drake, A.F., 2011. *Adolescents with tracheostomies—Communications in cyberspace*. International Journal of Pediatric Otorhinolaryngology, 75(5), pp.678-680.

McMullan, M., 2006. *Patients using the Internet to obtain health information: how this affects the patient–health professional relationship.* Patient education and counselling, 63(1), pp.24-28..

Moffatt, F; Martin, P; Timmons, S. 2014. *Constructing notions of healthcare productivity: the call for a new professionalism?* Sociology of Health & Illness, 36, 5, (686–702) URL: http://onlinelibrary.wiley.com/doi/10.1111/1467-9566.12093/pdf

Nash D.B. Health 3.0. *Pharmacy and Therapeutics.* 2008;33(2):69-75.) URL: http://www.ncbi.nlm.nih.gov/pmc/articles/PMC2730068/

P@H, 2015. *Why Patient@Home?*, Patient@Home, University of Southern Denmark, Accessed 20-01-2016, URL: http://www.en.patientathome.dk/

Pols, J. 2013. *The Patient 2. Many: About diseases that remain and the different forms of knowledge to live with them.* Science and Technology Studies, 26, 2, (80-97). URL: http://www.sciencetechnologystudies.org/system/files/v26n2Reader.pdf#page=82

Randeree, E., 2009. *Exploring technology impacts of Healthcare 2.0 initiatives*. Telemedicine and e-Health, *15*(3), pp.255-260.

Santana S, Lausen B., Bujnowska-Fedak M., Chronaki CE., Prokosch HU., Wynn R. *Informed citizen and empowered citizen in health: results from an European survey*. BioMed Central. Family Practice 2011, 12:20.

Snow, R., Humphrey, C. and Sandall, J., 2013. *What happens when patients know more than their doctors? Experiences of health interactions after diabetes patient education: a qualitative patient-led study*. BMJ open, 3(11), p.e003583.

Sociale Medier 2016. *Sociale Medier 2015 i Danmark*, Accessed 20-01-2016, URL: http://www.socialemedier.dk/

Steele J. 2014. *Health 3.0 is Holistic*, Radar Online, 29-1-2014 Accessed 20-1-2016 URL: http://radar.oreilly.com/2014/01/health-3-0-is-holistic.html

Strauss, A. & Corbin, J. 1990. *Basics of qualitative research: Grounded theory procedures and techniques*, Newbury Park, CA, Sage.

Van Manen, M. 2006. *Writing qualitatively, or the demands of writing.* Qualitative Health Research, 16, 713-722

van Uden-Kraan C.F.; Drossaert C.H.; Taal, E.; Seydel E.R.; van de Laar M.A. 2009.

Participation in online patient support groups endorses patients' empowerment, Patient Education & Counselling. 2009 74,1, (61-9).

Weiner, S. J. & Schwartz, A. 2015. *Contextual Errors in Medical Decision Making: Overlooked and Understudied. Academic medicine: Journal of the Association of American Medical Colleges*.

Wentzer, H.S. and Bygholm, A., 2013. *Narratives of empowerment and compliance: Studies of communication in online patient support groups.* International Journal of Medical Informatics, 82(12), pp.e386-e394.

Winterton, J., Delamare-Le Deist, F. & Stringfellow, E. 2006. *Typology of knowledge, skills and competences: clarification of the concept and prototype.* Luxembourg: Cedefop Reference series; 64 Luxembourg: Office for Official Publications of the European Communities, Accessed 20-01-2014 URL: http://www.cedefop.europa.eu/EN/Files/3048_en.pdf

Zhang, Y., He, D. and Sang, Y., 2013. *Facebook as a platform for health information and communication: a case study of a diabetes group*. Journal of medical systems, *37*, 3, (1-12)

REQUEST: A RESPONSIVE AND FLEXIBLE SCREENING SERVICE FOR CLINIMETRICS

Jan-Willem van 't Klooster[1] and Miriam Vollenbroek[1, 2]
[1]*Roessingh Research And Development, Roessinghbleekweg 33b, 7522 AH Enschede*
[2]*University of Twente, Drienerlolaan 5, 7522 NB Enschede*

ABSTRACT

Though clinimetrics in general and obtaining and sharing health related indicators are important in many clinical disciplines, obtaining and accessing this information in a flexible manner is an issue because of difficult-to-use programs, fixed instruments and poor access to clinimetrics tools and results.
Hence, we developed and evaluated ReQuest, a novel Software-as-a-Service to create, share, conduct and manage clinical screenings and tests.
ReQuest works in such a way that a wide variety of screening instruments can be provided on a wide variety of ways: online and offline; on smartphone, tablet and computer; guided by a professional and self-assessed. Both tests and results can be managed and shared among peers and patients. New screenings can be designed within ReQuest, a.o. in English, German and Dutch.
ReQuest is successfully used in different EU research projects. Evaluation results show good acceptance in elderly asked to use the system for self-assessments. Researchers who use ReQuest to conduct and manage screenings, value (i) secure on-site storage, (ii) the flexibility in which they can create or reuse screenings and (iii) sharing screenings with others.

KEYWORDS

Clinimetrics, screenings, clinical research

1. INTRODUCTION

Clinimetrics focuses on the development and application of assessment methods and measurement instruments to describe or measure symptoms, physical signs and other clinical phenomena. Since the introduction of clinimetrics in 1987 by Alvan Feinstein, adoption and usage has increased significantly in clinical research and practice.

Nowadays, such clinimetrics systems should be well accessible, easy to use and adaptable towards increasing numbers of tests and patients. In their review, Terwee et al. (2010) demonstrate the importance of standardization in clinimetrics. De Vet et al. (2003) emphasize technical advancements in clinimetrics as crucial towards the future. Not only 'these data still have to be interpreted', also 'their origins should be evaluated and interpreted.' Tomba & Bech (2012) discuss the issue of clinical judgment analysis. To this end, the ability to share and compare *results* and the *screenings* themselves is key.

For individual health indicators, public health and policy, systematic screenings are commonsense (Giard, R. W. M., 2005). In the recent years, many hospitals and health centers saw themselves confronted with an ever-increasing demand of health figures, to be provided for reasons of quality control, benchmarking and comparison. The increasing number of tests, test formats and data points however, raise the demand for clinimetrics support that allow to manage both different screenings and screening results. Meanwhile, this process is facing barriers in logistics of acquisition, distribution, collection of forms; difficulty understanding and completing surveys by patients; the potential disruption of workflow; difficulty scoring and interpreting results; clinical relevance; and cost (Williams, Templins and Mosley-Williams, 2004).

This indicates the need for clinimetrics services that offer 24/7 availability, interoperability, means to access and compare patient results (both inter-subject and between subjects) and safely share them across stakeholders. Data security is a crucial factor in clinimetrics, i.e. knowing the medical data is in safe hands (Institute Of Medicine, 2009).

Research projects increasingly demand development of screenings and self-tests, sometimes including multimedia instructions to clarify the tasks at hand. Finally, we see increasing patient numbers in longitudinal multi-cohort studies requiring effort in screening management.

So to sum up, data security, adaptability, availability, easy-to-create screenings with rich content, interoperability, sharing of screenings, management of studies and flexibility for accommodation of screenings are important *criteria* in clinimetrics services.

Current services that fulfill these criteria to a certain extent are given in Table 1. In general, they lack functioning on tablet or smartphone (important for on-the-go usage), media support, connectivity, and sharing options. Besides, costs and data security can be an issue when general available services are used. So the question arises how to address these demands, especially in a reusable and flexible manner.

To be able to fulfill the mentioned criteria, it was necessary to develop our own clinimetrics service. This service is called ReQuest. It is a web-based program (SaaS service), available for clinical research and education free of charge.

The service is presented in this paper as follows. In Section 2, the design and implementation approach of ReQuest are discussed. The service is evaluated among users and patients. This is discussed in Section 3. Finally, Section 4 discusses conclusions and future work.

Table 1. Comparison of screening programs: RRD ReQuest, VitalHealth QuestManager, SurveyMonkey, Nedap Ons

Heading level	ReQuest	QuestManager	SurveyMonkey	Ons EHR
Target market	research, clinimetrics	surveys, care	surveys, general purpose	patient administration
Measurement instruments				
Number of questions	20	10	15	11
Media	video, photo	-	-	-
Standard database	yes	yes	no	yes
Screenings extensible	yes	yes	yes	yes
Instruction necessary	no	yes	no	yes
Measurement results				
Security	on-premise secured database	cloud database	cloud database	cloud database
Media	video, photo	-	-	-
Result fetching	all, per patient	all, per patient	all	all, per patient
Availability				
Online available	yes	only via EHR	yes	only via EHR
Works on tablet	yes, responsive	yes, limited	yes, responsive	yes, limited
Works on smartphone	yes, responsive	no	yes, responsive	no
Result fetching	all, per patient	all, per patient	all	all, per patient
Sharing				
Share/reuse instruments	on-premise secured database	cloud database	cloud database	cloud database
Personalize instruments	video, photo	-	-	-
API	yes	yes	yes	yes
Connect to EHR	yes	no	no	yes
Technology	very modern	modern	modern	very modern
License	Free for research / education	paid	paid (free trial)	paid
Validation	clinical research	practice	practice	practice

2. DESIGN

In this section the main design considerations are presented. It includes its data model, question types flexibility, sharing options, and security aspects. Then briefly the implementation is discussed.

2.1 Data Model

A *screening* consists of parts and questions. A *part* consists of questions. Each question has a label, optional explanation and uses a question type, such as a Likert scale, multiple choice, stop watch and different numeric / text / decimal types. In this design, many different screenings can be modelled. Currently, ReQuest has over 50 standardized and widely used screenings available, mainly in rehabilition, physical exercise, physiotherapy and evaluation domains. Descriptions and calculations can be added too.

Questions are stored within each answer, such that should a question definition change over time, the original question text answered is always available. Results are modeled as shown in Figure 1.

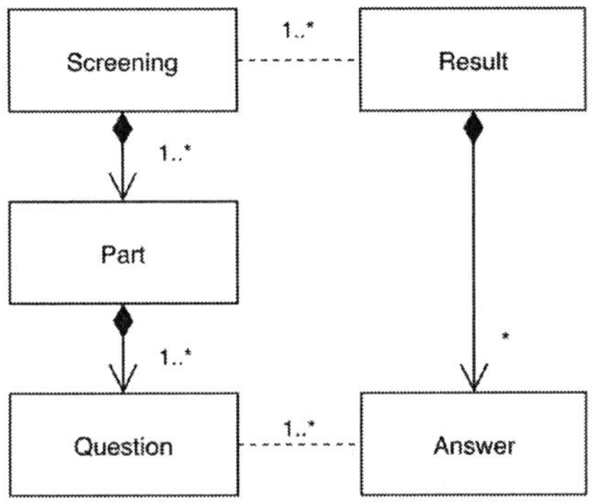

Figure 1. Nested modeling of screenings, screening parts and question, and results with answers. Left column: a Screening consists of one or more Part(s). A Part consists of one or more Question(s). Right column: whenever a Screening is answered it is stored as Result. A Result consists of zero or more Answer(s) to modeled Questions

2.2 Question Types

Question types refer to a kind of question. E.g., a *numerical question* is a question that should be answered by a number; a *boolean question* should be answered by either yes or no. In ReQuest, 20 question types are available, including numeric, decimal, date and multiple choice. The full listing is shown in Table 2. In each question, multimedia (images, movies) can be added to explain or illustrate the item. By defining multiple choice options, many discrete questions more can be formulated.

Table 2. Question types available in ReQuest

#	Question type	Explanation of rendered answer field(s)
1	**STRING_SMALL**	Small text
2	**STRING**	Text
3	**TEXT**	Large text
4	**NUMERIC**	Numeric
5	**DECIMAL**	Decimal
6	**LIKERT_1_5**	Likert 1-5
7	**LIKERT_1_7**	Likert 1-7
8	**SCORE_1_10_DECIMAL**	Score 1-10 (Decimal)
9	**SCORE_1_10_NUMERIC**	Score 1-10 (Numeric)
10	**BOOLEAN**	Boolean question
11	**MC_SINGLE**	Multiple choice, 1 answer
12	**MC_SINGLE_HORIZONTAL**	Multiple choice, 1 answer, horizontal
13	**MC_MULTI**	Multiple choice, multiple answers
14	**DATE**	Fillout with date (shows date picker)
15	**TIME**	Fillout with time (hh:mm)
16	**EMAIL**	Fillout with email address
17	**PICTURE_SINGLE**	Image, 1 point to be selected
18	**PICTURE_MULTI**	Image, multiple points to be selected
19	**EXPLANATION**	Just an explanation label, no answer
20	**STOPWATCH**	Stopwatch (shows a timer)

2.3 Sharing

Once a screening is created, ReQuest allows sharing of screenings in three different ways:

1. The screening can be added to a publicly available catalogue. This means each other user of ReQuest can reuse the screening. This is useful for standardized tests that are used frequently.

2. Besides that, the screening and its results can be shared with a specific number of other users. These other users can access the screening and its results.
3. The final way of sharing is to allow a specified number of other users to access the screening and its results, and also allow them to change the screening or manage it. Again, this full-access way of sharing can be set up regardless of whether the screening was added to the catalogue.

2.4 Security

As indicated by a.o. the Institute Of Medicine (2009), a major issue in using publicly available questionnaire services for clinical research, is the fact that the medical data collected is stored in places not controlled or operated by the responsible organization itself. ReQuest overcomes this issue by using on-site storage. In the case of the implementation at Roessingh Research and Development, it means that the ReQuest databases are protected in dedicated server areas owned and controlled in their own premises.

Moreover, in ReQuest only the owner of a screening is allowed to define who may access either the screening or the screening results, allowing for fine-grained access control to possibly sensitive information.

2.5 Implementation

Ruby on Rails (Ruby on Rails, 2015) was chosen as the framework to implement ReQuest due to its suitability for fast web application development. Moreover, it stimulates convention-over-configuration dictating a clear Model-View-Controller (MVC) separation, support for agile development and good support for automated code testing. Main highlights of the MVC levels are discussed below.

2.5.1 View

Vitannen (2011) concluded a survey stating that usability is among the large points for improvement in clinical software for physicians. ReQuest addresses this issue by using a well-known templating format. In this way, the user interface is very recognizable. Twitter Bootstrap (Bootstrap, 2016) is used for the View. Bootstrap enables responsive design and allows a wide-spread layout of the various screens of the service.

Rails' built-in localization functionality is used to provide versions in English, German and Dutch. New screenings can also be designed in other languages.

To be able to integrate ReQuest in other healthcare services or electronic patient records, an API is inevitable. The ReQuest API is currently used to access screenings, questions and results.

RABL (RABL, 2016) is used for API templating, as shown in e.g. Listing 1. It defines the JSON or XML structure and allows CRUD operations in this format consecutively.

Listing 1. RABL templating for retrieving screening by textual identifier

```
collection [@screening] => :screenings
attributes :id, :name, :description, :text_id, :created_at, :updated_at
child :parts do
  attributes :name
    child :questions do
      attributes :id, :name, :description, :question_type, :position, :greyed_out,
:required, :default_answer, :render_instructions, :condition, :regexp, :options,
:media_link, :image_link, :created_at, :updated_at, :multiple_choice, :text_id
          child :multiple_choice do
        attributes :name, :description, :inactive, :created_at, :text_id, :updated_at
        child :multiple_choice_options do
          attributes :name, :description, :score, :inactive, :created_at, :value,
:updated_at
        end
      end
    end
end
```

Figure 2 shows the most important interface screens of ReQuest.

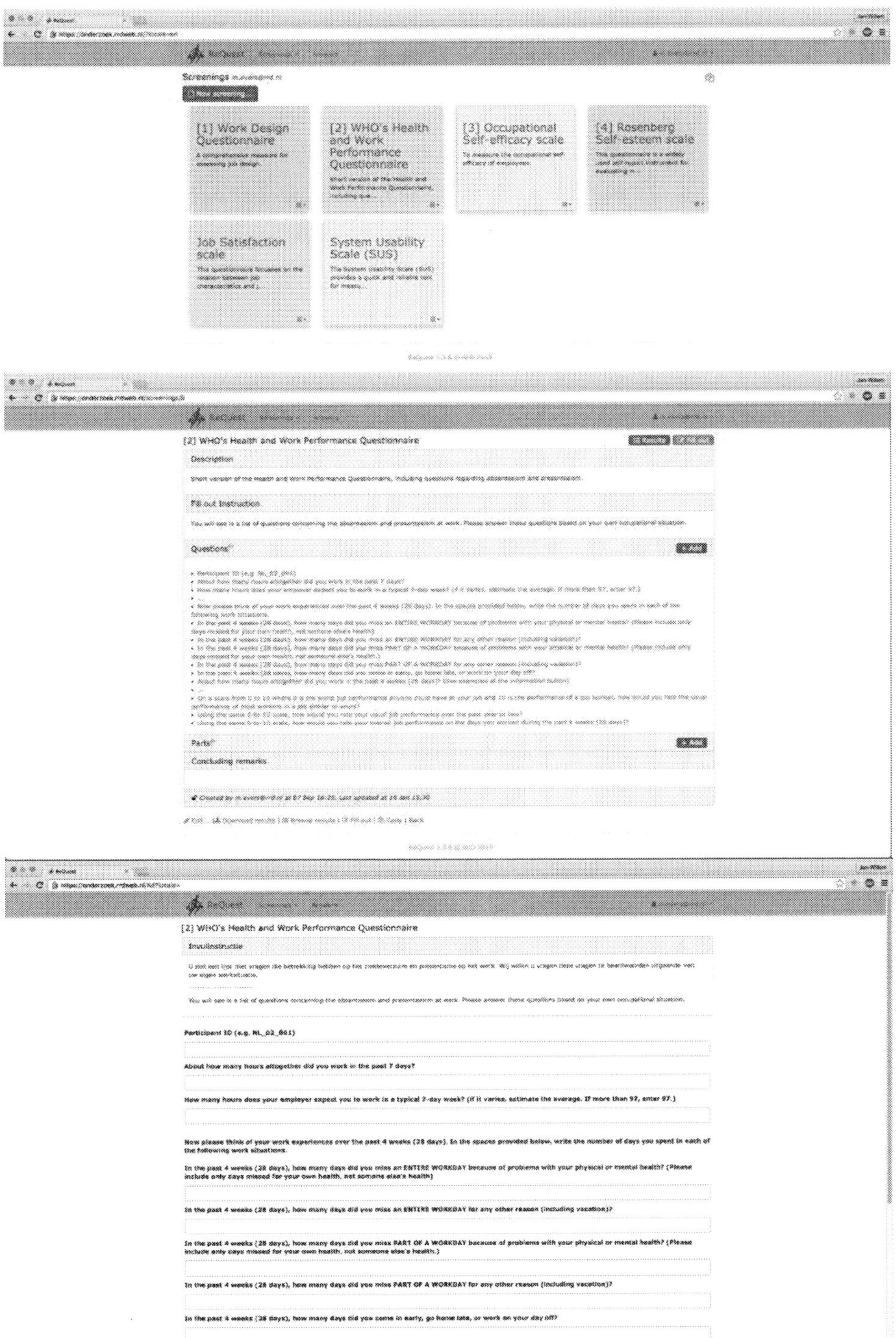

Figure 2. Sstarting from top: (i) a set of personal screenings, (ii) creating a new screening, (iii) filling out a screening

2.5.2 Model

The models shown in Figure 1 are all backed by means of the ActiveRecord Object-Relation Mapping (ORM). The library (in Ruby called gem) deep_cloneable was used for data manipulation tasks related to the nested models of the screenings. I.e., copying of screenings requires deep copies of items from different tables. For performance reasons, this is preferably done using native SQL queries. E.g.,

```
clazz = self.deep_clone include: [ :questions, {parts: :questions} ]
```

allows for this using the deep_cloneable gem. It consecutively takes care of referential integrity of the new instance. The model shown in Figure 1 is implemented and backed by MySQL. There are separate databases for test, staging and production deployments.

2.5.3 Controller

To prevent access of screenings patients don't have to do with, hashed id's are used to expose the fillout url to the user. E.g. https://screening.rrdweb.nl/eFHxaz. Json or Xml can be produced for a plethora of read and write operations. This can be used in the API to access screening definitions and to access screening results given a specific screening identifier. Also, complete screening results can be browsed through. Alternatively, it is downloaded in .csv file format for future data processing in e.g. R, SPSS or Matlab.

The software is hosted on GitHub and tested on CircleCI (code coverage 92,1%) using RSpec unit and integration tests. ReQuest uses the semantic versioning convention for new releases. At the webpage http://roessinghresearch.github.io/request/ technical documentation is given, the service itself is available at https://screening.rrdweb.nl.

3. EVALUATION

3.1 Method

Currently, researchers working on a.o. FP7 Perssillaa, AAL Pearl and ZonMW Life use ReQuest in longitudinal studies. We set up an evaluation for ReQuest to evaluate usage systematically in 3 ways:

1. ReQuest is designed to allow participants, especially elderly, to fill out screenings themselves. Hence, the System Usability Scale was used to find perceived usability among 8 elderly filling out a screening on a laptop in their own home. The screening-of-interest was the SF-36 (Ware et al., 1992) as it used in multiple European prevention projects RRD participates in.
2. Screenings can be created and managed with ease. To validate this, it is also evaluated with 7 different researchers that use ReQuest to create and conduct screenings for their daily work using the System Usability Scale (SUS) (Bangor, 2008). Then, a semi-structured interview, incorporating questions regarding the specific properties of ReQuest, is conducted to further assess perceived usefulness and their opinion regarding ReQuest as compared to other screening services.
3. The service should respond quickly and be 24/7 available. Thus, we monitored the performance of ReQuest using NewRelic application performance monitoring.

3.2 Results

By the end of April, 2016, over 70 screenings were made in ReQuest and over 10.000 questions were answered. ReQuest has about 3400 pageviews per month. ReQuest performs fast: 99,7% of requests are handled within 0,5 second; 0,3% within 0,5-2 seconds based on end-to-end measurement.

The SUS delivers a single number on a scale from 0 to 100. The elderly filling out the SF-36 assessment in their own home in ReQuest using a wifi-connected laptop. They reported a score between 75 and 97,5, resulting in an average benchmark score of 84,4 (sd 6,51).

The researchers also filled out the SUS based on their experience in creating and managing screening using ReQuest. They reported a score between 70 and 85, resulting in an average benchmark score of 78,9. Note that creating a screening is a more demanding task than filling out an already created screening.

The researchers consulted value the ease at which it is possible to create and conduct screenings (n=3).

'*ReQuest is just that what I need*', one researcher said. '*Commercial alternatives may provide more functionality but are often not as intuitive or as trustworthy.*' Another said: '*It's easy to create screenings in a fast way. It offers the necessary options that are needed when developing a screening. Furthermore, it's possible to add media files in the screening*'. This allows to design screenings with images and videos.

They valued the possibility to discuss program improvements with the designers of ReQuest (n=5) and access to technical support. One other researcher saw it as a must:

'*Without ReQuest it would not be possible for me to create self assessments*.' One researcher said. '*It was explained clearly to me and allows me to process results easier than on paper as I did beforehand.*'

'*Especially for healthcare, it is important that we know where the data resides and who has access to it.*' was mentioned regarding data security. Since ReQuest allows on-site storage, this crucial aspect can be guaranteed if needed. All interviewees agreed on this being an advantage.

Five out of 7 interviewees said their study participants did not face substantial problems in filling out screenings. Two did not know.

Sharing of screenings is not used by all. Those working in European projects valued screening sharing, as it allows them to share and conduct screenings in multiple centers. Sharing of screening is favored slightly over sharing results. '*If I want to share results, I may as well email the data file*' one researcher said, denying need for built-in support.

The responsiveness (it works on all screen sizes) was received positively as well as the user interface.

The number of question types is sufficient, yet 3 researchers would have been better off with even a few more. In most cases they work around missing types. E.g., selecting an answer from a dropdown list can be replaced by using a multiple choice question.

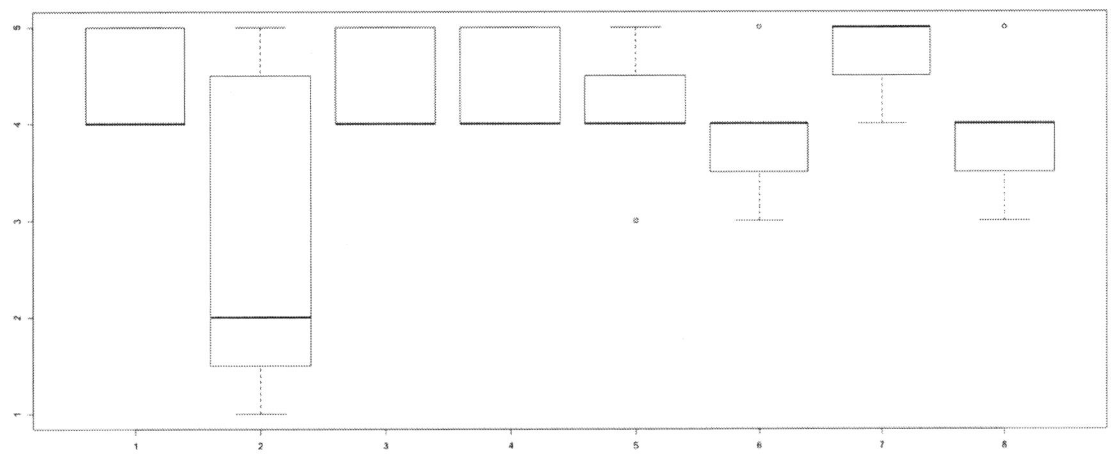

1: The number of different types of questions (e.g. open, closed, numeric...) ReQuest allows is sufficient.
2: I am missing question types to design my screenings.
3: it is positive that results obtained with ReQuest are stored in-house, not in the cloud.
4: ReQuest has a standard set of screenings. I value this property of ReQuest.
5: I value the ability to be able to share SCREENINGS with other users.
6: I value the ability to be able to share RESULTS with other users.
7: I appreciate that ReQuest works on different screen sizes.
8: The graphical user interface of ReQuest is clear.

Figure 3. Likert-based evaluation (1 – strongly disagree … 5 - strongly agree)

4. CONCLUSIONS & FUTURE WORK

In this article we presented ReQuest, a web-based screening and clinimetrics service. ReQuest is a flexible tool to support clinimetrics. Using this service, it is possible to conduct existing screenings as well as new screenings in an easy and manageable way. The service allows a variety of assessment to be conducted on different screen sizes such as smartphone, tablet or laptop. Currently, ReQuest is used in national and European research projects FP7 Perssillaa and DECI, AAL Pearl and ZonMW Life. The current work learns us that safeguarding of data, sharing of screenings and results for cooperation purposes, and the ability to manage a study process are key factors for researchers adopting such a tool.

Seven screening designers were interviewed. They are satisfied with the process of offering and improving ReQuest continuously. They value on-site data storage that safeguards their measurements and the possibility to share and reuse screenings. New screenings can be created quickly according to the users. SUS indicated good usability in elderly filling out self-assessments as well in researchers using ReQuest.

A remark is that the number of users inquired prevents bold statements about the outcomes. Another note to be made when evaluating usage of such a flexible tool in patients, is that the usability of the system itself is blurred with the usability of the screening created with it. I.e., end users evaluate the overall perceived screening, and not only the program that allow to conduct a to-be-configured screening. Of course, such scoring is dependent on the actual screening, that is why we choose a common standardized test.

Currently, ReQuest is used in different national and European projects. A larger evaluation is planned to study longitudinal clinimetrics service usage in these large research projects. In future, the screening managing functionalities are expanded to ease management of longitudinal studies.

ACKNOWLEDGEMENT

This work is supported by the Life project, granted by the Dutch healthcare research agency ZonMW under number 509.80219. Danielle Koning is acknowledged for her support in carrying out measurements.

REFERENCES

Bangor, A., Kortum, P. T., & Miller, J. T., 2008. An empirical evaluation of the system usability scale. *Intl. Journal of Human–Computer Interaction*, 24(6), 574-594.

A.R. Feinstein, 1987. *Clinimetrics,* Yale University Press, New Haven, CT.

Fette, A.M., 2006. A clinimetric analysis of wound measurement tools. *World Wide Wounds*. January 2006. http://www.worldwidewounds.com/2006/january/Fette/Clinimetric-Analysis-Wound-Measurement-Tools.html. Last visited: 4 May, 2016.

Fava et al., 2012. Clinimetrics: the science of clinical measurements. *International Journal of Clinical Practice*. Volume 66, Issue 1, pages 11–15, January 2012.

Giard, R. W. M. (2005). Prestatie-indicatoren als maat voor de kwaliteit van medische zorg: retoriek en realiteit. *Nederlands tijdschrift voor geneeskunde*,149(49), 2715-2719.

IOM (Institute of Medicine), 2009. *Race, Ethnicity, and Language Data: Standardization for Health Care Quality Improvement*. Washington, DC: The National Academies Press.

Tomba, E., & Bech, P., 2012. Clinimetrics and clinical psychometrics: macro-and micro-analysis. *Psychotherapy and psychosomatics*, 81(6), 333-343.

Ruby API Builder Language (RABL), 2016. https://github.com/nesquena/rabl (last visited 21 01 2016).

Ruby on Rails, 2015, 2016. http://www.rails.org (last visited 21 01 2016).

Terwee C.B. et al., 2010. Mind the MIC: large variation among populations and methods. *J Clin Epid,* 2010; 63: 524-34.

H.C.W. de Vet et al., 2003. Current challenges in clinimetrics. *Journal of Clinical Epidemiology* 56 (2003) 1137–1141.

Viitanen, J.; Hyppönen, H.; Lääveri, T.; Vänskä, J.; Reponen, J. & Winblad, I., 2011. National questionnaire study on clinical ICT systems proofs: physicians suffer from poor usability. *Int J Med informatics*, Elsevier, 2011, 80, 708-725

Ware Jr, John E et al, 1992. "The MOS 36-item short-form health survey (SF-36): I. Conceptual framework and item selection." *Medical care* (1992) 473-483.

Williams, C. A. et al., 2004. Usability of a computer-assisted interview system for the unaided self-entry of patient data in an urban rheumatology clinic. *Journal of the American Medical Informatics Association*, 11(4), 249-259.

A CALL FOR KNOWLEDGE: ASSESSING THE EFFECTIVENESS OF TEXT MESSAGING FOR SEXUAL AND REPRODUCTIVE HEALTH IN NORTHERN GHANA

Johannes G. de Kruijf
Department of Cultural Anthropology
Utrecht University, the Netherlands

ABSTRACT

In the context of a massive increase in mobile phone use, text messaging has recently become a popular focus of projects that aim to improve the health status of people in Sub-Saharan Africa in cost-effective and rapid ways. In many cases, these SMS programs concern HIV/AIDS and/or the encompassing theme of sexual and reproductive health. Primary objective of most interventions is disease prevention. Despite the apparent proliferation of SMS projects, review articles observe a lack of critical impact assessments and claim stronger evidence is needed to justify the widespread optimism concerning the potential of text messaging as behavior change communication. This paper offers such an assessment. It reports a mixed method study of the effectiveness of an SMS program that was initiated by the Dutch consortium Connect 4 Change to teach teenagers in remote areas of Northern Ghana about adolescence, sexually transmitted diseases, abortion and contraception. Research data show the impact of the intervention was severely undermined by technological challenges and inadequacies concerning the design of text message content. Despite some indications of general sensitization and subsequent behavioral change in qualitative data, no evidence of relevant knowledge increase was found. Such absence of impact highlights the crucial importance of 'formative research'; preliminary (preferably participatory) assessments of the context of implementation and the target population in order to design appropriate applications and socio-culturally sensitive messages. Also, it urges a rethinking of the particular potential of mobile technology. The notion of soft power is thus introduced to include contemplation of the immaterial property of (high) technology in marginalized settings.

KEYWORDS

Behavior Change Communication, Ghana, mHealth, Sexual and Reproductive Health, SMS

1. INTRODUCTION

Over the past fifteen years, Sub-Saharan Africa has experienced the greatest increase in mobile phones users in the world, reaching almost one billion in 2016 (Ericson Mobility Report, 2016). The rapid spread of (so far predominantly) feature phones, which allow for short message service (SMS), is thought to offer new and innovative opportunities for disease prevention efforts in the region. A wide range of projects involving SMS applications have been described in (grey) literature. According to an assessment of such literature by Déglise et al (2012), modes of intervention vary between one-or-two way communication, with or without incentives, and with educative games. Proponents of such applications suggest that advantages of SMS include its cost effectiveness, as well as its ability to disseminate information immediately, assure an adequate level of confidentiality, and have delivery confirmed (Atun et al., 2006). Besides HIV/AIDS, the theme of sexual and reproductive health (SRH) appears to be the most prominent focus of SMS based interventions. Text messaging is thus for instance employed to convey health tips on SRH (e.g. AppLab, Uganda), to connect recipients to counsellors (e.g. Learning about Living, Nigeria), to organize contests (e.g. Beat it, South Africa) or quizzes (e.g. Text to Change, Uganda), and to refer to alternative sources of information such as websites (e.g. Health on Line, Mali).

Despite the popularity of SMS projects and the expanding body of literature on texting as mHealth, reviews observe solid impact studies remain scarce and are urgently needed to establish stronger evidence, both of the successful application of SMS to improve sexual health (Lim et al., 2008) and of the potential of mHealth as a tool to foster behaviour change (Gurman et al., 2012). This paper concerns such an impact study. It provides a critical assessment of an SMS intervention aimed at the improvement of sexual and reproductive health of adolescents in Northern Ghana. The intervention was part of the 'ICTs in community health project'. This project was designed by a consortium of Dutch NGOs called Connect 4 Change (C4C) in collaboration with the Ghanaian Association of Church-based Development NGOs (ACDEP). It aimed to contribute to the improvement of healthcare delivery in ten communities in remote rural areas in Ghana's Northern Region and Upper East Region. The project's ultimate objective was to 'integrate appropriate and sustainable use of information, communication and supporting technologies into ten ACDEP member health institutions'. Specific target areas were sexual and reproductive health and maternal care. Capacity increase and behavioral change was aimed for by means of voice and text message programs, the training of traditional birth attendants and student members of local health clubs (so called peer educators), knowledge sharing events, and improvements in communication (mobile phones) and information access in Tele-centers. Additionally, interventions were implemented to automate and improve the efficiency of record keeping practices at ten ACDEP member healthcare delivery institutions.

Data and analysis below are part of a more comprehensive project impact assessment that was conducted by researchers attached to the Utrecht University department of Cultural Anthropology. It consisted of two field visits (October '14 and '15) and combined qualitative and quantitative explorations to map and explain success and failure in context. The ultimate goal of the study was to draw lessons from the project that enable ACDEP and the C4C partners to contemplate options for the post-project trajectory and identify areas of improvement to enhance impact and secure the sustainability of change.

2. BACKGROUND

The ICTs in community health project consisted of interventions in ten health clinics in Ghana's Northern Region (Salaga, WaleWale, Langbinsi and Nalerigu) and the Upper East Region (Garu, Zorko, Woriyanga, Kongo, Namolgo and Widana). It involved the implementation of an electronic patient record system (HMIS), the establishment of Community Health Information Centers, and the distribution of feature phones among traditional birth attendants (TBAs), traditional medical practitioners (TMPs) and peer educators. The distribution of these phones was part of the effort to facilitate communication (with clinic staff) and improve information and work towards capacity building. Concerning the issue of information, the mHealth component involved a SMS program (addressing peer educators on sexual and reproductive health) and a voice message program (addressing TBAs and TMPs). This intervention analysis focuses on the SMS program alone.

The SMS (sub-)program targeted peer educators; teenagers selected at schools (because of outstanding academic performance) and trained to inform and instruct their peers on issues concerning sexual and reproductive health. The SMS program was designed to boost the knowledge of these peer educators on some crucial themes (adolescence, HIV/AIDS, abortion, and contraception). The objective of this sub-program was to increase the kind of awareness that would produce (sexually restrained) behavior change among local adolescents and thus decrease the incidence of particularly teenage pregnancies (and abortions) and STDs.

Between August and October 2013, a sum total of 172 feature phones were distributed among peer educators in five (out of the ten C4C) communities. Table 1 provides an overview of the messaging system and the implementation format:

Table 1. Mobile Phones

Distribution		**Phones distributed among peer educators (Q3 2013)**				
Community	*Langbinsi*	30	Wale Wale	30	*Salaga-Loloto*	45
	Nalerigu	47	*Kongo*	-	*Widana*	-
	Garu	20	*Namolgo*	-		
	Woriyanga	-	*Zorko*	-	**TOTAL**	**172**
Text messages						
Frequency	Once a week					
Repetition	Maximum of 7 attempts (every 30 minutes) to deliver the message to the phone inbox.					
Moment	Mondays, in the afternoon					
Content I	146 different messages were composed and sent					
Content II	The last time a new message was composed was in January 2014					
Implementation trajectory						
Stage 1 *(pre-implementation)*	*Stakeholders* ***consultative meeting*** *on mobile phone program*					
Stage 2 *(implementation)*	*Distribution and* ***initial training*** *for trainer-of-trainers (once)* *Messages scheduled and sent out to recipients.*					
Stage 3 *(post-implementation)*	*One residential* ***refresher training*** *for 10 focal persons from the clinics (once)* *Continued to send scheduled messages to recipients*					

There was no general protocol for the further dissemination of information. Recipients belonged to various local secondary schools with different systems of peer education. Often, peer educators with an ACDEP phone would transfer the message to their peers in plenary sessions scheduled in/by school for the sake of SRH information sharing. These meetings are frequently organized by teachers or a school coordinator. As such, text messaging complemented other information strategies (e.g. leaflets, software/offline applications, posters).

3. METHODS

As indicated, the impact assessment consisted of two field stays; the researchers have visited the area for two weeks in October 2014 and for two weeks in October 2015. Because of logistical considerations and time constraints, four communities were selected for the impact study. Given their size, the composition of their populations, the presence (or absence) of facilities, and their project participation, these four were considered representative of the entire cluster of communities involved in the C4C project. The researchers have spent four days (2x2) in each of the communities. Eight focus group sessions with peer educators were scheduled in 2014; four focus groups with only female students and four other focus groups with only male students. Also, a mixed focus group interview was conducted. The size of the groups varied between five and nine (one consisted of nineteen adolescents). The age of the participants ran from fourteen to twenty-three. In 2015, a survey was conducted among thirty-three peer educators from fourteen different schools. Eight of these peer educators are from Woriyanga, a community *with* a peer educator program were *no* phones were distributed among peer educators. Comparison between the Woriyanga (and Garu) data and the data from the other communities will help to isolate the effect of the intervention. The survey was meant to test the knowledge on SRH. It was divided in three sections; free recall of text messages, cued recall, and a knowledge test.

Along with the focus groups and the survey, semi structured interviews were organized at all four sites in 2014 as well as 2015. Besides peer educators, teachers and local NGO staff were interviewed. The researchers have also visited secondary schools and Community Health Information Centers. The latter was to assess the availability of (alternative) information on SRH.

4. FINDINGS

Research data suggest knowledge-increase was limited (at the most) and provide no substantial evidence of behavioral change. Malfunctioning phones proved a crucial obstacle to success. Interview data and observations reveal large-scale technological failure. It seems that more than 95% of all phones (including the 406 phones distributed among TBAs and TMPs) were not functioning by October 2015. This percentage coincides with estimations by ACDEP and clinic staff about the total number of ACDEP phones still in use.

The main reason for phone failure was a malfunctioning battery. Phone batteries did not last, or spoiled while charging. Informants stress that the particular type of phone that was opted for was inadequate, especially considering the rural context of use. It was a feature phone (brand: *Q7*) that was cheaper than popular competitors (e.g. *Nokia*, *Techno*) but proved far less robust. Qualitative data shows the selection process was flawed and insufficiently organized/monitored. As the excerpt from an interview with an ACDEP staff member indicates, poor communication and the delegation of tasks partially explain the selection problems:

> [...] Initially we were like can we look for phones...there was a criteria, like long lasting battery, maybe a torch and a radio, a very simple one that TBAs who are not literate and TMPs who are not literate can easily use. So, I was in Tamale and I went around. I went to the *Tecno* office because I was using a *Tecno* at that time that I could guarantee that the battery would take more than three days. So we went there and we got a similar [one] that was much cheaper than what I was using and it could do this thing. Then we went to other places, then we looked for *Nokia*, then we looked for *Samsung*, which was much durable and...in Tamale everyone that was like *Tecno* is good. Because of the wide use of *Tecno*, which was cheap so...we communicated it to [the funder] and I think that [C4C] has a consultant, a local consultant in Tamale. So the idea was that [the funder] can contact the local consultant, he is their representative to check the phones, the cheap one; they use their own criteria... We were not previewed to what went on with the consultant. So they agreed and the consultant came and communicated to us that this is what we find, we...the cost of the phone is this, the number is 540 or so, and this is the type, QC or [Q7]. So [this brand] is something that is very new to us in Tamale. *(Interview 2014)*

In an e-mail, the consultant explains how circumstances caused the switch from Tecno to Q7:

> The original choice of phones was *Tecno T209* [...]. However, there was a delay in the release of funds, between the submission of invoice (17th May 2013) and the release of funds (11th August) there was a new 20% duty on all mobile phones. This meant we had to look for other brands, (at least *G-Tide*, and *Nokia*) but could not get any phone within budget. Total budget available was GHS34 per phone plus activated SIM card. Apart from the lack of *Tecno T209* at the Tecno offices at the time, (most importers had stopped importing phones with the hope that the tax would soon be repealed) the few shops that had a few available were selling them between GHS38.00 and GHS42.00. In the meantime, we had deadlines to meet, a fixed number of phones to acquire and a fixed budget. We finally, in consultation with ACDEP and [the other organization], opted for the model that came within budget – Q7. The team negotiated a 3 month warranty [...]. Unfortunately, the complaints only started coming in after the 3 month warranty period. I concede, the rate of spoiling is beyond the normal and has on at least 2 occasions, voluntarily repaired and replaced phone components in order to ensure a smooth implementation of the project at ACDEP. But that was also very soon seen not to be sustainable. *(E-mail, December 2015)*

Notwithstanding the explanatory value of misfortune, the fact that such poor luck had profound consequences indicates limited project robustness and exposes organizational weaknesses pertaining communication, decision-making and flexible adjustment.

4.1 Knowledge as an Indicator of Effectiveness

Despite the eventual failure of phones, text messages were initially received and read by all the peer educators who had received a phone. A survey was designed to test the knowledge of the content of these messages. A control group was included to be able to discern the ICT contribution (ex post facto).

Thirty-three peer educators (aged 13-18) have completed the questionnaire: sixteen participants (Langbinsi, Nalerigu and Garu) and seventeen non-participants (Garu and Woriyanga). Some important observations and insights can be drawn from this modest survey.

Response to the *free recall assignment* – the first in the questionnaire – was minimal. Participants were asked to write down all the information from text messages they had remembered. Most of the participants did not mention more than just one theme. Answers were brief and thematic, sometimes factually incorrect. They did not (ever!) provide any factual information that clearly connected to the specific jargon/content of the messages.

Following the free recall, the participants (with ACCEP phones) were asked to recall specific text messages concerning the five SMS themes: HIV/AIDS, contraception, STIs, teenage pregnancy and general information adolescence/menstruation. The participants without phones were asked to 'write down everything you know' about these same themes. The themes were provided (hence, cued recall). Interestingly, the control group gave twice as many correct answers as the intervention group [4 versus 2.1 per participant]. The control group gave only slightly more false answers [1.9 versus 1.7]. Furthermore, the control group gave more answers that were partially correct yet clearly not deriving from intervention-related information [1.9 versus 1.1]. Finally, the control group shows a lower non-response than the intervention group [0.2 vs. 0.9]. A possible explanation is that, whereas the intervention group was restricted to recalling information from the text messages only, the control group was encouraged to 'just write down everything' concerning the themes provided. However, the relatively high number of false responses from the intervention group – [1.7] compared to the control group [1.9] – suggests that **the intervention did not contribute to the eradication of incorrect knowledge** (e.g. the belief HIV is transmitted through eating from the same bowl or sharing one towel with an infected person). Results from the knowledge test underline this observation.

In the final part of the survey, we tested knowledge deriving from information distributed via text messages. We used the words and phrases from the text messages. Again, the control group provides more correct answers than the intervention group, but the difference is less striking [1.9 versus 1.5 correct responses per participant]. Remarkable is the difference in false responses [7.2 control group versus 4.9 from intervention group]. In contrast, the level of non-response was higher within the intervention group [5.9 versus 3.2]. Both groups gave a similar number of 'partially correct yet different source' answers, and almost the same number of 'incongruous' responses.

It can be argued that although the intervention did not result in higher levels of correct knowledge, it has made beneficiaries more cautious and probably more insecure regarding the precision of their answers (see the relative high number of non-responses [5.9, which is about 4 times the number of correct responses]. Certainly, the control group gives the most 'fantastic' responses. Examples are 'STIs came and replaced STDs', 'abortion is legal in both boys and girls', or 'contraception are tomatoes (soup) and vitamins'. However, if we count the non-responses as false responses, the intervention group has not only less correct responses [1.4 versus 1.9] but also more false/non-responses [10.8 versus 10.4]. They thus scored slightly worse than the control group. In general though, the amount of false and 'incongruous' responses and 'non-responses' is unacceptably high: more than four times the number of correct responses for the intervention group, and more than 3.5 times for the control group.

All in all, these survey results provide no clear indication that the intervention has contributed to adolescent's improved knowledge of sexual and reproductive health. In fact, it can be questioned if even the correct responses relate to information from the intervention or simply accidentally coincide with it. For instance, we randomly selected 27 key words from the messages (e.g. 'birth control', 'testing', 'breastfeeding', 'safe sex') of which we could detect only 3 among the responses. On the other hand, there are clear signs of the presence of shared knowledge that does *not* result from the intervention. Examples are the statements across all four locations that during menstruation a girl should 'bath [sic] regularly', 'keep one-self clean', or 'bath twice/three times a day'. This was mentioned by 22 out of 28 respondents but does not originate from the messages.

5. DISCUSSION

Data on information awareness and comprehension indicates there exists a wide and crucial **gap in vocabulary and phrasing** between the words/phrases used in the C4C text messages and those used by the adolescents. Employed phrasings, like 'pre-disposing factors' and 'Intra-uterine device' prove incomprehensible to the adolescents (and might not be the most relevant bits of knowledge either). This suggests that **the content of the messages has not been determined/ designed on the basis of a solid understanding of the beneficiaries and their context**.

In fact, the impact assessment reveals a variety of complications and flaws in the process of composing messages. For instance, some messages had a clear moral (Christian) or **normative undertone**. For instances, according to one text message: '*the best way to prevent teenage pregnancy is to practice abstinence. This is the only way to prevent unwanted pregnancy*'. Although this message suits the Christian character of the local partner, it does not suit the character and intention of the C4C intervention. Furthermore, some information provided in the messages was **factually incorrect**. An example is a message in which Candidiasis was referred to as a STD. Though fungal infections can be spread through sexual intercourse, it is *not* an STD as *Candida* species are considered part of the normal vaginal flora (and it thus occurs in celibate women). Obviously such an error can result in moral stress resulting from 'catching' this infection.

All in all, findings indicate (a) large scale technological failure; an unacceptably high number of malfunctioning phones and (b) the absence of evidence that phones/messages have contributed to improved sexual and reproductive health knowledge of participating adolescents. Despite these disillusioning results, the project evaluation of participants themselves shows high levels of appreciation. Notwithstanding malfunctioning phones, adolescents highly value the project and argue information from the text messages has not only increased their sexual and reproductive health knowledge but also contributed to a change in attitude, practices and behavior:

> *(PE 1) It has made us to know our health; It made us to stay away for STIs; Students are [now] able to manage themselves and abstain from sex until they complete or get married.*
> *(PE 2) It [the project] help [sic] me to prevent teenage pregnancy. It help me to avoid sexually transmitted diseases. It help me to avoid naming and shaming.*
> *(PE 3) This [the project] has made us, the teenagers, in the school, stop having casual sex and it has reduce teenage pregnancy in the community and helped to reduce STIs including HIV/AIDS*

Although indications of such behavioral change cannot be found in (inaccurate and incomplete) clinic statistics (e.g. on HIV testing, adolescent counseling, adolescent treatment of STI, and distribution of condoms), data do suggest increases in awareness with positive implications for adolescent practice. A definite strength of the intervention is the integration of mobile phones into pre-existing networks of information exchange, such as pre-existing adolescent health clubs. Consequently, new information can effectively be shared, both with peers during frequent (pre-existing) peer education meetings as well as with outsiders (family, community members).

A next strength of the intervention concerns the 'soft power' of technology. Qualitative data shows that phones are symbolic instruments of inclusion and recognition. Being selected a phone recipient enhances the peer educator's status as an endorsed educator. Furthermore, in contexts of information scarcity, the phone can be seen as a legitimizing device that, because of its technological advancement, possesses an inherent authority and is generally expected to convey only superior knowledge (that cannot be questioned) [1].

---

[1] An important observation considering the status component concerns the importance of gender differences. Both boys and girls appear equally motivated. However, girls' access to ICTs is more restricted because of sociocultural reasons. For instance, girls expressed their concerns that phone ownership could give them a bad reputation. As one girl said: *'If they see you with a phone, they will usually say you are a bad girl. Because if they see you with a phone at your ear like this, they will say you are calling your boyfriend [...]'.*

6. CONCLUSION AND RECOMMENDATIONS

Considering the survey results as well as the extremely short lifespan of the distributed phones it is justified to conclude that the intervention itself did not result in the improvement of the health status of adolescents in Ghana's Northern and Upper East Region. As indicated, there is neither statistical indication of any immediate effects of phone use for communication and information. However, participants claim that information on SRH has affected their perceptions and behavior. According to the interviewees, both peer educators and medical staff, changes are undeniable. Irrespective of the (lack of) factual/technical knowledge increase, the SMS program seems to have served to enhance the sensitizing effect of existing/ broader programs (e.g. SRH education and health clubs). This suggests that **a particular strength of the mobile phone as ICT4D is that of supportive device**, one that is strategically included in a multifaceted quest for positive change. This observation coincides with the conclusion of other SMS program analyses. For instance Chib et al. (2012), in their assessment of a HIV/AIDS text message campaign in Uganda, argue that realizing the potential of SMS programs requires making text messaging part of an integrated media communication strategy.

Whether the phone as a supportive device should be prioritized over other interventions in the realm of health care should depend on the outcomes of **a comprehensive (preferably participatory) need and possibility assessment in a specific locality/context**. Gurman et al. (2012) mention the need for 'formative research' in this respect. In Northern Ghana, mobile phones were in theory a cost-efficient and appropriate collaborative tool for communication and information sharing. Yet, the absence of a thorough preliminary understanding of the program setting – both in terms of technology and the topic of SRH – undermined the design of a technologically appropriate and culturally sensitive intervention. First and foremost, effective communication and information was thus hampered by the unfortunate selection of an inadequate phone (as well as poor connectivity together with the erratic power supply in the isolated communities). Second, if/while the phones functioned, the process of information and content creation proved too ineffective to produce a noticeable increase of knowledge.

Most peculiar is the lack of communicative strategy, didactical/pedagogical focus, and insight in socio-cultural convention that shows in the formulation of the text messages. Of course, **knowledge-based vision should always inform content creation** in the context of behavior change communication. As Briscoe and Aboud (2012) also argue, messages should be carefully tailored and land in an environment that is supportive of motivating individuals and communities to make positive health behavior changes. It seems rather ironic that in Northern Ghana, the complex process of (behavior change) communication was taken for granted in a quest for education. Messages did just not match the audience's level of comprehension. The unexpected observation that such flawed communication – along with the technological challenges – did not prove completely detrimental indicates a factor rarely mentioned in literature on mHealth; the 'soft power' of technology. Although more research on this immaterial property of (high) technology in marginalized settings is required, the distribution of phones among selected individuals appears to have aided the authority of peer educators as experts and helped to prioritize the theme of SRH.

REFERENCES

Atun, R., Y. Samyshkin, L. Thomas, S. Morris, M. McKee. S. Sittampalam & R. Coker. 2006. The Role of Mobile Phones in Increasing Accessibility and Efficiency in Healthcare. Retrieved (last accessed June 19, 2016) from https://www.vodafone.com/content/dam/vodafone/about/public_policy/policy_papers/public_policy_series_4.pdf

Briscoe, C. & F. Aboud. 2012. Behaviour Change Communication Targeting Four Health Behaviours in Developing Countries: A Review of Change Techniques. *Social Science & Medicine*, 75(4):612-621.

Chib, A., H. Wilkin, L.X. Ling & B. Hoefman. 2012. You have an Important Message! Evaluating the Effectiveness of a Text Message HIV/AIDS Campaign in Northwest Uganda. *Journal of Health Communication: International perspectives*, 17(1):146-157.

Déglise, C, L.S. Suggs & P. Odermatt. 2012. Short Message Service (SMS) Applications for Disease Prevention in Developing Countries. *Journal of Medical Internet Research*, 4(1):1-20.

Gurman, T., S. Rubin & A. Roess. 2012. Effectiveness of mHealth Behavior Change Communication Interventions in Developing Countries: A Systematic Review of the Literature. *Journal of Health Communication*, 17(1):82-104.

Lim, M., J. Hocking & C. Aitken. 2008. SMS STI: A Review of the Uses of Mobile Phone Text Messaging in Sexual Health. *International Journal of STD and AIDS*, 19:287-290.

Ericson Mobility Report. 2016. Retrieved from: https://www.ericsson.com/mobility-report/

Short Papers

A WEB ARCHITECTURE OF A FRAMEWORK TO EVALUATE SMARTPHONE-BASED FALL DETECTION SYSTEMS

Eduardo Casilari and Pablo Ortuño
Universidad de Málaga,
ETSI Telecomunicación, Universidad de Malaga, Andalucia Tech, 29071 Malaga, Spain

ABSTRACT

Smartphones have become an interesting alternative to deploy wearable and cost-efficient Fall Detection Systems (FDSs). During last five years, many research works have proposed a variety of algorithms for Smartphone-based FDSs. These systems are typically evaluated by testing the detection decision with a set of predefined movements (emulated falls and Activities of Daily Living or ADLs). However, up to date, there is an almost complete lack of a 'benchmarking' procedure to compare the actual performance of these algorithms in a rigorous way. This paper proposes an architecture that enables investigators to create and share with the research community their own datasets with falls and ADLs. The architecture, which consists of a simple downloadable smartphone application and a database server that can be accessed from a conventional web browser, is aimed at configuring a global framework to assess the effectiveness of FDSs based on commonly accepted evaluation patterns.

KEYWORDS

Fall Detection Systems; smartphone; mHealth; Android; accelerometer.

1. INTRODUCTION

Falls have become an important concern for national health systems as long as they are a leading and recurrent cause of unintentional injuries amongst adults aged over 65 (Yoshida 2007). Aiming at minimizing the morbidity linked to falls as well as the psychological anxiety known as FOF (Fear of Falling), diverse types of automatic Fall Detection Systems (FDS) have been proposed to promote a rapid medical intervention after an accident. Due to the worldwide popularity and plummeting costs of smartphones, during the last decade many research efforts have been devoted to develop FDSs that can be inexpensively integrated in the smartphones just by installing a simple application. In contrast with other previous solutions (such as context-aware systems and vendor-specific wearable devices), smartphone-based FDSs straightforwardly benefit from the computing capacities, wireless interfaces and embedded inertial sensors (basically an accelerometer and in some cases, a gyroscope and a magnetometer) that are natively supported by current smartphones.

FDSs can be contemplated as binary decision systems targeted at discriminating human movements provoked by falls from those produced by any ordinary ADL (Activity of Daily Living). The validity of the evaluation testbeds that employ simulated falls has been mistrusted and long discussed (Klenk et al. 2011) (Bagalà et al. 2012). Nonetheless, owing to the inherent difficulty of assessing the performance of a FDS with actual accidents, most works in the literature employ emulated falls and ADLs to evaluate the efficacy of the detectors. These falls and ADLs are systematically mimicked by a prearranged set of experimental subjects (normally a small group of 1 to 8 persons recruited among healthy and young adults). In this regard, a noteworthy initiative in this field has been promoted by the European Commission funded FARSEEING (FAll Repository for the design of Smart and sElf-adaptive Environments prolonging INdependent livinG) project (FARSEEING 2015), whose goals include to define a meta-database of real-world falls. Under this project, a panel of 40 experts from different disciplines collaborated to recommend a set of procedures and a format that a clinical dataset describing falls should follow (Klenk et al. 2013). Conversely, in spite of these recommendations, the analysis of the state-of-the-art clearly reveals the absence of an universally accepted database of 'traces' or mobility patterns that can be adopted as a common basis to compare the effectiveness of the different proposals in the literature (Casilari et al. 2015). This lack of clinical databases of realistic fall signals, also highlighted in (Schwickert et al. 2013), hampers a methodical and reliable comparison of the diverse detection algorithms that have been suggested for smartphone-based FDS. As a consequence, the reproducibility and review of the results and the conclusions of most related studies are virtually unfeasible. In fact, whenever a new FDS is presented, authors usually design and execute their own validation experiments by emulating falls and ADLs with their own groups of experimental subjects. In addition, authors arbitrarily choose the typologies of the ADLs and falls to be mimicked as well as the location of the smartphone and the way in which it is transported by the subjects during the execution of the movements. Moreover, in most cases, the mobility traces (mainly the signal from the accelerometer) captured by the smartphones in the course of the tests are neither stored nor offered to the research community for further investigation.

Only a few works on FDSs have made their 'clinical' databases available. For example, a database with ADLs and falls simulated by 11 volunteers is available at the web page of the Università Politecnica delle Marche (Italy) and described in (Gasparrini et al. 2014). Another similar database (obtained with 10 subjects and containing more than 800 ADLs and 500 falls), accessible via Web, is investigated in (Medrano et al. 2014). The paper in (Vilarinho et al. 2015) presents a hybrid FDS that combines a smartwatch and a smartphone. Authors state that the acceleration data of the 171 tests (ADLs and falls) conducted for the evaluation of the prototype and even some videos illustrating the fall cases will be available in the Github repository of the project (although no URL link is provided). The work in (Vavoulas et al. 2013) also offers the records of the built-sensors of a smartphone resulting from monitoring the human emulation of 4 types of falls and 9 different ADLs. The Human-Computer Interaction Lab of South China University of Technology also published a dataset, commented in (Xue & Jin 2010), with different acceleration-based activity samples from 44 individuals executing 10 ADLs and wearing tri-axis accelerometer in three different positions (no fall signals are considered).

Another interesting database has been recently released by the research group on Embedded Systems and Computational Intelligence in the University of Antioquia (Colombia) (SISTEMIC 2006). The dataset comprises the signals captured from two accelerometers and one gyroscope integrated in a device attached to the waist. The movements (19 types of ADLs and 15 types of falls) were carried out by 38 volunteers (including 15 people aged above 60). The traces are accompanied by video records of the emulated movements.

The dataset provided by the Crowdsignals initiative (CrowdSignals 2016) consists of 5 hours of data collected by monitoring two experimental subjects wearing a smartwatch and smartphone with 19 embedded sensors (including an accelerometer, a gyroscope and an accelerometer). 8 different ADLs (sitting, washing hands, driving, etc.) but no falls were executed by the subjects during the monitoring period.

As it refers to commercial Smartphone-based FDS (currently available as Android or iOS apps, such as Spantec Fall Detector, Fall Monitor, T3LAB, Fade Fall Detector, etc.) none of them gives any insight into the detection algorithm that they implement or about the conditions in which the app was validated (see Google Play Store (Google 2016) for the specifics of these applications). In the particular case of FallSafetyApp, the web page of the vendor includes some video clips illustrating that the application was evaluated with mannequins (FallSafetyApp 2016). In any case, we can state that, nowadays, there is not any mature smartphone-based FDS.

In this work, we introduce an ongoing project to deploy a Web architecture that enables researchers to generate and share the mobility traces captured by a smartphone during the evaluation tests of a FDS. The system consists of a simple Android app and a database server that can be accessed by any ordinary web browser. The app allows to fully describe the experiments and to upload the traces to the server, where they can be edited, filtered and downloaded by any remote user.

2. PROPOSED SYSTEM

Public databases planned to be used as a 'benchmark' to assess the accuracy of fall detection algorithms should thoroughly describe the way in which the traces were obtained and generated. In the case of employing the built-in inertial sensors of a smartphone to characterize falls and ADLs, the traces of an experiment should comprise one of several files containing the sequence of the timestamps and signals acquired by the embedded accelerometer, gyroscope and/or magnetometer. In addition, another extra file should inform (in a standardized way) about the conditions, instrumental and methodology that was employed to perform the experiment.

In order to ease the generation of these files, we propose the system sketched in figure 1. The operation and architecture of the proposed system is as follows. After installing the corresponding app in his/her own smartphone, the user can describe the evaluation movement that is going to be executed. For that purpose, the app offers a set of predefined drop-down menus that allows defining aspects such as the subject's characteristics (sex, age, weight, name, email and/or identification and institution), the nature of the movement to be performed (emulated or real life, ADL or fall), the typology of the ADL (walk, sit, stand, run, etc.) or fall (lateral, backwards, etc.), the selected sampling rate of the sensors (e.g. Android offers four possible configurable rates), the part of the body (wrist, chest, thigh) where the smartphone will be located and the employed mechanism (specific band, belt, pocket, etc.) to attach the phone to the user. Simultaneously, the app automatically reads the main characteristics of the smartphone (model, Android version) and the built-in mobility sensors (range, resolution, vendor and model) as well as other data such as the day, time and the GPS coordinates of the testbed location. The app also includes a form that permits the user to input an optional more 'verbose' text description of the movement conditions. This description could incorporate additional information commenting the scenario where the experiment will take place (e.g. indicating the health conditions of the subjects or informing if mattresses or any other protection pads were employed to prevent subjects from being injured while emulating the falls).

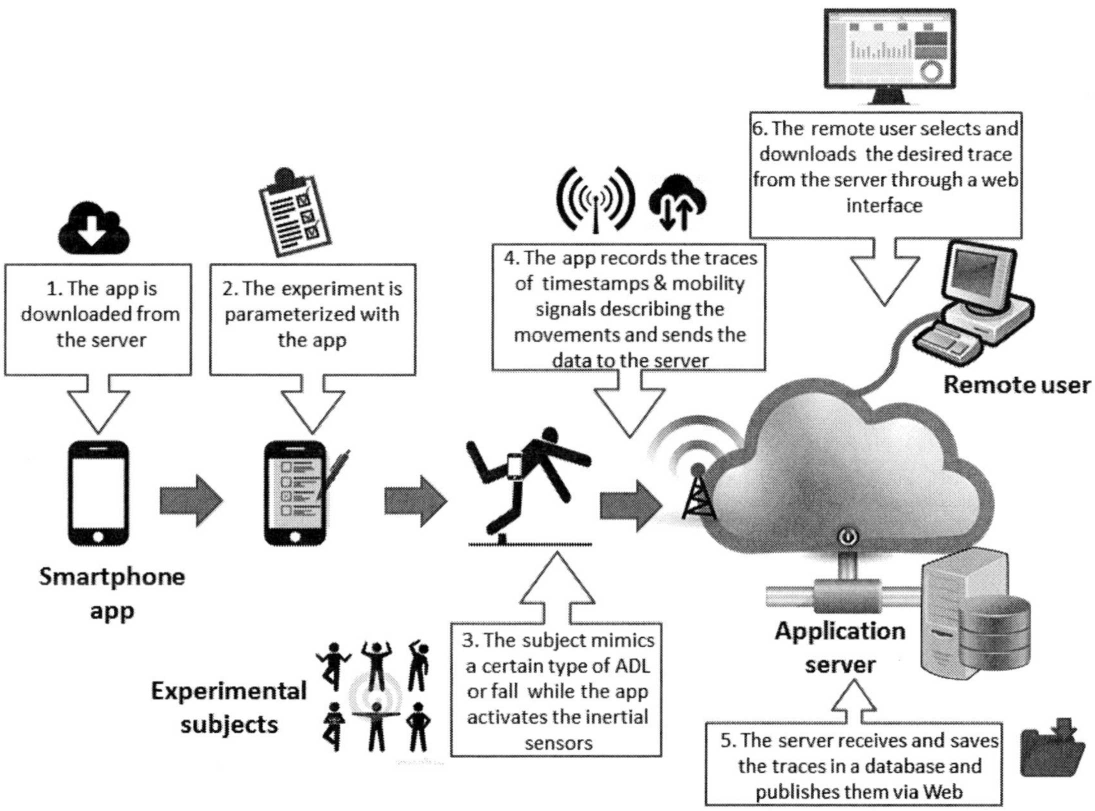

Figure 1. General architecture and dynamics of the proposed system

Once that the experiment has been properly characterized and the test subject begins to transport the smartphone, the monitoring phase can be triggered just by pushing a button in the app. So, the subject can emulate the movement that has been previously predefined while the smartphone stores the signals acquired by the mobility sensors in its internal memory (i.e. the data from the accelerometer and, if present in the device, from the gyroscope and the magnetometer). After the subject has executed the movement, the app sends (via Wi-Fi or 3G/4G connection) the captured signals (together with the whole associated data and the description of the experiment) to a remote database server. The experimental data can be accessed and browsed from any point in Internet through an ordinary Web interface. Thus, the interface enables any remote user to access and download the data depending on several search criteria (type of the movement, subject, smartphone model…). In addition, through the web interface, authorized remote users (such as those that create new datasets) can edit and correct the descriptors of the uploaded traces and even add complementary information (such as video clips of the performed movements) to help other users to identify the nature of the performed movements and replicate the experiments (if necessary). In the database, the traces associated to a particular experiment consist of two simple files with the same name and different extension: firstly, a text file describing the testbed and the conditions of the experiments, secondly, a CSV file containing the numerical series of the variables sampled from the sensors with their corresponding timestamps. The employed widespread CSV (Comma Separated Values) format allows the remote user to effortlessly process the traces through a wide set of software tools, from simple spreadsheet applications to high level mathematical analysis environments (such as Matlab). Consequently, the database will offer a set of traces describing a variety of emulated (or even real) movements, which could be used as a common framework by the research community to asses and compare in an offline and systematic way the performance of any existing fall detection algorithm. With respect to the previous (and scarce) publicly available datasets, the main advantage of our system is that traces are not limited to those generated by a single research team in a particular moment and a specific environment. Any researcher on the topic can potentially expand the database with new traces just by downloading and running an app while the movements are executed.

A pilot version of the developed system (with some example movement traces) is already available in the URL: http://falldetection.esy.es/.

3. CONCLUSION

Smartphones, which natively incorporate built-in inertial sensors, computing power and communication interfaces, provide a technological opportunity to implement Fall Detection Systems at practically no cost. Hence, smartphone-FDSs have recently turned into a topic of increasing interest in the field of telemedicine and mHealth (mobile Health). A still open issue in this domain is the evaluation of the fall detection algorithms with benchmark datasets. The review of the related literature reveals the complete absence of a reference framework to validate and compare the existing (and forthcoming) proposals. As a matter of fact, every new work proposing a new FDS envisages its own evaluation procedures to assess the effectiveness of the developed prototype. In most cases, whenever a new detection algorithm is designed, authors are obliged to recruit experimental subjects that emulate falls and Activities of Daily Living (ADLs) so that the performance of the system can be appraised.

This paper has presented an ongoing work to define an architecture that allows the researchers on this field to create and share their own databases of traces describing emulated falls and ADLs. The system includes two elements: a smartphone application (which characterizes and records the experiments) and a web server that receives the description of the experiments and the data traces captured by the smartphone. These traces can be easily accessed, downloaded and edited by any remote user from a browser.

The tool is aimed at providing a common reference framework to compare and validate algorithms for smartphone based-FDSs as long as it will help researchers to share with the research community their own samples and, in turn, to obtain the traces generated by other investigators.

In any case, we have to take into account that the assessment of a wearable FDS must face both technical and sociological aspects. Most analyses in the literature only evaluate the effectiveness of the detectors to differentiate falls from ADLs, neglecting other factors, such as the impact of the FDS on the autonomy, application coexistence or computing power of the smartphones. Moreover, a comprehensive evaluation of the FDS should prioritize crucial human aspects such as ergonomics or user acceptance (especially if we assume that older people are the target population for these products).

ACKNOWLEDGEMENT

This work was supported by European FEDER funds and the Spanish Ministry of Economy and Competitiveness [grant TEC2013-42711-R] (URL: http://www.mineco.gob.es/portal/site/mineco/)

REFERENCES

Bagalà, F. et al., 2012. Evaluation of accelerometer-based fall detection algorithms on real-world falls. *PLoS One*, 7(5), p.e37062.

Casilari, E., Luque, R. & Morón, M., 2015. Analysis of android device-based solutions for fall detection. *Sensors*, 15(8), pp.17827–17894. Available at: http://www.mdpi.com/1424-8220/15/8/17827/pdf [Accessed September 9, 2015].

CrowdSignals, 2016. AlgoSnap Sample Dataset. Available at: http://crowdsignals.io/ [Accessed April 19, 2016].

FallSafetyApp, 2016. Fall Safety App and Fall Safety Service for Organizations. Available at: https://www.fallsafetyapp.com/ [Accessed March 8, 2016].

FARSEEING, 2015. FARSEEING (FAll Repository for the design of Smart and sElf-adaptive Environments prolonging INdependent livinG) project. Available at: http://farseeingresearch.eu/ [Accessed March 8, 2016].

Gasparrini, S. et al., 2014. A depth-based fall detection system using a Kinect® sensor. *Sensors*, 14(2), pp.2756–2775.

Google, 2016. Google Play Store. Available at: https://play.google.com/store [Accessed March 10, 2016].

Klenk, J. et al., 2011. Comparison of acceleration signals of simulated and real-world backward falls. *Medical engineering & physics*, 33(3), pp.368–373.

Klenk, J. et al., 2013. Development of a standard fall data format for signals from body-worn sensors. *Zeitschrift für Gerontologie und Geriatrie*, 46(8), pp.720–726.

Medrano, C. et al., 2014. Detecting falls as novelties in acceleration patterns acquired with smartphones. *PloS one*, 9(4), p.e94811.

Schwickert, L. et al., 2013. Fall detection with body-worn sensors: a systematic review. *Zeitschrift für Gerontologie und Geriatrie*, 46(8), pp.706–719.

SISTEMIC, 2006. SisFall Dataset (SISTEMIC Research Group, University of Antioquia, Colombia). Available at: (http: 85 //sistemic.udea.edu.co/wp-content/uploads/2016/03/SisFall_dataset.zip), [Accessed April 19, 2016].

Vavoulas, G. et al., 2013. The MobiFall dataset: An initial evaluation of fall detection algorithms using smartphones. In *Proceedings of the IEEE 13th International Conference on Bioinformatics and Bioengineering (BIBE 2013)*. Chania, Greece: IEEE, pp. 1–4.

Vilarinho, T. et al., 2015. A Combined Smartphone and Smartwatch Fall Detection System. In *Proceedings of the 2015 IEEE International Conference on Computer and Information Technology; Ubiquitous Computing and Communications; Dependable, Autonomic and Secure Computing; Pervasive Intelligence and Computing (CIT/IUCC/DASC/PICOM),*. pp. 1443–1448.

Xue, Y. & Jin, L., 2010. A naturalistic 3D acceleration-based activity dataset & benchmark evaluations. In *Proceedings of the IEEE International Conference on Systems Man and Cybernetics (SMC' 2010)*. Istanbul, Turkey, pp. 4081–4085.

Yoshida, S., 2007. *A Global Report on Falls Prevention Epidemiology of Falls*, World Health Organization. Available at: http://www.who.int/ageing/projects/1.Epidemiology of falls in older age.pdf.

MANAGEMENT OF EMERGENCY DEPARTMENTS USING COMPLEX SYSTEMS MODELLING

Hao Bai[1], Mark Burkitt[1], Laura Smith[1], Gemma Poulter[2], Alan Kyffin[2], Chris Greenough[2] and Mike Holcombe[1]

[1] *Advanced Computing Research Centre, University of Sheffield, Sheffield, UK,*

[2] *Software Engineering Group, Rutherford Appleton Laboratory, STFC, UK*

ABSTRACT

Managing Emergency Department (ED) in the light of unpredictable patient demand, limited resources and the need to meet national targets is a major challenge. Launched by the UK's Department of Health since 2000, the National Health Service (NHS) is now facing a 4-hour maximum time target to meet, where failures will trigger considerable financial penalties. However, the target is often missed especially with an increasing demand year on year, and it's widely accepted that process management is one of the key reasons. In this work, a state-of-the-art approach using Agent-based Modelling (ABM) is introduced to provide a *decision support system* for the ED to obtain information ahead of time in order to optimise the management of human and material resources. By modelling each patient, member of staff and other related resources as agents, this model has the nature of flexibility and capability of providing any detail. With the benefit of modularised design, this platform can also be easily customised for any ED.

KEYWORDS

Emergency Department; management; modelling; prediction; decision support

1. INTRODUCTION

Managing Emergency Department (ED) has been a challenge for a long time [Derlet 2000]. Since the 4-hour time target was introduced by the UK's Department of Health in 2000, many EDs have struggled to meet it. According to the survey carried out by the British Medical Association, seven main reasons were listed for not reaching the target [BMA survey]: 1) Not enough in-patient beds; 2) Delayed discharges; 3) Delay in accessing specialist opinion; 4) Not enough nurses; 5) Not enough middle grade doctors; 6) Department too small; 7) Delay in accessing diagnostic services. Thus ED experiences a lack of human/physical resources, and also has bottlenecks in resource management. Over the next few years, the National Health Service (NHS) in the UK will continue experiencing budget pressure, which makes it difficult to improve the ED services by increasing the amount of staff and physical resources. Therefore, improving the efficiency of NHS systems, especially the ED, becomes necessary in order to provide a comprehensive and high quality care with limited funds, and to improve efficiency filling the gap between demand and funding [NHS five year forward view]. There is a predicted growth in attendance of 5-7% year on year [Health Building Note 15-01]. However, recent reports in the press have indicated a near 10% rise in attendance at ED. As shown in Fig. 1, the failure to meet the 4-hour target is likely to be amplified in coming years without an improved resources management strategy.

To help manage ED, simulation has been proposed as a possible solution [McGuire 1994]. Studies have been carried out using different methodologies, such as traditional Discrete Event Simulation (DES) [Connelly and Bair 2004], Agent-based Modelling(ABM) [Wang 2009 and Cabrera et.al. 2012] etc. So far, there is no successful system capable of simulating the whole emergency department with deep-level details. This paper proposes an ABM based approach that models the complete emergency department based on historical data and interaction rules, and provides a rich set of real-time and predictive information of the status of the department. Agent-based modeling is used in many areas for the analysis and management of complex systems, especially those involving humans. It is radically different to the discrete event approach adapted to hospitals from the world of factories.

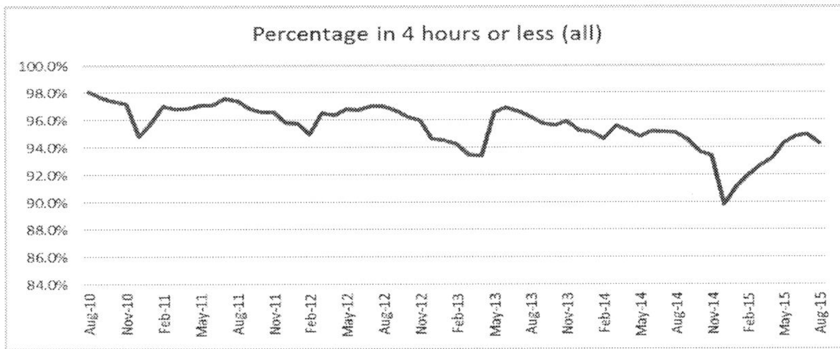

Figure 1. The number of patients seen within 4 hours over the past 5 years at all NHS EDs

2. THE EMERGENCY DEPARTMENT MANAGEMENT EMERGENCY

The growing demand on emergency services and the target regime that it has to satisfy has lead to significant problems for hospitals, both financial and reputational. Hospitals are fined for missing their ED targets, for example, this could amount to £500k a quarter. Hospitals are also increasingly judged through publicly available league tables.

Given this situation, a decision support system driven by an Agent-based Model (ABM) can help the managers in ED to improve their control through prediction of resources and patients. The system provides the benefits of overview of future potential bottlenecks within the department.

2.1 The Hospital Context

Table 1. Number of patients seen within 4 hours and penalties (05/04/2015 to 28/06/2015)

	Attendees to Department of ED	**Attendees seen > 4 hours**	**Percentage of attendees seen < 4 hours**	**Penalty Charges**
Total number of patients over 140 EDs	3,702,458	329,060	91.1%	£39,487,200
CMFT ED	37,220	3,542	90.5%	£425,040

The agent-based model was designed and developed in collaboration with the ED of Central Manchester Foundation Trust (CMFT) that is one of the busiest EDs in the UK. Data from the CMFT shows there were 5,628 attendees visiting the ED in the last week of July 2015. An average of 86% of attendees were seen to transfer or be discharged within 4 hours that is lower than the current target required by the NHS of 95%. ED staff are constantly under a high pressure due to the high demands and inevitably the quality of care provided is affected.

In order to urge EDs to improve their performance, a financial penalty of £120 per breach under the 95% target is enforced on a monthly basis. In the period of June 2015, ED of CMFT had 12,005 major patients, in which 7.68% (916) of those were not seen within 4 hours triggering a penalty charge of £109,920. A comparison is given in the Table 1 to show the how the 4-hour target was met and penalty charges across the country. The figures in the table indicate a large number of hospitals in the UK are struggling to meet the target, and the CMFT ED needs extra effort.

Under these circumstances, an ABM based decision support tool was developed to help CMFT ED to improve their management by simulating the operation of the whole department. The CMFT ED has three major emergency zones: *red* for patients delivered by ambulance who require rescue; *amber* for urgent medical treatment; *green* for stable patients. As a detailed simulation technology, the agent-based modelling is able to simulate each individual member of staff, patient, and other resources, e.g. room, cubicle etc. in each zone. Based on the knowledge of how patients flow within the department and the historical data, this model is capable of reproducing the accurate picture of the department.

2.2 The Model

The implementation is based on the Flexible Large-scale Agent Modelling Environment (FLAME) [www.flame.ac.uk, Coakley *et.al.* 2006 & 2012]. FLAME is a generic agent-based modelling system which can be used to develop applications on majority of computing systems ranging from laptops to HPC super computers. FLAME based ABM has been successfully used in many fields, such as economy and biology [Holcombe *et.al.* 2013, Bai *et.al.* 2014 and Gavin *et.al.* 2015].

ABMs represent complex systems as a collection of autonomous entities – *agents* – and set of rules of how these *agents* interact and behave. Among the agents in this model are patients, doctors, nurses, nurse practitioners, consultants, receptionists, porters. Each agent has integrated into it a dynamic data set that identifies it, its attributes and location, etc. There can be many different examples of each of these agent types. Each agent type can behave in a number of ways depending on what they are, what their attributes are, who they are interacting with and what options are open to them.

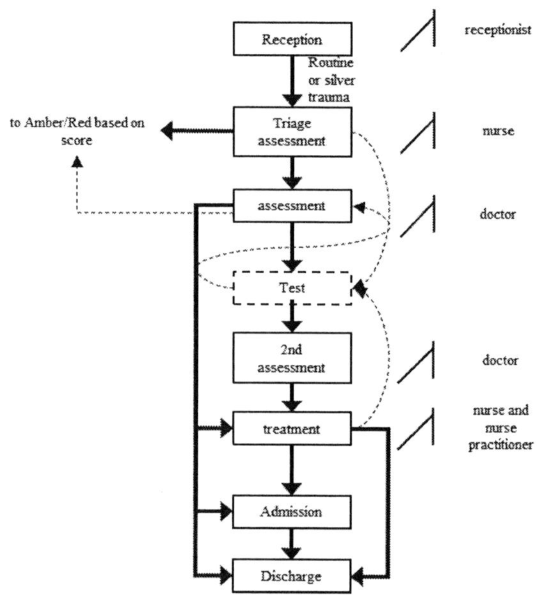

Figure 2. Patient flow in green zone of the CMFT ED department

Thus within CMFT, for example, on arrival in the green zone reception (self-admitting patients) the receptionist agent will collect information from the patient and when this happens the patient agent will be given an *id* and a basic set of details which are then 'attached' to it in the model as attributes. The patient then waits in the waiting area unless they are identified as an emergency. The next stage will be triage and for this a nurse, when available, will examine the patient and record the basic aspects of triage. This information is again attached to the patient but will also be broadcast for other staff to read when appropriate – e.g. when making decisions about who a doctor should see next or during further treatment. The patient flow in green zone is presented in Figure 2. The patient agents go through the flow and interact with staff agents at each process. Depending on the availability of staff and other resources, the performance of green zone is therefore simulated and the model naturally generates outcome based on randomness.

An example of model output is shown in Fig. 3. In this figure, the profile of patient arriving at and discharged from green zone is displayed, where the numbers were counted hourly. As a comparison, the usage of resources, such as reception, triage and cubicles within green zone, are presented. Starting a simulation from 8:30 in the morning, it can be observed that there are more patients present in the daytime. With the patient number increasing, more resources are used for patient registration, triaging, diagnosis and treatments. The model outputs correctly reflect the status of daily running of the department.

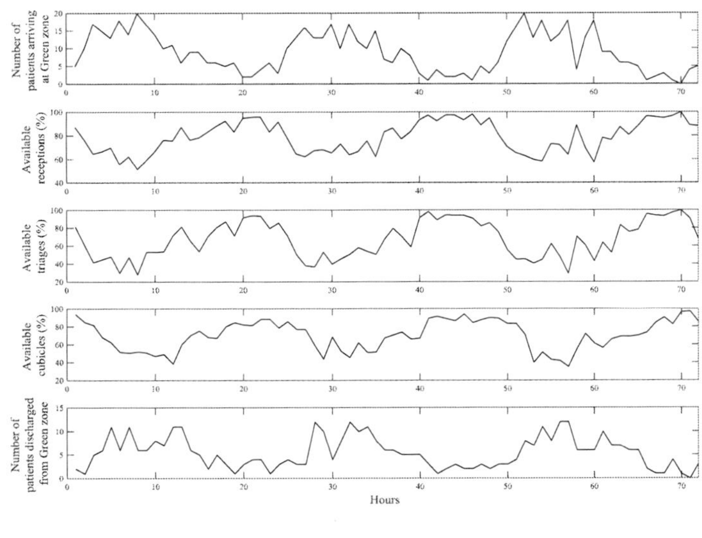

Figure 3. Patient flow in green zone *vs.* usage of resources

2.3 The Decision Support System

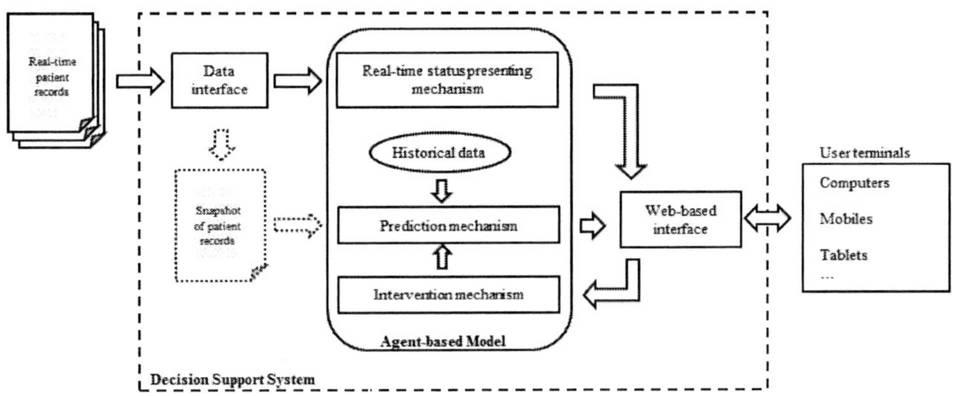

Figure 4. Diagram of ABM-based decision support system

The system model is based upon the detailed set of agent types which have been developed based on both observations at the hospital but also from the use of historical data going back 3 years. The decision support system is composed of a web-based interface and the ABM model running at the back-end. The architecture of proposed system is displayed in Fig. 4. The system could connect with the real-time patient record database for presenting the current status of the whole department. Upon user's request, when a prediction is needed, a snapshot could be generated, and from where a *faster than real time* predictive simulation would be run based on historical data. The predictive results would be provided to user for looking ahead of time. If an intervention is necessary, user would be able to apply the intervention to the model, and re-run the simulation to test its effectiveness.

3. CONCLUSION AND FUTURE WORK

As a state-of-the-art approach, the Agent-based Model captures the complexities of process of patient flow in ED department, and provides the opportunity to monitor the situation and to predict what is likely to happen in the next 4, 12, 24, 48, ... hours. This gives managers and clinicians a foresight view to understand where the blockages are and optimise their management accordingly. Future work will be carried out to collect more information and data from CMFT ED for the purpose of further model testing and validation, which will improve the realism and accuracy of this system. Real-time patient records and staff measurement data will be connected to the model to enable the model to present on-going status of the department and to predict from any chosen scenario.

ACKNOWLEDGEMENT

The authors thank colleagues at the Central Manchester Foundation Trust for their collaboration, and Mathew Hall for developing the web-based interface.

REFERENCES

Bai H, Rolfe MD, Jia W, Coakley S, Poole RK, Green J & Holcombe M., Agent-based modeling of oxygen-responsive transcription factors in Escherichia coli. PLoS Comput Biol, 10(4), e1003595, 2014

Cabrera E., Taboada M., Iglesias M., Epelde F. and Luque E., Simulation Optimization for Healthcare Emergency Departments. Proceedings of the International Conference on Computational Science (ICCS), 9 (0) (2012), pp. 1464–1473

Coakley S., Gheorghe M., Holcombe M., Chin S., Worth D., and Greenough C. Exploitation of HPC in the FLAME agent-based simulation framework. In Proceedings of the 2012 IEEE 14th Int. Conf. on HPC and Communication HPCC '12, pages 538–545, Washington, DC, USA, 2012

Coakley S., Smallwood R., and Holcombe M. Using X-machines as a formal basis for describing agents in agent-based modelling. SIMULATION SERIES, 38(2):33, 2006

Connelly L.G., Bair A.E., Discrete Event Simulation of Emergency Department Activity: A Platform for System-level Operations Research. Acad Emerg Med 11:1177-1185, 2004

Derlet R., Richards J., Overcrowding in the nation's emergency departments: complex causes and disturbing effects, Annals of Emergency Medicine, 35(1) 63-68, 2000

Gavin F., Jonathan W., Mike H. and Giuseppe B., Modelling the Transport of Nanoparticles under Blood Flow using an Agent-based Approach, Scientific Reports 5, Article number:10649,doi:10.1038/srep10649, 2015

Holcombe M, Coakley S, Kiran M, Chin S, Greenough C, Worth D, Cincotti S, Raberto M, Teglio A, Deissenberg C, van der Hoog S, Dawid H, Gemkow S, Harting P and Neugart M., Large-scale modeling of economic systems. Complex Systems, 22(2), 175-191, 2013

McGuire F., Using simulation to reduce lengths of stay in emergency departments, Proceedings Simulation Conference, IEEE Press, 861-867, 1994

Wang L., An agent-based simulation for workflow in emergency department, in: Systems and Information Engineering Design Symposium, 2009. SIEDS'09., 2009, pp. 19-23.

Web materials

[BMA survey] http://www.rcem.ac.uk/CEM/document?ID=3156&subID=1027

[Health Building Note 15-01] https://www.gov.uk/government/uploads/system/uploads/attachment_data/file/206267/15_01final3_v3.pdf

[NHS five year forward view] https://www.england.nhs.uk/wp-content/uploads/2014/10/5yfv-web.pdf

HOW EXISTING SOCIAL MEDIA ASSIST HEALTH AND CARE CO-PRODUCTION: A QUALITATIVE STUDY

Hadi Daneshvar and Stuart Anderson
The University of Edinburgh
10 Crichton Street, Edinburgh, EH89AB, UK

ABSTRACT

Future of health in EU faces the triple challenges of ageing, fiscal restriction and inclusion. Co-production offers ways to manage informal care resources to help them cater for the growing needs of aged people. Social media (SM) are seen as critical enabler for co-production. We investigate the uses of SM as an enabler of co-production in health and care. We conducted a qualitative study using interviews and online observations. We found that particular types of SM are currently used for coordination and communication across boundaries to enable co-production. Particularly Twitter and Facebook are used to connect carers, disseminate information and invite volunteers. However many other types of SM are still not in use in this sector because of shortcoming of systems and lack of user awareness. Nevertheless, carers showed interest in using systems which help them to engage people in shaping of services, sharing of experiences and encouraging care activities.

KEYWORDS

Social Media; e-Health; Co-production.

1. BACKGROUND

Health in EU faces the triple challenge of ageing, fiscal restriction and inclusion(Carretero et al., 2012). In the UK the number of elderly people will increase to 6.6 million over the next 25 years. In Scotland, by 2035 the 65+ years group is projected to have grown by 82%. It is the most elderly age-groups of the population that are projected to increase most dramatically. Between 2010 and 2035 those aged 65+ are projected to increase by 82%. .

As a result of the matters described, this study intends to focus on three important types of problems caused by aging population in Scotland, and the wider UK. These problems can be highly categorized as: 1) Increasing population of elderly people; 2) Unavailability of enough resources; 3) Exclusion from society. These lead to increased governments` need for expenditure to provide and deliver health and care services, increased need for expenditure by older people while their income has often fallen, and finally exclusion from society due to limited resources.

The statistics show, the needs of elderly people are growing and there is an increase of requirement for carers. Now the population of informal carers are around 10% of 62M population of the UK. It is projected that this number will increase to 9 million(Carers, 2012) of 73.2M (around 12% of population) in the next 25 years. The current value of care is worth an estimated £119bn per year – almost the spending as the NHS England's budget. So an important challenge is how to resource for care and health of elderly people in future. In this respect we have a big pool of informal carers which need to be managed to achieve care needs.

Currently many of the public services are poorly equipped with potential social economy of family and neighbors. Depending solely on economic growth to fulfil the finance needs of public services, could lead to less efficient quality of service and outcomes. Hence new ways of overcoming this shortage are needed (Boyle and Harris, 2009). In order to reshape service delivery we need to equip different resources. Examining the concept of co-production is an initial step in reforming the service delivery. Boyle and Harris from NEF(the New Economics Foundation) give a definition for co-production:

"Co-production means delivering public services in an equal and reciprocal relationship between professionals, people using services, their families and their neighbors. Where activities are co-produced in this way, both services and neighborhoods become far more effective agents of change." (Boyle and Harris, 2009).

Co-production offers ways to manage informal care resources. The full participation of informal carers in the co-production of health and care has the potential to play a significant role in the sustainability of health and care delivery. A pressing issue is how to coordinate this massive resource with the formal health and care system to enable true co-production of health and care. Increasingly e-health, is seen as the tool to re-shape healthcare systems(Gaddi et al., 2014). In particular, social media (SM) are seen as critical enabler for co-production(Lin and Lu, 2011). Communications is a key element in co-production that enables coordinating across various bounders. SM cuts across boundaries and is well understood.

SM are online tools for the creation and sharing of digital content. They aim for widespread use and are capable of supporting an unlimited number of users. Kaplan and Haenlein (2010) defined SM as "a group of Internet-based applications that build on the ideological and technological foundations of Web 2.0, and that allow the creation and exchange of User Generated Content". Dissemination of content operates at Internet speeds. It has been argued that SM have caused a change in social action in many areas (Asur and Huberman, 2010). The term was coined by Chris Shipley after his research and reports on technology trends (Newson et al., 2009). SM have since become media for the creation and maintenance of connection and interaction amongst individuals (Kwai Fun IP and Wagner, 2008). They are currently used widely by diverse range of users and have become amongst the largest number of most visited sites worldwide (Alexa.com, February, 2016).

2. OBJECTIVE

Our study investigates the current and possible future for SM as an enabler of co-production in health and care for elderly people. To achieve these aims two main sets of questions are asked: what are the current uses of SM in health and social care? How can SM be reshaped to enable (and reshape) health and care co-production?

We consider a typology of opportunities and limitations of SM for health and care. The table 1 shows existing health and care service bundles with (1) existing or (2) new SM tools. New health and care service bundles with (3) existing or (4) emerging SM tools. In this part of the research we focused on the first part type "Existing health and care service with existing SM tools".

Table 1. SM and Health and Care

		Health and Care	
Social Media		Current Service	Change Services
	Current App.	Use of current SM in existing situation (1)	Current SM help to change health and care service (2)
	Redesign	New SM to cater for the current health and care services (3)	New Social Media with New Health and Care Services (4)

3. METHODOLOGY

We conducted a qualitative study investigating the sociotechnical aspects of the current and possible future uses of SM by different organizations and groups as an enabler of co-production. We used purposeful sampling to select organizations and group who were providing care services to elderly people. We used a combination of interviews and online observations of groups and activities to collect data. Combining different methods enabled us to triangulate the data sources to validate our findings.

For the wider study, we conducted 20 semi-structured interviews which focused on the services offered, the types of online applications (particularly SM) used, their challenges and future possibilities. For this paper we focused on interviews conducted with voluntarily and professional organizations that provided care service to elderly people or private support to carers. The role of interviewees were directors, care supervisor, business manager or digital and marketing manager of the organizations. We also observed the activities of 10 groups/organizations in different SM environments (e.g some closed groups which used Facebook or other types of public SM for their purpose and activity). For coding and analyzing data we used Nvivo software.

Table 2. Social media use by different groups

Existing SM	Professional Carers	Voluntary organizations	Private Sectors
Social Networking (FB.)	• Companies have a Facebook page but it is not widely used by professional carers • Some carer supervisors use SN informally to communicate with carers to • Arrange meetings between them and carers • Provide feedback on carers work • But there are some difficulties in using SN such as FB in this regard: • Many carers have low computer skills • Care data is very sensitive so they must be cautious in using SM as a communication medium for such data.	• Voluntary organizations use SN to broadcast information to people (carers and families and sometimes patients) • Information about new rules and regulations • Research findings • Promote activities of government which are for the benefit of carers • Promote Services for carers and families • Offer replacement care service • Courses • Promote activities for carers and patients • Reading groups • Group walks • Invite volunteers (carers) to contribute to activities for elderly people • Sponsored walk • Connect with other organizations and promote their services	• Private groups exist for patients and family carers which are very close in nature (not even happy to be interviewed) • These groups are managed by volunteer carers who may have a very different backgrounds • E.g. COPD group in Scotland, only people can join if they have the COPD condition or they are carers of people with this condition
Micro-Blog (Twitter)	• Rarely used (Mainly used for broadcasting of information) • E.g. activities or new job opportunities	• Similar use to SN, however with higher frequency, also more communication and connection with other organizations. • Also used to receive updates about contents of events when they are not present in an event. • Additionally the nature of twitter allows them to communicate with wider people and organizations by posing questions. Enables them to receive feedback from the wider community. • Microblogs are also used for fundraising	• Not observed
Blog	• Not familiar and not used widely.	• Not used widely, because its time consuming to keep it updated with long posts • Only very large volunteer organizations use blogging. Patients and families share stories and experiences. Other patients and families respond	• Not observed
Forum	• Not familiar and not used widely.	• Online forum not used, because information need to be controlled and its time consuming to do this control	• Not observed
Website	• Website used to promote and present the company with very limited interaction with audience.	• Website used as a news page with very limited interaction with audience.	• Not observed

3. RESULTS

Our findings showed very limited use of SM in elderly healthcare services. There were two interrelated causes for this limited use: 1) users were unaware of the ways they could use the available systems; 2) systems did not cater for the needs of the users. In this regard we found there are no (or very limited) particular forums or on-line communities formed for detailed discussions around elderly healthcare matters. The same applies to blogs, where use is very limited (or non-existent) by professional carers and volunteer organizations. The very few specialized groups that exist were closed groups, so much that they did not even accept to be interviewed or observed.

The main reason for not using SM were said to be:

- Low skills and lack of awareness of users particularly professional carers;
- High costs and effort needed to update and monitor the contents of these SMs;
- Professional carers are not familiar with professional use of technology. Although many carers showed interest and some were even using SM for personal communication, they had very limited knowledge of how can these systems be used to achieve benefits in the course of giving care for elderly people;
- Problems in accessing to the online resources because of low level of skills and confidence, in particular for elderly people.
- Carers and professional carers have security concerns in using SM to share information

We also found that the majority of websites used by care organizations use web 1 technologies with very limited interaction functionalities. Furthermore, there was no specialized SM designed to be specifically used for elderly health and care services. The reason for this is that there is no business case to see the benefits of having web 2 technologies and using SM in organizations as well as by professional carers (knowledge and resources are limited).

We also found that there are certain types of SM used in co-production of healthcare services for elderly people. Particularly Twitter and Facebook are amongst the commonly used types. Twitter is used by service organizations as a tools for communication between organizations and their followers and receive feedback from them. It is used to invite volunteers to contribute in different activities. Facebook on the other hand was used to create informal connection with carers in order to appreciate their work and encourage them in taking on care activities. The main characteristic of Twitter that promoted its wide use were that it was easy to use, very quick in dissemination of information and flat in nature. These reasons led to higher use of Twitter in comparison to Facebook functionalities such as re-tweeting and hashtag (e.g. for fundraising purposes and information dissemination) are widely used by organizations to collaborate with each other in achieving their goals. Facebook on the other hand was used commonly by carers and private sector (patients and families) because groups (more private) can be formed to discuss matters in a more private manner.

In conclusion, so far in this research, after characterizing uses of SM and identifying their benefits and risks of SM for health and care, we have identified coordination tools and social sense-making as key emerging trends in the use of SM. Also we have found that only particular types of SM are used (rarely) by particular carers/ carers organizations. Finally, we have identified health and care integration as a key driver for service redesign.

In the next steps of this project we aim to find on the new services which can be offered with existing SM. Then we will be exploring the possibilities of designing a new SM to cater for the existing need and new services required.

REFERENCES

National Population Projections 2010-based Statistical Bulletin, 2011. Office for National Statistic.

Asur, S. & Huberman, B. A. Predicting the future with social media. Web Intelligence and Intelligent Agent Technology (WI-IAT), 2010 IEEE/WIC/ACM International Conference on, 2010. IEEE, 492-499.

Boyle, D. & Harris, M. 2009. The challenge of co-production. *London, New Economics Foundation.*

Carers, U. 2012. Facts about carers 2012. *Policy briefing. London: Carers UK. Available at: www. carersuk. org/media/k2/attachments/Facts_about_carers_Dec_2012. pdf (accessed on 7 May 2013).*

Carretero, S., Stewart, J., Centeno, C., Barbabella, F., Schmidt, A., Lamontagne-Godwin, F. & Lamura, G. 2012. Can Technology-based Services support Long-term Care Challenges in Home Care? *Analysis of evidence from social innovation good practices across the EU CARICT Project Summary Report*.

Gaddi, A., Capello, F. & Manca, M. 2014. eHealth, care and quality of life. Milan, Heidelberg, New York, Dordrecht, London: Springer.

Kaplan, A. M. & Haenlein, M. 2010. Users of the world, unite! The challenges and opportunities of Social Media. *Business Horizons,* 53, 59-68.

Kwai Fun Ip, R. & Wagner, C. 2008. Weblogging: A study of social computing and its impact on organizations. *Decision Support Systems,* 45, 242-250.

Lin, K.-Y. & Lu, H.-P. 2011. Why people use social networking sites: An empirical study integrating network externalities and motivation theory. *Computers in Human Behavior,* 27, 1152-1161.

Newson, A., Houghton, D. & Patten, J. 2009. *Blogging and other social media: Exploiting the technology and protecting the enterprise*, Gower Publishing, Ltd.

AN INFORMAL MULTI-STAKEHOLDER NETWORK AS A COLLABORATIVE STRATEGY TOOL IN E-HEALTH

Øyvind Heimset Larsen[1] and Hilde G. Corneliussen[2]
[1]*Researcher,*
[2]*Senior Researcher,*
Western Norway Research Institute
PO Box 163, NO-6851 Sogndal, Norway

ABSTRACT

The county of Sogn & Fjordane is one of 19 counties in Norway, with approximately 109 000 inhabitants in 26 municipalities. This county has achieved outstanding results in the field of e-Health, as one of the first counties in Norway that completed the process of moving message in the health sector to an electronic platform, by June 2014. What has made this challenging region become a spearhead and the 26 municipalities able, not only to engage in, but to achieve technological innovative and organizational results and success in health services? The analysis takes the theory of Triple Helix systems as a starting point, and shows that an informal multi-stakeholder network has been a vital driving force, operating across the small and diverse units without challenging existing power structures.

KEYWORDS

Community e-Health Services, e-Health Infrastructure, Organizational change, System Integration, Multi-stakeholder network

1. INTRODUCTION

The county of Sogn and Fjordane (S&Fj),[1] one of Norway's 19 counties, is a challenging region for public services, owing to the great distances and demanding topography, including long fjords and high mountains dividing the landscape. Despite the regional characteristics constantly putting local infrastructure and communication to the test for the 26 municipalities in the county, the region has shown outstanding results in the field of e-Health and new technology in health care. S&Fj was one of the first counties to complete the process of establishing electronic messaging between local health services, general practitioners and specialist care in 2014, and has since illustrated a unique collective and ambitious engagement in welfare technology (Ambient Assisted Living). The aim of this paper is to explore which mechanisms that have been active in making this challenging region become a spearhead in Norway. What has made the 26 independent municipalities able, not only to engage in, but to achieve innovative and organizational results and success in implementing technology in health services?

In order to explore this, we will revisit documentation from and evaluation of the processes that introduced e-messaging, and point at mechanisms working as driving forces in the processes. In order to understand innovation in the health care sector in this region it is necessary to include a perspective of cross-disciplinary and even cross-sectorial cooperation, and we will connect this discussion to Bergek et al's theory of Triple Helix networks (2008).

In the next section we will present the empirical case of e-messaging, before we present the Triple Helix systems theory and the analysis with our findings.

[1] The county of Sogn & Fjordane covers 18 623 km2 of mountains and fjords, with approximately 109 000 inhabitants. The municipalities has populations between 800 and 12 800 (Statistics Norway, 2015, http://bit.ly/20Nd61T).

2. THE CASE: TECHNOLOGY INNOVATION IN HEALTH CARE

Health care services are the single largest expense for the municipalities, and it is a sector heavily relying on personnel to achieve its goals. Over the last decades we have seen a growing emphasis on increased efficiency and reduced costs in the health care sector, and during the last decade also increased emphasis on innovative use of technology as tool to achieve these goals (NOU 2011:11). Our aim here is to explain why the county of Sogn & Fjordane and its 26 otherwise independent municipalities have managed to achieve common goals of establishing technological solutions in the health care services.

2.1 Contextualizing Health Care Services in Sogn & Fjordane

The many small municipalities in S&Fj have a tradition of working together to develop infrastructure and services. The regional health trust, which is the specialist health care, controls the hospitals and medical clinics in the county. The municipalities' health care service is organized in health centres across the 26 municipalities, with each centre employing a handful of people who complete a range of tasks. The 26 municipalities have been early adopters of technology in health care, being among the first to connect to the secure telecommunications network for health care, developed and managed by the government enterprise Norwegian Health Network in 2007, and they were also among the first to complete a full scale regional transition to e-messaging in 2014. Implementing new technologies in health care services requires changes that go deep in the institutions, involving training, new procedures, the reallocation of resources and new organizational structures. How did this process unfold in S&Fj?

2.2 Electronic Messages in the Health Care Sector

Since the mid-1990s there have been national plans for involving IT in health care services. In the early phases, between 1997 and 2007, the landscape was dominated by enthusiasts and local projects more than nationally coordinated projects. The national goal was to move towards electronic solutions, and to see the IT development in relation to organisational development and a holistic perspective on health care. The national secure telecommunications network for health care was part of this initiative, and a second phase was the initiative to move all communication between local health services, general practitioners and specialist care to an electronic platform, also involving a development towards electronic patient journal, all of this rooted in governmental reports and white papers in the period.[2] Simultaneously, a reform in the health care sector – The Coordination Reform, directed more responsibility to the municipalities (HOD 2008-2009), thus increasing the need for precise and quick information and communication between the regional specialist care and municipalities, as the goal was to deliver health care as close to the patients' home as possible.[3]

The national goal set for electronic messages in the health care sector was that it should be "the norm for exchange of patient information" between all involved parties by the end of 2015, supporting an increase in "information safety and accessibility."[4] Verbal communication will always be important in a therapeutic setting and in the personal meeting between patient and clinician. But there is also a need to exchange information between different parties in the treatment chain, a need increasing in accordance with a more specialized health service. And this exchange of information has to be secure and robust.

A series of projects were initiated in the region in the period between 2005 and 2014, starting with a project in 2005 aiming to put a greater emphasis on effective use of ICT and broadband technology within health care services operating in the intersection between municipalities and the health trust. This was followed by several projects under headings of "IT for health care services in the county of Sogn & Fjordane", and later "Electronic messaging in health care services in Sogn & Fjordane".

From the start in 2005 to 2012 it was a long and winding road that dominated (Larsen and Skogseid 2012). The e-messaging project between local health services and general practitioners (GPs) was completed

[2] The first National Action Plan for the ICT sector in health "More health for each bIT", 1997-2000. The second was "bIT for Bit" 2001-2003. The third was "Interaction S@mspill 2007", 2004-2007. The fourth was "Interaction 2.0" 2008-2013 that included "the national message boost." Then followed the current National Action Plan for eHealth 2014-2016), https://ehelse.no/english.
[3] The Coordination Reform, https://www.regjeringen.no/globalassets/upload/hod/vedlegg/bhh-ppt-trondheim-en-24-09-08.pdf.
[4] https://www.nhn.no/english/Pages/electronic-messaging.aspx.

in 2009, with new secure infrastructure established on all GPs offices and new work procedures developed, also this as one of the first counties (HDIR 2009). The following rollout initiatives were successful, although they to some extent where delayed and hampered by external factors. Between 2012 and 2014, however, the e-messaging project realised the task of establishing electronic messaging between local health services, GPs and specialist care in Sogn & Fjordane. Thus, by June 2014 – one and a half year before the national goal – all municipalities had established e-messaging as one of the first counties in the country.

2.2.1 What Made the Municipalities Able to Achieve this Goal?

Throughout all the phases of establishing e-messaging, IT-Forum was a central driving force. IT-Forum Sogn & Fjordane is a multi-stakeholder network, established in 1995. The County Governor of S&Fj was together with Western Norway Research Institute initiator to IT-Forum, and a number of public and private organizations were invited to participate. Although members of the group have shifted over years it has always included a cross-sectoral group, with representatives from a selection of municipalities, the Health Trust, practitioners, county governor, research and industry. The purpose of IT-Forum is to initiate and coordinate "activities that promote efficient and development-oriented use of information and communication technology in the private and public sectors", and to work "to strengthen and develop cooperation and coordination between agencies, organizations and businesses in Sogn & Fjordane."[5] Despite its informal structure and voluntary participation, IT-Forum has, with its goal-oriented and long-term visions, contributed to moving public positions from cities to rural locations in S&Fj, and it has secured an infrastructure investment of 300 million NOK in broadband in the county.

In the process reviewed here, IT-Forum acted as a tool for collaboration between the municipalities, developing a checklist with detailed instructions for establishing electronic messages, and securing communication and information between the actors. All 26 municipalities enrolled in the project in the first phase. Cross disciplinary project groups were established in every one of the municipalities, ensuring that both health care personnel and IT people were involved in the work. Within the project structure, a couple of municipalities acted as test bed for practical implementation of the e-messages, and for finding and solving challenges concerning the integration between electronic messaging systems and a total of three different electronic journal systems. In this way, challenging circumstances for implementation and enforcement, such as lack of national standards, could be dealt with by the main project group and the municipalities acting as test bed. Incentives for the municipalities to participate were found in the checklist and support system provided by the IT-Forum project, as well as in an economic contribution from the project to participating municipalities.

In the evaluation the project was described as a success story, not only because it made the county a spearhead, but also due to the qualitative effects: it secured written messages where there had been only oral messages, it made messages efficient and securely sent between the different units involved with a patient, which was in particular important for municipalities' ability to comply with requirements established by the health reform, effective from 2012. It also appeared to be a success story due to the attitudes it created. From different actors we heard phrases like: "It has been incredibly fun to work with this project because people can see the benefits!" We found an impressive documentation of health care workers' will to participate as well as their satisfaction and pride from what they had achieved through the projects: "People are proud of working here" (Corneliussen 2014b).

2.3 Analysis with Network Theory

The Triple Helix systems theory, described by Etzkowitz and Leydesdorff (2000), focuses on the relations between research institutions, industry, and government with regard to innovation. This theory will help to illustrate the driving forces in the case described above. Etzkowitz and Leydesdorff's theory suggests that innovation is happening in the intersection between industry, knowledge producing institutions and government, rather than in the industry alone (Etzkowitz 2008). The metaphor of the helix illustrates that knowledge, action and funding interact and affect each other in a way that supports innovation. In a business development context, the three helixes *researchers*, *companies* and *authorities* interact in processes of innovation. In a peripheral region there is an organizational thinness, according to Tödtling and Trippl

[5] http://www.it-forum.no/.

(2005). This will influence the helixes in a system like S&Fj. The helixes are rarely equal, not even in regions with organizational thickness (Etzkowitz 2008). The public sector in S&Fj is relatively larger and influences the system more heavily than industry and research institutions compared to other counties in Norway.

In a previous research project (Nesse, Skogseid et al. 2014), we explored what makes heterogeneous networks like IT-Forum work well, based on a small adjustment of Bergek et al. (2008). Firstly, it must be attractive to join. Secondly, the network must be able to develop new knowledge and share this within the network. Third, innovation needs to come out of the network, in other words: something new that would not be possible without the network - for example, product, procedures or organizational structures, should be the outcome of the cooperation in the network. A fourth feature is that the network must have legitimacy. Both contestants and others outside the network must feel that the network is credible and positive. Fifth, the network should be able to mobilize resources, raise funds, establish contacts and build knowledge in a more efficient manner than the individual participants would be able to. And finally, knowledge and product developed within and through the network should be able to benefit others outside the network. Thus, if you spot a free rider, you have a clear indication that you have succeeded with the network.

2.4 IT-Forum as a Collaborative Strategy Tool

When we analyse the success story of establishing e-messaging in S&Fj[6] with the help of Bergek et al.'s theory, we find that the qualities and values of the multi-stakeholder network of IT-Forum is vital for the case.

Gaining legitimacy and mobilizing: IT-Forum gained legitimacy through its open and inclusive structure together with its long history as a local task force network. The trust towards this network has developed over 20 years of deep and long-lasting engagement in issues that have affected all or most of the municipalities, and in particular in cases where the municipalities have been too small to take on these issues on their own.

The funding raised for the projects was achievable as a result of the IT-Forum network, and it was of high importance as economical help for the municipalities as well as the associated institutions. The funding made it possible both to establish individual and externally funded projects in the municipalities, as well as to involve knowledge providers and researchers in the process.

While the funding contributed with legitimizing the structure, the main and dominating mobilizing force seems to be the ability to work on behalf of all the municipalities in the county. Thus, challenges of participating with (competitive) neighbour communities seem to be eliminated through this overarching structure where every one of the municipalities was invited to participate.

Knowledge and innovation: The main value of the network activities in the e-messaging projects is related to knowledge produced in and through the network. The multi-stakeholder network included knowledge and expertise from health care sector, county as well as municipality leader level, as well as knowledge providers and technology experts. This cross-disciplinary, cross-sectoral network structure thus contributed in creating a knowledge platform targeting the municipalities in this particular region, with its distinct set of challenges. It is important to emphasise that IT-Forum was not operating in isolation, and the e-messaging projects run by the regional Health trust Helse Vest were tightly coupled to and supported by the projects coordinated by IT-Forum. In addition, the IT-Forum network also provided input to national level government and organizations based on project experiences of working with e-messages in the municipalities.

Free riders as a sign of success: The municipalities acting as pilots helped by solving technical issues and developing routines, as well as acting as motivators and supporters for the other municipalities. The evaluation of this project illustrated the importance of the network in making the transfer to e-messaging becoming achievable for the municipalities (Corneliussen 2014a): "Several municipalities must unite, make a network so that they don't have to work alone – they have to collaborate. … It isn't necessary that every municipality invent everything on their own." And some of the municipalities clearly enjoyed this as a free ride: "We were last, and therefore we learnt from the other municipalities. They had dealt with the initial problems."

---

[6] The analysis is based on several project reports including case studies in the region: Nesse et al. 2014, Corneliussen 2014a & 2014b.

3. CONCLUSION: THE EFFICIENCY OF AN INFORMAL NETWORK STRUCTURE FOR SMALL INDEPENDENT UNITS

There was more than one driving force behind the process of establishing e-messaging, however, the cooperation across municipalities proved to be one of the most important single factors. As illustrated here, the multi-stakeholder network IT-Forum has been vital in initiating, planning, organizing and running the innovation projects for the public health care services in the region.[7] The coordinated efforts, and particularly the tight coordination with the regional Health Trust is unique also in a national setting.

The trust and ability to recruit very different actors, from public to private, communities of practice to research, created a structure that made it possible to support actions across the boundaries of the municipalities. The network was also a key to secure necessary funding for the municipalities as well as supporting actors. In short, the network has strengthened a region with many small municipalities that would otherwise not have access to the necessary resources to support all aspects of these innovation processes.

Some of the challenges that health care services in small units like the municipalities in this region meet, are different from challenges found in more densely populated areas – typically also the areas where more centrally organized piloting happens. In this perspective, the multi-stakeholder network has been vital in creating action force across the small units, while still allowing each unit to operate as an independent unit.

Limitations include that the network has no formal authority, and it has been voluntary to participate in all activities initiated and driven by the network. Consequently, the network does not make more formal structures obsolete. However, this informal structure might make it easier for the independent units to participate without having to negotiate in power structures. This also opens for applying this model in other regions with many small units, as it does not threaten existing power structures or the units' independence.

ACKNOWLEDGEMENT

We would like to thank the dedicated IT-Forum task force members and all the local project participants for their enthusiasm and efforts. The endeavour made by all the different actors made possible the successful innovation projects for the public health care services in the region.

REFERENCES

Corneliussen, H. G., 2014a. *Meldingsløftet Sogn og Fjordane - Små og tett på!* Vestlandsforsking-rapport nr. 8/2014.

Corneliussen, H. G., 2014b. A Success Story about a Communication Reform in the County of Sogn & Fjordane'. *Proceedings from Scandinavian Conference on Health Informatics*. C. E. Moe and M. Fossum. August 21-22, 2014, Grimstad, University of Agder, Norway.

Etzkowitz, H., 2008. *The Triple Helix: University-Industry-Government / Innovation in Action*. New York and London, Routledge.

Etzkowitz, H. and L. Leydesdorff, 2000. "The dynamics of innovation: from National Systems and "Mode 2" to a Triple Helix of university-industry-government relations." *Research Policy* 29 (2000): 109-123.

HDIR, 2009. "Samspill – IT. Nasjonalt meldingsløft - helseforetakenes 30 på topp liste pr 1.4.09 (30 top

listed health trusts by April 2009) ." (The Directorate of Health, Oslo, Norway).

HOD, 2008-2009. "Helse- og omsorgsdepartementet (2008–2009) St.meld. nr. 47: Samhandlingsreformen. Rett behandling – på rett sted – til rett tid.".

Larsen, Ø. H. and I. Skogseid, 2012. "lectronic Interaction in Health Services and Organizational Change – Lessons From the Norwegian County of Sogn & Fjordane." *IADIS International Journal on WWW/Internet* 10(1): 65-78.

Nesse, J. G., I. Skogseid, K. Skarbø and Ø. H. Larsen, 2014. *Innovasjon i Sogn og Fjordane – vilkår og barrierar*, Dokumentasjonsrapport Publisher: Vestlandsforsking VF rapport 3/2014.

NOU, 2011:11. "Innovasjon i omsorg."

Tödtling, F. and M. Trippl, 2005. "One size fits all? Towards a differentiated regional innovation policy approach. ." *Research Policy* 34(1203-1219).

[7] Cf. the County Governor of Sogn & Fjordane http://blogg.fylkesmannen.no/fmsf/ (8 December 2014) and the Sogn & Fjordane County Municipality Director of Finance, Heggheim http://www.porten.no/artikler/meiningar/it-forum-eit-foredome.

DISTANCE COURSE APPLIED TO MEDICAL DOCTORS WORKING IN THE PUBLIC HEALTH SYSTEM IN BRAZIL: STUDENT PROFILE ANALYSIS AND PERFORMANCE EVALUATION

Rosália Morais Torres[1], Humberto José Alves[1], Maria do Carmo Barros de Melo[1], Bruna Matias[1], Luciana Magalhães Braga[1], Alaneir de Fátima Santos[1] and Simone Ferreira dos Santos[2]
[1]*Health Technological Center of the Medical School of the Federal University of Minas Gerais (UFMG)*
[2]*Department of Health of the State of Minas Gerais, Brazil*

ABSTRACT

Distance medical education has proven to be an attractive alternative for updating medical professionals, although little is known about the profile of this target audience. The profile and performance of those public health system professionals in the State of Minas Gerais, Brazil, enrolled in a distance electrocardiography course structured into 3 modules with increasing levels of content complexity were evaluated in this article. It was verified that the approval, failure and discontinuance rates did not have significant variations relating to gender and the complexity of the subjects addressed, indicating that the difficulty level of the content may not be a determining factor for the performance and adherence of medical doctors in distance courses. Mostly professionals enrolled in the course were younger, recently graduated in medicine and still undergoing professional training. Given this characteristic of the sample, approval and failure rates were very similar in those groups, indicating great motivation for this education modality. However, difficulties to follow the course suggested little time availability from those professionals to dedicate to distance courses, despite their greater need and interest for them. Medical doctors in the age group above 35 years old seem to be less interested in enrolling in distance courses, however, when they enroll, they tend to get good approval rates, good performance and lower discontinuance rates. Those profile and performance aspects should be considered when planning distance courses for medical professionals.

KEYWORDS

e-learning, virtual learning environments and issues, medical profiles

1. INTRODUCTION

Medical Distance Education is a growing modality due to the fast advancement of the medical area knowledge[7]. Medical doctors face the challenge of conciliating the need of continuous updating of their knowledge with difficulties, usually attributed to the little time available to attend professional updating and improvement courses[3]. Distance medical education can be the ideal methodology for the continuous education of the medical professional due to the freedom of access at any time and place.[1,6, 11] Despite that potential, little is known about the performance and profile of that target audience, that, obviously, may vary widely in terms of gender, age, gender, time since graduation, type of specialization, among other factors.[8,9] All those factors may influence, alone or together, the discontinuance, permanence and performance of that medical professional in distance education courses. Understanding those factors is important in order to establish strategies to adapt the distance courses offered in this very promising area for this education methodology[4,9]. This article aims to evaluate the profile and performance of the medical doctors enrolled in a electrocardiography modular course offered free of charge in the health public system of the State of Minas Gerais, Brazil, analyzing if this profile impacted the discontinuance, approval and failure in every module.

As the course comprised 3 increasing complexity modules, the profile/performance parameters were analyzed considering the level of complexity of every module. The relevance of this article relays on the outstanding lack of publications on the subject, aiming to contribute for the discussion of the professional profile and its importance in the distance education for the medical area.

2. METHODOLOGY

The electrocardiography course, a partnership between the Federal University of Minas Gerais and the State Health Department, was offered free of charge to approximately 4,000 medical doctors that are part of the Health Basic Units (UBS). Between August, 2013 and February, 2016, five groups were offered, each of them consisting of 3 learning modules. When enrolling in the learning platform, the professionals were invited to complete a questionnaire about their professional and personal data. The data analyzed in this study were extracted from those questionnaires and from the analysis of the grade in the evaluation activities of the education platform. For statistical purposes, all students that accessed and had activities in the course platform were defined as effective, excluding dropout students from that group. Students enrolled in the course who never entered the education platform were classified as dropout students, as well as those that formally manifested their wish to discontinue the program and also those whose grades were ≤ 6.0, indicating minimum activity in the platform. The effective students defined as approved were those who had grades $\geq$ 60, in every Module and those effective students who failed had < 60 points in every Module.

The course consisted of 3 independent modules, being the first one mandatory and prerequisite for the remaining ones. The difficulty level established for every module was, respectively: low (basic, mandatory Module), moderate (2nd Module) and high (3rd Module). Therefore, each student could sign up for 1, 2 or 3 modules. The decision for offering independent modules, with progressive escalation of the difficulty level, was based on the assumption that medical doctors could have no time or enough motivation to pass through all the modulus. That would allow the professionals to choose the most suitable level of electrocardiography knowledge appropriate to their daily practice and the time they could dedicate to the course.

3. RESULTS

The amount of 2,755 students enrolled in the Basic Module, corresponding to 69% of the initial target audience, estimated in 4.000 medical doctors of the Health Basic Units (UBS) of the State of Minas Gerais. A total of 1,496 students were considered effective (54.3% of the students enrolled), verifying an initial dropout of 45.7% of the students enrolled. That dropout was greatly due to students that enrolled but never entered the education platform, being attributed to causes not related to the course. A total of 905 students were approved in the basic module (60.49% of the effective students enrolled) and 591 failed (39.5%). This module was a prerequisite for the others ones, therefore, the students could only enroll in the intermediate and advanced modules after being approved in the basic module.

3.1 Students Performance Determinants in Basic Module

Among the students approved in the basic module, 49.28% were women and 50.72% were men, which showed that the approval was not influenced by gender (P=0.666). In this module, the age group was important for the approval, with a significant difference in the approval between the age groups (P=0.000), being higher in the age group between 26 and 30 years old, (33% approval) and 31 to 35 years old (20.67% approval). The time since the medicine graduation was also statistically significant for the approval in this module (P=0.00), being the approval higher (54%) in the group between 0 and 5 years old since graduation. Among the students approved this module, 59.67% obtained grade A (score higher than 90 points), 23.31% obtained grade B (score equal or higher than 80 and lower than 90 points), 8.62% obtained grade C (score equal or higher than 70 and lower than 80 points) and 8.40% obtained grade D (score equal or higher than 60 and lower than 70 points) with no difference between genders relating to the grade obtained. There was a predominance of grades A and B in the age groups between 26 and 30 years old and 31 and 35 years old.

At the same time, there was predominance of grades A and B in professional who had graduated between 0 and 5 years before. In the basic module the failure, that is, final score lower than 60 points, was higher for women (P=0.004) with 54.15% failure among women. There was also statistically significant difference in the failure in terms of age group, higher in the 26-30-year-old and 31-35–year-old groups (P=0.000). The same was verified in relation to the time since graduation, verifying a higher failure rate in the group of professional who had graduated between 0 and 5 years before (60.41% of the students failed). The discontinuance was not statistically different between genders (P=0.592), being 50.75% for women and 49.25% for men. The higher formal discontinuance rate was between 0 and 5 years since graduation (55.28%), much higher than the second highest rate of 14.22%, with P=0.000 between groups. Coinciding with the shorter time since medical graduation, it was found that, among the dropouts, most students were aged between 26 and 30 years old, being 32.10% and 22.28% aged between 26 and 30 years old and 31 and 35 years old, respectively.

3.2 Students Performance determinants in the Intermediate Module

The amount of effective students in this Module was 754 professionals, being 461 approved (61,44%) and 293 failed (38.46%). The approval was equal for both genders (P=0.666) and different for the time since graduation (P=0.00), with 50.78% approval in the group of professionals who had graduated between 0- 5 before. The age group was different between the groups, with 29.28% of the students approved in the age group between 26 and 30 years old (P=0.00). The failure, in this Module, was the same for both genders (P=0.683), but different within the age groups, with a higher failure rate between 26 and 30-year-old (32.08%) and 31 and 35–year-old (21 50%) ranges, being those two age groups accountable for about 54% of the students who failed in this Module. Students graduated between 0 and 5 years before accounted for 54.61% of failure, making the relationship between the time since graduation and failure significant (P=0.00). As for performance, 60% of the students approved in this Module obtained grade A, 17.14% grade B, 13.45% grade C and 9.11% grade D. The gender did not influence the performance of the students in any of the grade scales (A, B, C or D) in this module. Statistical findings regarding performance for age group and time since graduation are related among themselves, indicating that most of the grades A and B were obtained by students aged between 26 and 35 years old and being graduated between 0 and 5 years before.

3.3 Students Performance Determinants in the Advanced Module

The amount of effective students in this Module was 368 professionals, being 306 approved (83%) and 62 failed. The approval rate was the same for both genders (P=0.109). The approval was different (P=0.00) according to the age group, with 26.14% of approval in the age group between 26 and 30 years old (P=0.00). Regarding the time since graduation, most of the students approved had graduated between 0 and 5 years before (P=0.00). Failure in this module was not determined by gender (P=0.075), but was influenced by the age group and time since graduation (P=0.00) for both variables. Students aged between 26 and 30 years old accounted for 37.10% of failures and the students graduated between 0 and 5 years before accounted for 62% of failures in this advanced module. Relating to performance, 57.84% of the students approved this Module obtained grade A, 23.53% grade B, 13.73% grade C and 4.90% grade D. The gender did not influence the performance of students in the grade scales A and B, but there was more grade C among men and more grade D among women. Statistical findings regarding performance for age group and time since graduation showed a greater amount of students with grades A and B in the age group between 26 and 30 and between 0 and 5 years since graduation.

4. DISCUSSION

Gender distinguished the students of this course in only one aspect: the approval rate that was slightly higher for women in the Basic module. It is possible that the double professional and domestic workday influenced the availability of time for women to devote initially to the course. Most students accessed the course at night, outside working hours, time in which, considering the social structure in Brazil, women are more focused on taking care of their children.

It was found in all Modules that both the rates considered positive, such as approval and achievement of grades A and B, and negative, such as discontinuance and failure, were predominant in younger age groups, between 26 and 39 years old, being graduated between 0 and 5 years before. Apparently opposite, such statistical data was strongly influenced by the great amount of enrollments of young and recently graduated medical doctors, who were part of the "*Mais Médicos*" [More Medical Doctors] Program of the Brazilian government. A feature of this group is the high workload, associated to medical training activities which are mandatory within the program and the preparation for medical residency tests. Those factors mean that those professionals have less availability to dedicate for distance courses, but more interest and need of them. Therefore, this was the predominant group in the enrollment (higher interest or need), and the group with higher initial dropout (concrete impossibility to continue the course after enrollment) and higher failure rate (professionals who started the course, but were unable to continue it). The same way, as the predominant group, the approval rates were higher. Thus, it was verified that the largest percentage of younger students enrolled was reflected in the statistical analysis in all Modules and for all variables analyzed, except for gender.

5. CONCLUSION

Distance education shows great potential for medical training and updating [1,3,7]. However, the behavior of medical doctors face this education modality is unknown. In this work, greater interested in e-learning was verified - reflected in the amount of enrollments - by young and recent-graduated medical doctors of both genders. More familiar with technology, those professionals do not face difficulties relating to aspects involving ICT. Although showing high performance[5,2], this group is, simultaneously, more inclined to dropout and failure due to the lower availability of time (excess of professional activities and need to need to prepare to tests in the beginning of the career). However, in the moment those professionals manage to dedicate to the course, they are the ones presenting the best performance, reflected in the high approval rates and better concepts reached, regardless the complexity of the subject addressed. This is probably the group of professionals that most seeks for improvement courses, being a potential audience for distance education. It is possible that short term courses are more suitable for this group of medical doctors, aspect that should be considered when planning distance courses for this target audience. On the other hand, doctors older than 35 years old seem less interested in general for distance courses, which reflects in the lower amount of enrollments as of this age group. They probably need more professional motivation to seek, in distance courses, new training modalities. However, the moment they are motivated, they tend to have lower failure and dropout rates than the younger audience, concluding the course with good performance[5,10]. Those professionals may need more assistance related to the use of technology and, probably would feel more comfortable with courses in which they have more time to carry out the activities. An interesting aspect observed in this course is that the content complexity did not affect the performance parameters, something that needs more study. In view of the possibility of different profiles within the medical groups, it is important to adapt e-learning strategies for their specificities, seeking to make the learning methodology, monitoring and ideal time suitable for the offering of courses[4].

ACKNOWLEDGEMENT

This research was supported by the Department of Health of the State of Minas Gerais, Brazil. We are grateful to the entire and dedicated team who participated of this project.

REFERENCES

1- Barnett GO, 1995. Information technology and medical education. *J Am Med Inform Assoc*, vol. 2, no 5, pp. 285-295.

2- Bhargava, R, 2009. How doctors are using social media. Fresh influence, Ogilvy Public Relations, Available from: http://blog.ogilvypr.com/2009/09/how-doctors-are-using-social-media/

3- Cardoso, SH, 1998. Educação médica à distância pela Internet. *Revista de informática médica,* vol. 1, no 5. Available from: www.informaticamedica.org.br/informaticamedica/No0105/cardoso.htm

4- Christant EL et al, 2003. O papel do ensino a distância na educação médica continuada: uma análise critica. *Rev Assoc Med Bras*; Vol 49 No 3,pp 326-9. Available from: http://www.virtual.epm.br/material/tis/amb.pdf

5- Jason GJ & Clapham, M, 2014. Attitude to e–learning among newly qualified doctors. *The clinical teacher* 2014; vol.11, pp. 20–23.

6- MacKenzie JD & Greenes RA, 1997. The World Wide Web: redefining medical education. *JAMA,* vol 278, no. 21, pp. 1785-1786.

7- Neves-Junior, WFP et al, 2011. Educação a Distância (EaD) em Física Médica. *Comunicação saúde educação* vol. 15, no. 39, pp.1199-1206.

8- Sabbatini, R, 2001: O crescimento do conhecimento médico On-Line, *Revista Médico Repórter.* Available from: http://www.sabbatini.com/renato/papers/reporter-medico-14.htm

9- Sabbatini, RME; Cardoso, SH: Capítulo 24: O setor de saúde e a EAD. In: Litto, FM & Formiga, M org.): *Educação a Distância: O Estado da Arte*. São Paulo, Brazil: Editora Pearson and Associação Brasileira de Educação a Distância, pp.198-208.

10- Sklar, B, 2006. The current status of on-line continuing medical education. Master Dissertation, University of California. Rev. Available from: http://www.cmelist.com/mastersthesis/

11- Zucker J et al, 1996. A comprehensive strategy for designing a Web-based medical curriculum. *Proc AMIA Annu Fall Symp*; pp 41-45

EXPLORING SECURITY OF REMOTE PATIENT MONITORING DEVICES USING NFC TECHNOLOGY FOR IDENTIFICATION OF THE FRAIL ELDERLY

Brian Ondiege[1], Malcolm Clarke[1] and Glenford Mapp[2]
[1]*School of Information Science, Computing and Mathematics, Brunel University, Uxbridge*
[2]*School of Science and Technology Middlesex University, Hendon*

ABSTRACT

Security is a major issue for telehealth. With the increasing possibilities that telehealth is offering, security is becoming a greater concern. The ISO/IEEE 11073 Personal Health Device standard does not address the issue of security and authentication of medical devices which is vital in determining the integrity of a telehealth system. This study has developed selection criteria to be used in designing and selecting technology for the frail elderly. These criteria are developed from the perspective of the frail elderly to ensure that any proposed technology is easily accepted. This study proposes the use of NFC technology for use in identifying the person using the device. The approach was tested using a blood pressure monitor modified with an NFC scanner which is used together with an NFC card for identification purposes. Whenever a patient is to take their BP, they identify themselves first to the device and then are allowed to take their (blood pressure) BP. Identification is critical in providing quality care in the healthcare setting and the dangers associated misidentification can result in the patient getting the wrong care.

KEYWORDS

Frail Elderly Technology; Identification Technique, Remote Patient Monitoring Devices, NFC Technology

TERMINOLOGIES

Telecare is the term for offering remote care of elderly and physically less able people, providing the care and reassurance needed to allow them to remain living in their own homes [4].

Telehealth- "is the remote exchange of data between a patient at home and their clinician(s) to assist in diagnosis and monitoring typically used to support patients with Long Term Conditions, uses technology to provide services that assist in the management of long term health conditions, including Chronic Obstructive Pulmonary Disease (COPD), Chronic Heart Failure (CHF), Diabetes and Epilepsy [1]".

Remote Patient monitoring (RPM) devices- is technology that allows patients to use medical devices outside of conventional clinical settings, i.e. their homes, which may increase access to care and decrease healthcare delivery costs [5].

1. INTRODUCTION

The number of people aged 60 or over is expected to pass the 20 million mark in the UK by 2030. The number of people aged 65+ is projected to rise by nearly 50% (48.7%) in the next 17 years to over 16 million in the UK [12]. With the ageing population being the highest consumers of health care, these demographics present challenges to the healthcare sector. Furthermore, since telehealth is still in its infancy, it makes it an ideal scenario for opportunistic security attacks because the security issues have not been adequately addressed.

Security and identification are therefore issues of major concern in a telehealth system because patient information is very critical, which in the wrong hands can threaten patient care. There are few articles that have been published that address the problem of security of remote patient monitoring devices. During the initial stages of this research, a study was conducted with healthcare professionals to determine their perception and knowledge of security and the security of remote patient monitoring devices. It was found that the perception of healthcare professionals towards security was very low. One area in remote patient monitoring that has been overlooked and poses a serious threat to successful implementation of telehealth, is the identification and authentication of the patient using the remote patient monitoring device. Telehealth is frequently used for monitoring of the elderly and the critically ill. Telehealth is seen as beneficial to patients who have high levels of medical requirement. Telehealth techniques allow these patients and their physicians to closely monitor medical conditions and if an emergency occurs, they will then act on it.

2. RESEARCH OBJECTIVES

This research looks at the ISO/IEEE 11073 Personal Health Device (X73-PHD) which is a standard for RPM devices security. This research is guided by the following research questions:

- How do you know that the patient at the other end is whom they claim to be?
- How can multiple patients in the same household share a single device
- What is the best technology for identification of the frail elderly using remote patient monitoring devices

Dangers associated with patient misidentification

Failure to correctly identify patients can result in wrong diagnosis, transfusion errors, testing errors and many others [11]. Risks to patient safety occur when there is a mismatch between a given patient and components of their care, whether those devices are diagnostic, therapeutic or supportive [10].

2.1 Other Identification Techniques vs the Proposed Solution

Passwords have proved to be very insecure and burdensome to the user, especially for the frail and elderly, the thought of having to remember or change them frequently puts a strain on a patient who also has to worry about managing their disease which makes using passwords impractical [9].

Biometrics mechanisms attempt to provide a comprehensive security solution. Although with the possibilities that biometrics offers, the authors believe that it is not appropriate for identification of the frail elderly since most people might feel that it is intrusive because of its relation to criminal profiling. Some people reject fingerprint identification as repulsive and invasive since it is seen as a way of the government to monitor and control people. Biometric devices are expensive and the cheaper ones are not reliable, it would not make sense to put a scanner that costs more than the patient monitoring device because it will make the device costly and thus not be widely deployed[6], [7]. This is why the researchers believe that a more comprehensive solution that encompasses the needs of frail elderly is required and the main reason that researchers has chosen NFC technology for identification of the frail elderly using remote patient monitoring devices.

How Current RPM devices work

The current system works by the patient transmitting their readings that they have taken using a remote patient monitoring device. The readings are sent wirelessly from the device to the home gateway and then it sent via mobile communications to the hospital that is monitoring the patient and stored in the database where the clinicians can later retrieve them. Each device that is used in remote patient monitoring is assigned a unique ID, for example, a serial number which is used in identifying the transmitter. This only shows device identification but does not show user identification, which is determined by establishing the identity of the patient on the other end. The current system therefore is limited in its ability to verify the identity of the patient.

The first problem is verification of the person using the device as the measurement is not supervised in the home and the device does not have a mechanism for identification and authentication. If the patient cannot be verified and incorrect data is received then there is a possibility of incorrect diagnosis and incorrect treatment. The second problem is that current devices are limited in the number of people who can use each device at a given time. Only the person who is being monitored is allowed to use the device, and in case someone else in the house uses the same device the readings will not be accurate and can cause harm to the patient if they receive the wrong care.

Figure 1 shows how the current remote patient monitoring device works. A patient who is to be monitored at home, has their blood pressure is taken, which is sent wirelessly to a home gateway, and the gateway transmits the readings to a server which stores them. The doctor who is monitoring the patient then views the readings.

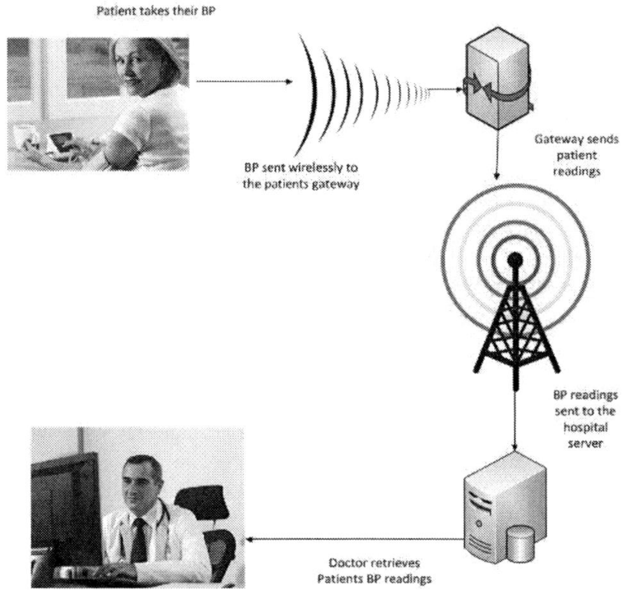

Figure 1. The Current System

3. DISCUSSION

Why NFC technology

Near field communication (**NFC**) is a set of communication protocols that allow two electronic devices, one of which is usually a portable device such as a smartphone, to start communication by bringing them within 4 cm of each other.

The following criteria were used:

1. **Usability**

Usability plays a vital role in technology for old people; it builds confidence and trust when using technology.

Patients should not have to think too hard when they are using a technology, nor should they have to refer to a manual when using it; this makes them look less intelligent and leads to time wastage.

Device prompts should be logical, sequential and effortless to ensure that the patient uses less time, enjoys using the device, makes a recommendation and looks forward to using it continuously. NFC technology provides an effortless and fast means of identification and authentication.

2. **Familiarity**

NFC technology is widely used in Great Britain and other developed countries. In London people are using it for public transport as part of the Transport for London (TFL) network and in making payments at the grocery stores. This study identified that NFC technology will elicit different reactions, with most of them positive due to ease of use and very few negatives for those who are not familiar with the technology.

3. **Cost**

Cost plays a vital role in implementation of any technology. NFC technology is affordable and secure; an NFC card cost less than 40c and can be re-used multiple times by different users, which makes them economically viable.

Proposed Solution

Figure 2 shows the proposed model that will be tested on a modified blood pressure monitor to incorporate NFC technology. This will be used by patients to identify them whenever they are taking their blood pressure:

4. A patient is issued with an NFC card
5. Patient scans the card on the NFC reader to identify themselves
6. If authenticated, the patient is prompted to take their blood pressure
7. BP and ID from the NFC card are sent wirelessly to the gateway in their homes
8. Readings from the gateway are sent to the hospital monitoring their readings
9. Data is stored in the database server in the hospital
10. Doctor views data from the server
11. Patient can log in to their portal to view their readings and previous history.

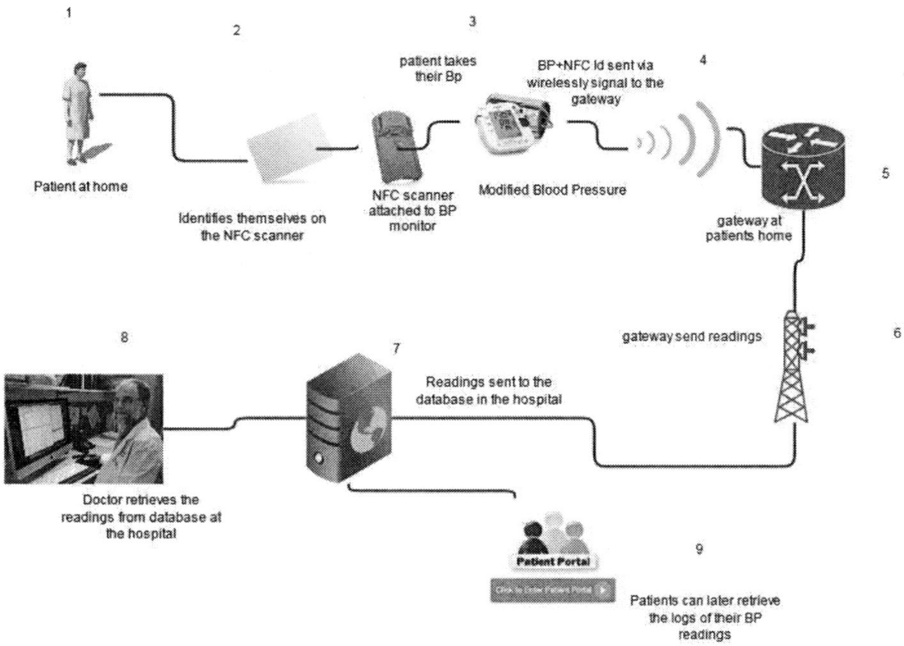

Figure 2. The Proposed New Model

This technology will be useful to the elderly who are living in care home and need to be monitored for chronic diseases since it will allow multiple people to share the same device.

A Capability-Based System for the secure use of NFC

Identification with NFC alone is not sufficient; therefore there is a need for a solution that will increase the security within the NFC framework. This paper proposes the use of a capability-based system to enhance security for tele-monitoring systems because the NFC_ID can be tampered with either while in storage or in use, thus resulting in misdiagnosis or fatality. It provides additional security by restricting access to data, people and devices.

Capabilities are therefore especially applicable in the context of eHealth as health data is very sensitive and hence must be protected from tampering and unauthorised access. Furthermore, capabilities allow us to run a role-based mechanism so restrictions can be based on the role of different people within the healthcare system such as doctors, nurses, technicians and administrators. Therefore in this system each entity must have a capability for example people, devices and infrastructure all must have capabilities. Capabilities can also be used to provide restrictions to access to data and resources to personnel based their roles. Figure 3 present the new format that will be used to present capabilities. More details on this capability format is found in [13].

The **Type** Field: This field is used to specify the type of object capability that is being used. Types could include Cloud Providers, Cloud platforms, users, applications, etc.

The **Property** Field: This field is used to define the properties of the object

The **Object Id**: This field is used to uniquely identify the object.

The **Random Bit** Field: This field helps to uniquely identify the object.

The **Hash** Field: The Hash field is used to prevent the casual tampering of capabilities.

TYPE	PROPERTY FIELD	OBJECT ID	RANDOM BIT FIELD	HASH FIELD

Figure 3. The New Capability Format

Therefore this paper proposes the use of capabilities to provide security for eHealth.

4. CONCLUSION

This study identifies the significance of patient identification in delivering quality care to the patients. Based on healthcare professional's perception, this study identified the issue of security as one area in remote patient monitoring devices that has been ignored and threatens the delivery of patient care. This study recognises that NFC alone for identification is not sufficient therefore this study proposes the use of capability-based system to enhance security of tele-monitoring devices and healthcare infrastructure.

REFERENCES

[1] Tsa (2015): What is Telehealth, Available at: http://www.telecare.org.uk/consumer-services/what-is-telehealth

[2] Ko, J.K.J. et al., 2010: Wireless Sensor Networks for Healthcare. *Proceedings of the IEEE*, 98(11).

[3] Independent Living (2015) Telecare Available at: http://www.independentliving.co.uk/cp-editorials/telecare-community-alarms/

[4] Huang, Ding-Long, Pei-Luen Patrick Rau, Gavriel Salvendy, Fei Gao, and Jia Zhou. 2011: Factors affecting perception of information security and their impacts on IT adoption and security practices. International Journal of Human-Computer Studies 69 (12) (12): 870-83.

[5] Sylla, I.T., 2012: Technologies for remote patient monitoring. *Electronic Products*, 54(9).

[6] Malone, C.S (2015): DNA Fingerprinting. Available at: http://www.csun.edu/~cmalone/pdf360/Ch16,17rDNA.pdf.

[7] A.K. Jain, Ajay Kumar, (2010): Biometrics of Next Generation an Overview, Second generation biometrics Springer.

[8] Technical Specifications (2011) NFC Forum. Available at: http://members.nfc-forum.org/specs/spec_list/

[9] B.B Zhu. et. Al (2014): "Captcha as Graphical Passwords—A New Security Primitive Based on Hard AI Problems," in *IEEE Transactions on Information Forensics and Security*, vol. 9, no. 6, pp. 891-904

[10] Australian Commission on Safety and Quality in Healthcare (ACSQHC), 2012. *Safety and Quality Improvement Guide Standard 6: Clinical Handover*,

[11] WHO (2011): Patient Identification Policy 1(May), pp.1-26

[12] National population's projections (2012) Available at: http://webarchive.nationalarchives.gov.uk/

[13] Mapp et al (2014): Exploring a New Security Framework for Cloud Storage Using Capabilities. 1st International Workshop on Cyber Security and Cloud Computing, Oxford UK.

Reflection Papers

TOPIC: FACTORS AFFECTING THE ACCEPTANCE OF MHEALTH TECHNOLOGY BY SAUDI DIABETICS AND DOCTORS

Bander Alkhudairi[1] and Dr Lyn Pemberton[2]
University of Brighton
[1]*Office: 616, Watts Building, Lewes Rd, Brighton BN2 4GJ, East Sussex, UK*
[2]*Office: 537, Cockcroft Building, Lewes Rd, Brighton BN2 4GJ, East Sussex, UK*

ABSTRACT

Healthcare technologies serve varying purposes for different users such as patients and doctors. Patients faced with problems accessing healthcare institutions may readily accept mHealth technology as a strategy for resolving their access limitations. In addition, patients faced with strict cultural limitations would also have a higher use for mHealth technology and therefore have good reasons to accept it. However, poor, uneducated, and elderly members of the society have poor perceptions of mHealth technology as compared to the wealthy, educated, and younger populations who have been exposed to the technology. Additionally, practitioners do not believe the current mHealth apps are secure enough to maintain a reliable confidentiality level with patients. Patients on the on the other hand are skeptical about the security of their personal information shared through the mHealth application.

KEYWORDS

Privacy, technology acceptance, education, exposure, support systems, Ehealth

1. INTRODUCTION

Acceptance of mobile health technologies varies by gender, age group, education level, exposure level, and availability of supporting systems. Gender differences in the use of health applications indicate that female nurses have embraced the use of reminders unlike the male practitioners on the same case. Age groups also differ in terms of technology use and their perceptions of its relevance. For instance, the younger generations with exposure to technologies such as Smartphones are more willing to use these same technologies to keep their medical conditions in check than older generations who often report lack of confidence in technology. Patients and doctors that have been exposed to the technology and have a good knowledge of it register higher acceptance levels than those without any exposure or with poor knowledge. Education level is also a significant variable in determining the acceptance of mHealth technology. Education level is linked to knowledge about how mHealth can save time, be useful in emergencies, assist those living far away from medical institutions, and those whose cultural limitations prevent the access of healthcare. All these factors influence the attitudes towards the acceptance of mHealth application as a medical technology solution. However, other significant findings show that patients and practitioners disagree about the level of privacy afforded by this technology while at the same time a majority of practitioners and patients do not have sufficient experience of using technological innovations (Study Findings).

2. BODY OF THE PAPER

2.1 mHealth Security Issues

Medical records for patients are stored in databases that have historically suffered threats of unauthorized access as well as revelation by practitioners. These reasons have largely contributed to the perception that the use of technology in medical settings suffers from potential risks of breaching patients' privacy. Just like the patients, doctors perceive the use of mHealth applications as a strategic solution to health care issues and needs facing diabetics and other chronic disease sufferers. However, they worry about the confidentiality factor of using technology to follow up medical conditions, consume and disburse personal information, and administer medical interventions remotely. Therefore, there are challenges to the acceptance of mobile technology in health care ranging from the ethical use of information systems to the practical application of such technologies on both sides. Patients stating that they welcomed the adoption of such technologies also said that they would not use the application if the medical institutions failed to install support systems and the practitioners are not efficiently trained how to use the applications (Venkatesh et al., 2003).

In addition to the aforementioned confidentiality issues, the very nature of information systems that enable the technologies does not appeal to every patient and doctor. With respect to information security, information systems have fallen victims to cyber-attacks targeting personal information of patients and doctors as well. As a result, real-time communications between patients and practitioners, which is interruptible by the potential cyber threats, is seen to be a risk, hence a valid reason for refusal to accept the technology in some cases. Although some channels of communication can be considered safe for the exchange of personal information, some users are deemed careless that they are capable of instigating identify theft among other cyber-related risks. Hence, knowledge management regarding safety and how to guarantee safety with mHealth is found to be an influencing factor in the adaption of mHealth health application.

2.2 Experience of Use

Education is a factor in determining the use of a technology in addressing a variety of medical conditions. The interaction between patients and doctors needs to follow known channels and be available for both sides. However, informed doctors and informed patients require costly training of the patients, a perspective sensitive to resources utilization and can be time-consuming. Likewise, informed patients and uninformed doctors render medical information systems irrelevant in the efforts to adapt mHealth as a contemporary medical approach for minimizing unattended medical cases and managing conditions. For patients who have experienced loss of information or indentity theft, cite insecurity as a factor influencing their resistance and acceptance levels of the mHealth. Thus, patients with first-hand experience of information loss and identity theft have a higher resistance level to mHealth technologies. Doctors who have not used mHealth applications before are similarly resistant to integrating technology into their practice (The University of Cambridge, 2011).

Practitioners without any experience of mHealth and who have worked many years in their profession are often more difficult to convince about using this technology and have a higher level of resistance to it; whereas younger practitioners and patients who have been exposed to the technology are more accepting (Holtz, 2008). However, older generations of patients also report that, regardless of their knowledge and experience in using mobile technology, some interfaces they have experienced are not user-friendly as they are difficult to navigate and involve lengthy processes. Although the older generation often fares poorly in the acceptance of mobile applications, it is also observed that the younger generation suffers from problems such as finding the applications boring and forgetting to use them (IMS Health, 2014). Therefore, it is observed that factors relating to personal behaviour and attitudes can hinder the use of available applications. The speed, reliability, interface and usability of the application and the personal behaviour of the user therefore influence the acceptance of medical technology (Holtz, 2008).

Combined, the lack of knowledge among the practitioners on how to use technological solutions make the adaption of the solutions very slow and at times non-existent. As a result, if a considerable number of professional practitioners are seen to admit that they have not used technological inputs such as mHealth in their practice, it is expected that patients managing chronic conditions may not benefit from a recommendation from these practitioners because their lack of knowledge would lead to the avoidance of such recommendations. Additionally, given that practitioners are more likely to offer solutions to patients on how to manage their conditions, those without the knowledge on the existence of mHealth application would not drive higher levels of use. As the results (from the study) suggests, even if more patients had mHealth application as a way of managing their conditions, their ability to use them and the overall of poor adaption and understanding by the practitioners would contribute to poor adaption of the technology. For practitioners looking forward to achieving better success rates with patients, adaption of such technologies and disbursing the same knowledge to the patients would increase the adaptability of mHealth. However, the observed poor adaptability is majorly contributed by factors related to poor training and development of practitioners.

2.3 Performance Rates

Performance rate relates to the number of times the technologies are used in addressing a real medical concern such as following up on patients, consultation with practitioners, updating health information, as well as the reliability of the technology. An example of a mobile health application is the Glucose Buddy which is portable as a Smartphone application. The performance rate of the app is influenced by the knowledge regarding how to install it. Patients and doctors with an idea of the app would have issues accepting it if they lacked the information about how to get the app working (Chomutare, et al. 2011). Crucially, the interfaces of the application consist of how to access and input information. These tasks rely on the default language or the languages information is rendered in. Thus, all patients with the knowledge about mobile health applications such as Glucose Buddy may be constrained to use it especially since the only available languages may not feature Arabic language options.

Faced with the above challenges, it is difficulty for patients and doctors to accept technologies which are perceived to work but which have been developed without regard for certain needs such as familiar language. However, solutions to the challenges are feasible considering that training is an option. As suggested by doctors, applications such as Glucose Buddy, if deployed to patients' Smart phones can serve as personal assistants reminding patients to regularly check their blood glucose or alternatively to get drug administration (Chomutare, et al. 2011).

2.4 Division in Acceptance

Division in acceptance relates to populations' willingness to consider the technologies proposed. Elderly doctors are seen to be at variance with younger doctors in acceptance levels. Older doctors are more likely to believe that information security and patient confidentiality is problematic as they believe mHealth applications to be insecure (Illiger, et al. 2014). On the other hand, the younger generation of doctors does not generally consider information security a significant issue hindering their adoption of mHealth since they have experience regarding the benefits and advantages of using such technologies (Venkatesh, et al. 2003).

Male patients also are observed to not have as many reasons to use mHealth applications compared to women in the KSA. As a result, women prevented by cultural bias from driving, being treated by men, or seeking healthcare intervention from a remote healthcare facility have a higher potential acceptance level for mobile health technologies. On the other hand, economically stable society members more readily accept the application as result of educational elevation over poor society members who do not have means of financing the enabling technologies such as Smartphones or subscribing to healthcare networks.

3. CONCLUSION

Technologies in health care serve varying purposes for different healthcare sector stakeholders. MHealth applications appear as promising solutions for diabetes self-management, access to appropriate healthcare and communication with medical practitioners for many Saudi diabetics. There are many scenarios where cultural, economic and geographical reasons make it difficult for certain Saudi diabetics to access healthcare and get the information they need. If such a solution is to be accepted by these populations, it is essential to understand how to address the issues of technology acceptance, especially by those sections of the population whose perceptions of mHealth are limited by not being knowledgeable about the technology or simply not being exposed to it like their richer and more educated counterparts. Additionally, the supporting frameworks are observed to be areas requiring improvement in order to guarantee accessibility to remote assistance through the mHealth applications.

REFERENCES

Venkatesh, V., Morris, M. G., Davis, G. B., & Davis, F. D. (2003). User acceptance of information technology: Toward a unified view. MIS quarterly, 425-478.

The University of Cambridge, (2011).*Mobile Communications for Medical Care a study of current and future healthcare and health promotion applications and their use in China and elsewhere*. Final Report.

IMS Health (2014*). IMS Health Identifies Opportunities for Mobile Healthcare Apps to Drive Patient Engagement, Enhance Delivery of Care | News*. [online]

Holtz, C. (2008). *Global health care: issues and policies*. Sudbury, Mass, Jones and Bartlett Publishers.

Illiger, K. et al. (2014). Mobile Technologies: Expectancy, Usage, and Acceptance of Clinical Staff and Patients at a University Medical Center. JMIR mHealth and uHealth, 2 (4), e42

Chomutare, E. et al. (2011). Features of Mobile Diabetes Applications: Review of the Literature and Analysis of Current Applications Compared Against Evidence-Based Guidelines. *J Med Internet Res*, 13 (3), 1-16

THE STATUS OF MHEALTH IN EUROPE AND A REVIEW OF REGULATIVE CHALLENGES

Maria Emmanouilidou
Kent Business School
The University of Kent, Canterbury, Kent, CT2 7NZ, UK

ABSTRACT

The rapid uptake of mobile access globally and the continuous technological developments in the field are opening up significant opportunities for healthcare. Mobile health (mHealth) represents the state-of-the-art of healthcare and is expected to play a fundamental role in the restructuring of healthcare systems. In particular, mHealth can support innovative models of care by placing the patient at the center and contribute to the sustainability of healthcare systems. The real value of mHealth is that it can overcome temporal, locational as well as organizational constraints thus providing a flexible and efficient continuum of healthcare. The European market is offering a vast potential for mHealth and consequently a number of related initiatives are taking place. Nevertheless, the full unlock of mHealth in Europe is impeded by regulatory challenges. The aim of this paper is to reflect on the current state of mHealth in Europe and on open and challenging regulatory issues that need to be addressed. The end goal of this reflection is to act as a forum for an ongoing dialogue and ultimately a coordinated roadmap for the ecosystem of mHealth.

KEYWORDS

mHealth, Europe, regulation, framework, review

1. INTRODUCTION

In the developed world healthcare systems face significant challenges such as the shrinking of budgets and the ageing of the population that has given rise to the growth of chronic diseases (GSMA, 2013). To counteract these challenges, governments at a worldwide level have set at the center of their healthcare strategic agenda the need for the fundamental reform of the status quo of healthcare systems (Emmanouilidou and Burke, 2013). In this reform process the role of information technology is instrumental and eHealth is at the forefront. The European Commission (EC) in 2004 submitted to its Member States the eHealth Action Plan that comprised a challenging list of implementation actions extending until 2010 (EC, 2007). In 2012, the EC unveiled a second eHealth Action Plan covering the period 2012-2020 that focused on addressing the barriers to the deployment of eHealth systems (EC, 2012).

mHealth represents the new edge of eHealth and aims to address the emerging problems of health systems whilst delivering innovative healthcare services. More specifically, mHealth can improve decision making, support the effective management of chronic conditions, provide suitable healthcare in case of emergencies and extend the coverage of healthcare provision. Although the potential of mHealth is widely recognized regulatory challenges constitute a barrier to the growth and uptake of mHealth solutions. This paper provides a reflection of the state of mHealth in Europe and aims to act as a forum for discussion of the key regulatory aspects that will facilitate the development of the mHealth ecosystem.

2. THE EDGE OF MHEALTH

The continuous advances in Information and Communication Technologies (ICT) and the rise of mobile Internet have accelerated the development of various mobile applications. In the context of healthcare, mobile-enabled health services have received the attention of application developers, network service providers, healthcare professionals and patients as well as the overall research community (Varshney, 2014).

The proliferation of the mobile health market is clearly depicted in the number of downloaded mobile health applications which at the end of 2010 surpassed 200 million and by 2017 this number is predicted to exceed 1.7 billion (Silva et al., 2015).

The World Health Organization (WHO, 2011, p. 6) considers mobile health (mHealth) a component of eHealth and defines it as "medical and public health practice supported by mobile devices such as mobile phones, patient monitoring devices, personal digital assistants (PDAs) and other wireless devices". This definition opens a broad spectrum of sophisticated mHealth applications that can involve among others disease prevention and wellness, monitoring and remote care, mobile decision making and emergency interventions (Varshney, 2014). The added value of mHealth is that it can support new models of care by surpassing geographical, temporal and organizational barriers (Silva et al., 2015) and can contribute to the sustainability of healthcare systems (GSMA, 2013). In particular, the main perceived benefits of mHealth include a reduction of healthcare costs, patient-centered care, greater empowerment of individuals to manage their health more effectively and proactively, an opportunity to deliver more consistent, coordinated and efficient healthcare when and where is needed and increased access to health services to remote or under-served communities (GSMA, 2013; Treacy et al., 2015). The early signs for the potential of mHealth are positive with the UK Department of Health reporting that remote care, if used correctly, can lead to a 20% reduction in emergency admissions, a 14% reduction in bed days and a 45% reduction in mortality rates (GSMA, 2013).

2.1 The State of mHealth in Europe

In April 2014, the European Commission (EC) launched a public consultation on mHealth, the Green Paper on mobile health (mHealth) inviting interested stakeholders namely administrations, health organizations, professionals, patients and the industry to provide their views in order to discover barriers and issues related to the use of mHealth (EC, 2014a; Treacy et al., 2015). The Green Paper sets the priorities in regulating mHealth which involve patient safety, patient data security and interoperability between mHealth applications and medical devices (Shapiro et al., 2014). Alongside the Green Paper, the Commission also published a Staff Working Document on the existing EU legal framework applicable to lifestyle and wellbeing apps (EC, 2014b). The aim of this document was to provide software developers and medical device manufacturers with legal guidance on the EU legislation and to determine if their products are subject to the European Medical Devices Directive (MDD) 93/42/EEC and 2007/47/EC or the In Vitro Diagnostic Device Directive (IVDD) 98/79/EC (Treacy et al., 2015). Nevertheless, mHealth in Europe is insufficiently widespread and remains largely in an experimental phase mainly due to the inadequacy of the national and European legal framework (ITU, 2014). This legal challenge encompasses three types of issues. First, applicable rules in Europe do not appropriately account for the specific requirements associated with the development of mHealth (ITU, 2014). Second, the European legal landscape is fragmented, dominated by individual State jurisdictions with no attempt to harmonize the rules (ITU, 2014). Third, the state of the European mHealth regulatory framework is lagging behind compared to the United States or Japan (ITU, 2014). The responses of the EC public consultation have indicated privacy, security, patient safety, transparency of information, the need for a clear legal framework and further evidence on the cost-effectiveness of mHealth as essential prerequisites (EC, 2015a). The public consultation findings were followed up in the mHealth stakeholder meeting at eHealth Week 2015 during which the EC presented the industry-led Code of Conduct on mobile health apps and planned initiatives to support web entrepreneurs' access to the market (EC, 2015b).

2.2 Examples of mHealth in Europe

Four EU-funded projects have been recognized as making the most of the mHealth potential. NEPHRON+ is a personalized treatment and management solution for patients with kidney failure that allows patients to consult and manage data and doctors to monitor the treatment and remotely adapt it (EC, 2014c). REACTION supports self-management and life-style changes for diabetes patients and includes GlucoTab, a mobile system that enhances medical information flow in hospitals (EC, 2014c). MOBIGUIDE is an intelligent decision-support system for patients with chronic diseases where biosignals are monitored and recommendations regarding lifestyle changes are made and INTERSTRESS is an ICT-based solution for the assessment and treatment of psychological stress (EC, 2014c).

Under the European Union Horizon 2020 research and innovation framework program further mHealth projects will be implemented including m-RESIST, HEARTEN, myAirCoach and EmERGE. m-RESIST is the first EU mHealth application that aims to develop a therapeutic program for patients with treatment-resistant schizophrenia (EC, 2015c). HEARTEN will develop biosensors that monitor breath and saliva heart failure biomarkers and send alerts to the patient's smartphone when his/her life is in a critical situation (EC, 2015d). myAirCoach will provide a holistic mHealth personalized asthma monitoring system (EC, 2015e) and EmERGE will develop a mHealth platform to enable self-management of HIV in patients with stable disease (EC, 2015f).

2.3 The European Regulatory Terrain of mHealth

To fuel the development and deployment of mHealth solutions the EU regulatory framework governing mHealth requires a clear roadmap. Currently, within the existing framework there are regulatory barriers that need to be urgently addressed. For example, mHealth devices are governed, though not explicitly, by two regulatory frameworks, the Radio Equipment and Telecommunications Terminal Equipment (RTTE) and the EU Medical Devices Directive (MDD) (Shapiro et al., 2014). This causes queries regarding the application and coordination between these two regulations (GSMA, 2013). Given the broad range of mobile medical devices that can vary in both their technical aspects as well as the type of service they offer it is essential to develop an appropriate classification scheme. This is also relevant for mobile medical applications as these can vary widely depending on the purpose and context of use. In the case of medical applications the boundaries between mobile medical apps and mobile wellness apps can become blurred as healthcare models become more patient-oriented (GSMA, 2013) and should therefore be clearly refined. In this respect, if the purpose and context of use of a mobile wellness app is placed within the regime of healthcare provision and supports a patient's diagnosis and treatment then it should be classified accordingly.

Patient safety is a key priority in healthcare and should therefore be at the core of mHealth solutions. This leads to the need for a thorough risk assessment of medical devices and apps. In order to successfully address this challenging element a broader engagement is required with the representation of relevant stakeholders. To prevent potential patient harm, rules and standards should be determined per medical condition and for the corresponding medical device or app. In relation to this the issue of liability will need to be considered especially when a critical condition occurs. The issue of liability is rather complicated in mHealth solutions as in addition to product liability or potential liability of healthcare professionals and third party service providers (ITU, 2014) there are also the actual patients who are heavily encouraged to self-manage their medical conditions. The transmission of sensitive health data that is performed via wireless networks in mHealth services highlights the issue of data privacy and security. To deal with potential confidentiality breaches patients should be provided with disclosures about the type of data collected and its use (Varshney, 2014) and be able to provide when needed their consent.

The regulatory framework of mHealth in Europe needs to overcome existing burdens and be tailored with the ongoing advances in mobile technology for mHealth innovative solutions to flourish. An open dialogue and a broader corporation and coordination at a pan-European level will enhance the accomplishment of the targeted objectives of mHealth solutions and lead to timely marketed mHealth products that can effectively and efficiently deliver the anticipated benefits to all healthcare stakeholder groups.

3. CONCLUSION

Mobile health, the new edge on healthcare innovation, is an emerging research field with mHealth solutions covering a broad spectrum of different types of mobile medical devices and medical applications. mHealth is expected to effectively tackle the challenges healthcare systems are confronted with thus leading to sustainable healthcare systems. The anticipated benefits of mHealth services highlight the central role of patient in the provision of healthcare and the ability to overcome temporal, geographical and organizational barriers. In Europe, mHealth remains to a large extent in an experimental phase mainly because of the inadequacy of the European legal framework. This paper aimed to provide a brief overview of the current state and progress on mHealth in Europe and a starting point for a comprehensive reflection of the challenges that are at the center of the mHealth regulatory agenda. Since this paper is restricted to a broad discussion of relevant themes and high level proposals for the associated regulatory challenges future studies could expand this reflection and provide an in-depth analysis of the topic.

REFERENCES

EC, 2007. eHealth priorities and strategies in European countries – eHealth ERA report. Available: https://ec.europa.eu/digital-agenda/en/news/ehealth-priorities-and-strategies-european-countries [07 Feb 2016]

EC, 2012. eHealth Action Plan 2012-2020 – Innovative healthcare for the 21st century. Available: http://ec.europa.eu/digital-agenda/en/news/ehealth-action-plan-2012-2020-innovative-healthcare-21st-century [07 Feb 2016]

EC, 2014a. Green Paper on mobile Health (mHealth). Available: https://ec.europa.eu/digital-agenda/en/news/green-paper-mobile-health-mhealth [07 Feb 2016]

EC, 2014b. Commission Staff Working Document on the existing EU legal framework applicable to lifestyle and wellbeing apps. Available: https://ec.europa.eu/ [07 Feb 2016]

EC, 2014c. What mHealth can do for you. Available: http://europa.eu/rapid/press-release_MEMO-14-266_en.htm [28 Mar 2016]

EC, 2015a. mHealth in Europe: Preparing the ground – consultation results published. Available: https://ec.europa.eu/digital-single-market/en/news/mhealth-europe-preparing-ground-consultation-results-published-today [28 Mar 2016]

EC, 2015b. mHealth in Europe: Next steps. Available: https://ec.europa.eu/digital-single-market/en/news/mhealth-green-paper-next-steps [28 Mar 2016]

EC, 2015c. m-RESIST: The first EU mHealth project for patients with treatment-resistant schizophrenia. Available: https://ec.europa.eu/digital-agenda/en/news/m-resist-first-eu-mhealth-project-patients-treatment-resistant-schizophrenia [07 Feb 2016]

EC, 2015d. HEARTEN A co-operative mHealth environment targeting adherence and management of patients suffering from Heart Failure. Available: http://cordis.europa.eu/project/rcn/194092_en.html [07 Feb 2016]

EC, 2015e. Analysis, modelling and sensing of both physiological and environmental factors for the customized and predictive self-management of Asthma. Available: http://cordis.europa.eu/project/rcn/194083_en.html [28 Mar 2016]

EC, 2015f. Evaluating mHealth technology in HIV to improve Empowerment and healthcare utilization: Research and innovation to Generate Evidence for personalized care. Available: http://cordis.europa.eu/project/rcn/194097_en.html [28 Mar 2016]

Emmanouilidou, M. and Burke, M., 2013. A thematic review and a policy-analysis agenda of Electronic Health Records in the Greek National Health System. *Health Policy*, Vol., 109, pp. 31-37.

GSMA, 2013. mHealth and the EU regulatory framework for medical devices. Available: http://www.gsma.com/ [07 Feb 2016]

ITU, 2014. Discussion Paper Filling the Gap: Legal and Regulatory Challenges of Mobile Health (mHealth) in Europe. Available: www.itu.int [07 Feb 2016]

Shapiro, J. et al., 2014. Review of 2013-2014 Mobile Medical Applications: Regulatory Challenges and Precedents. *Intersect*, Vol. 8, No. 1, pp. 1-26.

Silva B.M.C. et al., 2015. Mobile-health: A review of current state in 2015. *Journal of Biomedical Informatics*, Vol. 56, pp. 265-272.

Treacy, C. et al., 2015. Mobile Health & Medical Apps: Possible Impediments to Healthcare Adoption. *eTELEMED 2015: The Seventh International Conference on eHealth, Telemedicine and Social Medicine.* Lisbon, Portugal, pp. 199-202.

Varshney, U., 2014. Mobile health: Four emerging themes of research. *Decision Support Systems*, Vol. 66, pp. 20-35.

WHO, 2011. mHealth New horizons for health through mobile technologies: Based on the findings of the second global survey on eHealth (Global Observatory for eHealth Series, volume 3). Available: http://www.who.int/goe/publications/goe_mhealth_web.pdf [28 Mar 2016]

A REVIEW OF HUMAN BODY SHAPE DETECTION TECHNIQUES AND THEIR APPLICATION TO THE PREDICTION OF HEALTH RISKS

Shellyann Sooklal, Patrick Hosein and Surujpal Teelucksingh
The University of the West Indies,
St. Augustine Campus, Trinidad and Tobago

ABSTRACT

Human body shape detection techniques have applications across fields such as surveillance for security purposes and design and clothes fitting estimations in the fashion industry. We here propose the use of body shape detection for the classification and estimation of health risks since body shapes are directly related to fat distribution. With the phenomenal increase in prevalence of obesity worldwide and the ensuring epidemic of diabetes and other weight related sequelae, the monitoring of body shapes has assumed great relevance. This is especially so among children where earlier identification and timely interventions may impact future health outcomes. Body shape assessment by traditional manual methods can be tedious, time consuming and may exaggerate psychological effects on obese students. Therefore non-invasive approaches employing images has potential to both screen larger numbers as well as avoid direct contact. Although such techniques are available, the equipment necessary can be expensive thereby limiting access to those from poorer countries whose populations are most vulnerable to epidemiologic transition to obesity. In this paper we provide a review of body shape classification techniques and discuss possible low-cost alternatives that can be used to monitor body shapes of school-aged children using camera photographs.

KEYWORDS

Body shape, feature detection, image processing

1. INTRODUCTION

Human detection in images has been explored for many years but the interest and development of the field continues to grow due to the vast range of areas to which it can be applied such as medical analysis, video games and surveillance (Kart et al. 2011, Shotton et al. 2013, Lee & Nevatia 2007). However, studying humans from images and extracting body features are not straightforward. Challenges arise due to the wide variations in appearances (height, weight, shape), clothing and pose (Brandao et al. 2014).

The analysis of a person's body shape can be used to determine current and even potential health risks. Wang et al. (2015) pointed out that body health is not only determined by body fat percentage but also by how that fat is distributed throughout the body. They also explained that studies have shown that persons with more fat around their abdominal region have a higher chance of developing cardiovascular diseases and other associated diseases, such as diabetes and hypertension, than those who store fat around their hips. The distribution of fat around different areas of a person's body leads to different body shapes such as apple, pear, triangle, hourglass and rectangle. Identifying high-risk body shapes in a timely manner allows for timely intervention.

This paper gives a review of techniques that have been used to determine a person's body shape from images. The paper is organized as follows. Section 2 presents some applications which have been created to categorize a person's body shape. Section 3 discusses the findings from the reviewed work and proposes a method for determining the body shape of school children. Finally, the paper concludes in Section 4 with a summary of the findings.

2. APPLICATIONS TO CATEGORIZE BODY SHAPES

Most systems that have been developed to categorize body shape have been built for the specific purpose of virtual clothing fitting applications. One of these applications is presented by Simmons et al. (2004). They selected some basic female body shapes and used both visual and recorded data (3D body scan measurement data of the abdomen, stomach, hip and waist) to derive mathematical formulas which can categorize persons into the various shapes such as hourglass, rectangle, oval, triangle and diamond. Subsequently, Neophytou et al. (2013) built an online fashion application that uses the classification formulas developed by Simmons et al. However, instead of using data from 3D body scan data, measurement details were extracted from images uploaded by the user. Their system applied a GrabCut method (Rother et al. 2004), in order to subtract the background. The models' shapes were learnt by applying principal component analysis to vertex coordinates from the scans. Energy functions were used to match a user's silhouette to a model. The methods used by Simmons et al. were then applied to determine the user's body shape. A similar system built by Kart et al. (2011) required the user to enter personal information, such as weight, height and age, and also a 2D photograph of themselves taken in front of a plain, light color background and dressed in dark colored clothing. The user's information was processed using functions from MATLAB's Image Processing Toolbox. The RGB (red, green and blue) image was first converted to an HSV (hue, saturation, value) image, then noise reduction algorithms were applied. The negative of the image was taken and, finally, the pixels were counted in order to determine measurements for different parts of the body, such as the bust, waist and hip. The measurements, as well as calculated body mass index (BMI), were then passed to a decision algorithm which determined the body shape of the person. Sekine et al. (2014) presented a virtual clothes fitting application which overlaid images of clothing, worn by a model with a similar body shape, onto the user's image. They used depth images in order to easily obtain the locations of a person's joints and trained their system by matching depth images to 3D human body models of varying body shapes and poses. At run-time, the system matched the user's depth image with one of the pre-trained depth images. Characteristic measurements, such as height and waist width, were used to select the shape model whose shape best matched that of the user. After the body shape model was selected, the image of the clothing was scaled and overlaid onto the user's image. Likewise, Chang and Wang (2015) implemented a technique for estimating body shape in order to improve the experience of shopping for clothing online. Part-based feature extraction was first used on training images of persons in skin-tight clothing. Descriptive features of the person's body shape, such as height and leg length, were extracted from the semantic parts. Principal component analysis was performed on the features in order to reconstruct the person's shape and also to aid with modelling variation in shapes. Multiple camera views were used to improve the accuracy of the estimated shape. The previously selected features were used as constants across the various views. Linear regression was applied to match measurements across the different views of the person. Finally, optimal coefficients were calculated for each view in order to reconstruct the body shape of the person. The reconstruction coefficients were allowed to be different across multiple views. In order to evaluate the proposed technique, the DC Suite, a 'try-on' software, was applied on the SHREC' 14 dataset.

Apart from the use in clothing fitting applications, there are a few methods developed solely to estimate a person's body shape. For example, Balan and Black (2008) proposed a method for estimating a 3D body shape of a person using multiple images of the person under varied poses. They applied a generalized notion of a visual hull where the body shape of the person lay within the visual hull and belonged to a family of parametric 3D body shapes. They also identified skin regions that allowed the deduction of certain constraints, such as the overall weight of the person. They also used gender-specific models to refine the shape estimation process. In addition, Hasler et al. (2010) proposed a multilinear method for estimating pose and body shape. The method was independent of the type of clothing worn by the subject and worked for both single and multiple camera views. They first derived bilinear, analytical and independent models for pose and body shape using a database of 3D scans of 114 undressed subjects each of which were scanned in various poses. In order to determine the pose and body shape of a silhouette in an image, an initial 3D shape was chosen and similarities between the shape and the silhouette were calculated.

Although our application differs from those discussed above (i.e. clothing fitting), the techniques involved are quite similar since our objective is to identify the outline of the person's body in order to determine if the person's dimensions and shape indicate potential health issues. Another difference with our work is that we require a low cost solution since the budget available in schools for such programs is typically quite small.

3. DISCUSSION

Human shape detection and feature extraction in images is still a challenging task. In addition, little work has been done on body shape classification. The body shape classification applications described in the previous section were mainly for virtual clothing fitting purposes but techniques used by these applications can be applied to body shape detection methods for health-risk detection purposes. The latter applications first extracted the user from the image, applied various techniques to extract measurements of different body parts, such as shoulders, bust, waist and hip, and then classify the user into a specific body shape. These applications employed a general set of procedures. These include background subtraction, segmentation of body in order to extract useful details such as shape features, training and use of a classifier to determine body shape and testing of the proposed system with a popular dataset or one which was self-obtained. The results are compared against manually acquired values or with the results of similar methods in order to assess the method's accuracy and efficiency. However, typically less information is required for body shape categorization since precise values of dimensions are not needed but rather ratios of these dimensions. These modified techniques may now be put to a novel use of stratifying the population into differential risk categories for the purposes of intervening to preserve health. Furthermore, our interest is to create low cost tools and applications that would find widespread use in low income and resource-constrained settings.

3.1 Preliminary Experiments

Initial experimentations were done in order to build a system that can easily and efficiently categorize the body shapes of secondary school students in their usual school uniforms, without the use of expensive equipment such as 3D scanners. Images of two students (1 male and 1 female), dressed in full uniforms, were captured with a color camera and also with a Seek Thermal Imaging Camera. It was initially hypothesized that infrared images will give a more accurate outline of a person's shape, rather than color images. However, as seen in Figure 1, thin materials absorb body heat, whilst thick materials prevented the IR camera from detecting heat in those regions. Hence we concluded that the analysis of color images would provide better accuracy than infra-red images in detecting body shapes of school students.

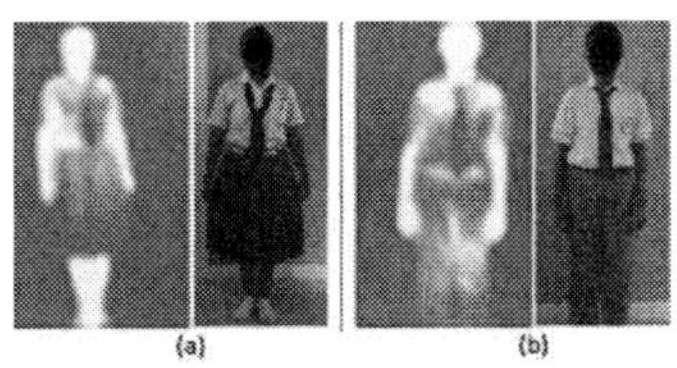

Figure 1. Infrared images of students along with their corresponding color images

3.2 Work in Progress

A 360 degree health application is being developed for secondary schools in Trinidad and Tobago. Heart disease is the leading cause of death in the country. It is well-established that pear-shaped individuals are at far less risk of heart attacks than those who are apple-shaped. Therefore, tracking the risks of developing heart disease at a young age and taking measures to prevent heart disease will be a useful public health intervention. The application collects data from students, processes it, derives a risk index and alerts the school's administration and parents/guardians of potential health risks. Once a month, students are asked to walk past a pair of cameras connected to a client application running on a tablet or laptop. One camera takes a frontal photo while the other takes a profile shot. These photos are automatically triggered since the system is meant to have minimal human intervention. Face recognition is used to identify the student. The frontal view of the person is captured when the user stands in front of a specially marked background. The marked background contains cues which can aid in automatically calculating a person's height. The students also stand on a weight scale that captures weight and body fat percentage via bioelectrical impedance analysis. The readings are sent to the client application via Wi-Fi or Bluetooth. Initial processing of the image is done on site before being transmitted via a Wi-Fi or cellular connection to a central database. This pre-processing includes edge detection in order to extract the important characteristics of the image. The resulting image is significantly smaller in size and hence can more easily (and cheaply) uploaded to the central server.

Further processing is done on the central server. Various dimensions of the image are measured and these are discounted to take into account clothing since school uniforms can alter the perception of a person's actual body size and shape. However, an advantage is that all students are similarly dressed thereby minimizing the effect of outerwear. These discount factors are being determined based on initial experiments whereby image dimensions and actual dimensions of children dressed in school uniforms are compared to determine the appropriate discount factor. Various ratios and measurements are calculated in order to aid in the categorization of a person's body shape. These include height, waist circumference, waist to hip ratio, waist to height ratio, torso length and the difference in volume between the top half of the torso and the lower half. Body Mass Index (BMI) is also calculated using a special chart for children between the ages of 5 and 17. Ethnicity detection is performed on the central server in order to incorporate an ethnic-specific criteria for BMI calculations. This is due to the fact that Caucasians are allowed the largest measures and Indians and Chinese have the least capacity and the lowest thresholds for decompensation as they grow fat. A combination of all of these dimensions and calculations are used as characteristics to determine which body shape category is most probable for the student. We also perform comparisons with historical data for that student to detect any trends. This information together with the body shape information and other data that may have been collected at the site (such as weight) or calculated by the central server (such as waist circumference) is then used to determine what action, if any, should be taken by the school.

4. CONCLUSION

The classification of human body shapes from images is a useful technique for predicting health risks. Most of the research in this area has been done in the fashion industry for virtual clothes fitting. However, we believe that we can identify children at high risk for adverse consequences of obesity using low cost equipment to non-invasively determine weight-related health issues of school-aged children. We are developing a low cost, automated framework for doing this and are presently running experiments.

REFERENCES

Balan, A. and Black, M., 2008. The Naked Truth: Estimating Body Shape Under Clothing. *In Computer Vision ECCV 2008, ser. Lecture Notes in Computer Science, D. Forsyth, P. Torr, and A. Zisserman, Eds. Springer Berlin Heidelberg, 2008*, Vol. 5303, pp. 15–29.

Brandao, A. et al, 2014. M5AIE - a Method for Body Part Detection and Tracking Using RGB-D Images. *In 2014 International Conference on Theory and Applications (VISAPP)*. Lisbon, Portugal, pp. 367–377.

Chang, W. and Wang, Y. F., 2015. Seeing Through the Appearance: Body Shape Estimation Using Multi-View Clothing Images. In 2015 IEEE International Conference on Multimedia and Expo (ICME 2015), Turin, Italy, pp. 1–6.

Hasler, N. et al, 2010. Multilinear Pose and Body Shape Estimation of Dressed Subjects from Image Sets. *2010 IEEE Conference on Computer Vision and Pattern Recognition (CVPR)*, San Francisco, California, pp. 1823–1830.

Kart, O. et al, 2011. Web Based Digital Image Processing Tool for Body Shape Detection. *2011 Web Proceedings in ICT Innovations*. pp. 139–147.

Lee, M. W. and Nevatia, R., 2007. Body Part Detection for Human Pose Estimation and Tracking. *IEEE Workshop on Motion and Video Computing, 2007 (WMVC '07)*. Austin, Texas, pp. 23.

Neophytou, A. et al, 2013. 3D Human Body Shape Estimation and Classification for Online Fashion. *Digital Fashion.*

Rother, C. et al, 2004. GrabCut: Interactive Foreground Extraction Using Iterated Graph Cuts. *In ACM Transactions on Graphics (TOG)*, Vol. 23, No. 3, pp. 309–314.

Sekine, M. et al, 2014. Virtual Fitting by Single-Shot Body Shape Estimation. *Proceedings of 5th International Conference on 3D Body Scanning Technologies*. Lugano, Switzerland, pp. 406–413.

Shotton, J. et al, 2013. Real-Time Human Pose Recognition in Parts from Single Depth Images. *In Commun. ACM*, Vol. 56, No. 1, pp. 116-124.

Simmons, K. et al, 2004. Female Figure Identification Technique (FFIT) for Apparel Part I: Describing Female Shapes. *In Journal of Textile and Apparel Technology and Management*, Vol. 4, No. 1.

Wang, S. et al, 2015. A Novel Quantitative Body Shape Score for Detecting Association Between Obesity and Hypertension in China. *In BMC Public Health*, Vol. 15, No. 7.

Poster

STRUCTURED AND INDEPENDENT DATA QUALITY ASSESSMENT OF AMBIENT ASSISTED LIVING CARE TECHNOLOGY IS NEEDED

Stefan Wagner and Femina Hassan Aysha Beevi
Aarhus University, Department of Engineering
Aarhus, Denmark

ABSTRACT

Background: Ambient assisted living care products are used for the monitoring of the well-being and rehabilitation efforts of patients and senior citizens. These technologies require a high data quality level in order to be valid. Objectives: To present the findings from a structured and independent data quality assessment method applied to five AAL and care products and provide recommendations for the field. Results: We found a lack of relevant independent effectiveness and efficacy studies and other technology assessments in the literature with regard to the five AAL products that were evaluated in the study. Furthermore, there was a lack of interest on behalf of the vendors to deliver their equipment for efficacy testing at our laboratory facilities. Conclusion: We suggest that all AAL care products should be duly evaluated for their efficacy levels in a structured and standardized manner by an independent organization before being implementation in the clinic.

KEYWORDS

Health technology assessment, ambient assisted living, care, data quality

1. INTRODUCTION

An increasing number of ambient assisted living care products are becoming available as part of an e-health trend of monitoring the well-being and rehabilitation efforts of patients and senior citizens. Working in close collaboration with local public sector care facilities and hospitals, we have carried out a number of efficacy evaluations of a range of ambient assisted living (AAL) care technology products. As a result of these evaluations, we have found several challenges including a lack of proper evidence for the clinical effectiveness and even a lack of properly structured and independent efficacy testing of the products. While effectiveness studies (which are studies that has focus on the effect of an intervention in a real world setting) are often difficult, time consuming, and expensive to carry out, an efficacy study may on the other hand easily be carried out by independent organizations in a laboratory setting for basic quality assurance at a low cost. However, many vendors and product buyers fail to arrange for such independent efficacy studies before acquiring and implementing new AAL technologies. This lack of proper data quality assessment of AAL products poses a potential threat to the safety of patients. The AAL care technology assessment method (CTAM) aims at addressing these challenges. It is based on the generalized method of health technology assessment (HTA), but with a focus on evaluating AAL technologies. The method is intended to provide evidence to policy makers for decision making on choosing AAL technologies.

2. OBJECTIVES

The aim of this poster is to introduce a structured and independent data quality assessment method for AAL and care products and provide recommendations for the field. Specifically, the poster reports on the specific results of an assessment of five AAL products as a case study on the quality assessment problems in the field.

3. METHODS

We evaluated five AAL products for a range of selected capabilities including fall detection (Wagner and Beevi 2015). For this the CTAM method was used with its six assessment dimensions of user requirements, technology characteristics, user perspectives, efficacy safety and transferability. The five assessed products are: MariCare's Elsi Smart Floor, Tunstall's Intelligent Ceiling, Compliant Concepts' Mobility Monitor, Any Group's MultiVagt and KR's Opus 5 Intelligent Hospital Bed. The main evaluation focus was on the support of a range of modalities or activities of a resident in a single home apartment, e.g. an assisted living care facility (nursing home), rehabilitation facility, private home, or in other types of assisted living facilities. Of these, the detection of falls, especially during the nighttime, is the most important modality and event. As part of the CTAM process a literature search for efficacy and effectiveness evaluations was performed. Also, we invited the vendors to provide access to their products for efficacy evaluation at our facilities.

4. RESULTS

We found there is a lack of relevant independent effectiveness and efficacy studies and other technology assessments in the literature. Furthermore, there was a lack of interest on behalf of the vendors to deliver their equipment for efficacy testing at our laboratory facilities, except for Any Group, who willingly delivered and installed their product series at our lab facilities. The efficacy evaluation of Any Groups products failed to live up to the specifications provided by the company on several points, indicating the difficulties of structured efficacy testing at a laboratory facility, and the need to perform additional assessments.

5. CONCLUSIONS

We recommend further independent evaluations of all five assessed products to be carried out with regard to their efficacy, clinical effectiveness, and total cost of ownership. A quantitative comparison of the products may be done by validating them in actual care settings or with simulated events as efficacy evaluation in the laboratory setting. Furthermore, based on the experiences of this case study, we suggest that all AAL care products should be duly evaluated for their efficacy in a structured and standardized manner by an independent organization as a mandatory task before marketing the product, and that these reports be made freely available to potential end-user organizations. Later, as the products mature further, the clinical effectiveness should be evaluated in real life care facilities and hospitals before they should be endorsed by clinical and academic societies. CTAM, along with other HTA methods, could be a relevant method for achieving these objectives.

REFERENCES

Wagner, S., Beevi, F.H.A. 2015. Healthcare Technology Assessment of Ambient Assisted Living Products for Fall Detection and Activity Tracking of Senior Citizens. *Journal of Pervasive Systems Engineering*, Vol. 4, No. 1, 2015.

AUTHOR INDEX